Animal and human calorimetry

J. A. McLEAN

Hannah Research Institute, Ayr, Scotland

G. TOBIN

Department of Physiology, University of Leeds

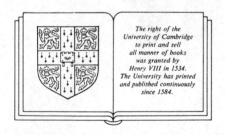

The right of the
University of Cambridge
to print and sell
all manner of books
was granted by
Henry VIII in 1534.
The University has printed
and published continuously
since 1584.

CAMBRIDGE UNIVERSITY PRESS

Cambridge

New York   New Rochelle   Melbourne   Sydney

Published by the Press Syndicate of the University of Cambridge
The Pitt Building, Trumpington Street, Cambridge CB2 1RP
32 East 57th Street, New York, NY 10022, USA
10 Stamford Road, Oakleigh, Melbourne 3166, Australia

First published 1987

Printed in Great Britain by the University Press, Cambridge

Library of Congress cataloguing in publication data

McLean, J. A.
Animal and human calorimetry.

1. Calorimeters and calorimetry. 2. Biological
chemistry–Technique. I. Tobin, G. II. Title.
QP135.M37 1987   591.19′283   86-34311

British Library cataloguing in publication data

McLean, J. A.
Animal and human calorimetry.

1. Calorimeters and calorimetry
2. Chemistry, Analytic – Quantitative
I. Title II. Tobin, G.
545′.4   QD117.T4

ISBN 0 521 30905 0

UP

Contents

Principal symbols and abbreviations

A Area

a Oxygen consumed per gram of substance oxidised (subscripts identify the substances as fat, F; carbohydrate, K; protein, P; urine, U; nitrogen of protein, N) (see Table 3.4. p. 36)

$\mathbf{C}$ Carbon retained in the body

Ch Chamber output (GLC)

c Specific heat (of gas or water vapour at constant pressure, c_p; liquid water, c_w; dry air, c_a)

c Excess isotopic concentration (DLWT)

D Thickness of insulating layer (GLC)

D Dose of labelled water (DLWT)

e Error or % error

F Fraction by volume of gas in dry air

F' Fraction in CO_2-free air

$F\text{I}$ Fraction in inlet or inspired air

$F\text{E}$ Fraction in exhaust or expired air

ΔF $= (F\text{E} - F\text{I})$

(All with subscripts specifying the gases, e.g. $F\text{I}_{O_2}$, $F\text{E}_{CO_2}$, ΔF_{CH_4}, etc.)

f Fraction by weight of an element in a compound or mixture (Subscripts specify the element, e.g., O, C, H, N; a second subscript specifies the compound or mixture as urine, U; or protein, P)

$\left.\begin{array}{l} f_1 \\ f_2 \\ f_3 \end{array}\right\}$ Isotopic fractionation factors (DLWT)

G Calibration factor of an oxygen analyser (initial value, G_i; mean value, $\overline{G}$)

I Zero-time intercept of isotopic concentration plot (DLWT)

i Current

K Thermal conductivity

k Sensitivity constant (GLC) of chamber, k_{ch}; platemeters, k_{pm}; thermopile, k_{th}

k Fractional isotopic turnover rate (DLWT) of ^{18}O, k_{18}; ^{2}H, k_2
L Latent heat of vaporisation (at temperature T, L_T)
L Distance from a temperature measuring point (GLC)
l Characteristic length (GLC)
M Quantity of metabolic heat produced (kJ)
$\dot{M}$ Rate of metabolic heat production (kW)
m Characteristic factor of gradient layer winding (GLC)
$\dot{m}$ Mass rate of water loss by evaporation
$\dot{m}$ Mass rate of air flow
N Quantity of urinary nitrogen excreted ($\dot{N}$ rate of excretion)
N   Nitrogen retained in the body
n Number
P Barometric pressure (initial value, P_i; value at STPD. P_o)
P_w Partial pressure of water vapour
ΔP Change in pressure
Pm Platemeter output (GLC)
Δp Platemeter correction (GLC)
Q Quantity of heat
Q Heat of combustion (quantity of heat per gram of substance oxidised –
       for subscripts see a above and Table 3.4. p. 36)
$\dot{Q}$ Heat flux or heat flow rate (Subscripts specify type, source, etc. e.g.
       convective, $\dot{Q}_C$; evaporative, $\dot{Q}_E$; via heat exchanger, $\dot{Q}_{EX}$; generated by
       a fan, $\dot{Q}_F$; heat leakage, $\dot{Q}_L$; non-evaporative, $\dot{Q}_N$; through walls, $\dot{Q}_W$;
       removed by circulating water, $\dot{Q}_w$)
q Heat evolved per litre of oxygen consumed (for subscripts see a above
       and Table 3.4. p. 36)
R Universal gas constant
R Reading of a gas analyser (initial value, R_i)
R Electrical resistance
r Humidity ratio (weight of water per weight of dry air)
r Respiratory quotient (for subscripts see a above and Table 3.4. p. 36)
S Sensitivity of oxygen analyser to total pressure (of air, S_a; of pure
       nitrogen, S_n)
T Temperature (mean skin, T_s; mean body, T_b)
Th Thermopile output (GLC)
t Time
V Volume
$\dot{V}$ Air flowrate or gas consumption rate at STPD (e.g. $\dot{V}_{O_2}$, $\dot{V}_{CO_2}$, etc.)
$\dot{V}'$ CO_2-free air flowrate
$\dot{V}_I$ Inlet or inspired air flowrate
$\dot{V}_E$ Exhaust or expired air flowrate
v Velocity
$\dot{v}$ Platemeter ventilation rate (GLC)
W Total pool of body water (mols)

x	Distance
$\alpha, \beta,$	
γ, d	Coefficients for oxygen, carbon dioxide, urinary nitrogen and methane in equations (Brouwer type) for calculating heat production
a	Temperature coefficient of resistivity
γ	Specific heat ratio
ε	Thermoelectric power (V/°C)
θ	Thermopile to chamber sensitivity ratio (GLC)
κ	Thermal diffusivity
λ	Separation between temperature measuring points (GLC)
ρ	Density (of air at STPD, ρ_o; at measuring condition, ρ_m)
ρ	Electrical resistivity
σ	Magnetic susceptibility
τ	Time constant of an exponential function
φ	Platemeter to chamber sensitivity ratio (GLC)
ψ	Error factor of chamber sensitivity (GLC)

A-D	Analogue to digital (converter)
CERT	CO_2-entry rate technique
DLWT	Doubly labelled water technique
DVM	Digital voltmeter
GLC	Gradient-layer calorimeter
IMP	Integrating motor pneumotachograph
K–M	Kofranyi–Michaelis respirometer
MIC	Mobile indirect calorimeter
RE	Retained energy in the body
STP	Standard temperature and pressure (273 C, 760 mm Hg)
STPD	Standard temperature and pressure for dry air

Preface

The impetus for writing this book was born out of a realisation that expertise in calorimetric techniques relied for its distribution largely on chance conversations between a limited 'club' of practitioners. Many research workers in medicine and other fields of biology have embarked on calorimetry in the mistaken belief that measurement of oxygen consumption and hence energy expenditure is a very straightforward process. It is almost certain that results and conclusions which have been reported from some calorimetric systems are incorrect. There has been no comprehensive text-book, and newcomers to the field have usually acquired knowledge by visiting establishments where calorimeters already exist. Subsequently, they have based their own equipment on what they have seen, sometimes unaware that a quite different form of calorimeter might be better suited to their particular needs. Furthermore, practitioners in human calorimetry and farm animal calorimetry seldom meet and have tended to develop their ideas along different lines. Complete systems for calorimetry were, and indeed still are, rarely available commercially and some commercial instruments are of poor accuracy; some indeed are based on incorrect scientific principles. Whether buying a complete system or building one up from components, it is necessary to have a full understanding of the basic principles and of the many sources of error that must be guarded against. We hope that this book will be useful to teachers and students as well as those embarking on calorimetry. The extensive treatment that we have given to the measurement of gas concentrations and flow-rates should also be useful to respiratory physiologists.

We have limited the scope of the book to calorimetry of man and laboratory and farm animals. Calorimetry of insects and aquatic species is not included. Also we confine ourselves to the methodology of estimation of respiratory gaseous exchange, heat production and heat loss. Nor do we go into biochemical, physiological or nutritional aspects, except in so far as is necessary for the understanding of calorimetric methods. Certain sections, which give mathematical or technical details not essential to a complete understanding of the remaining text, are included either in small type or as appendices.

It is a pleasure to acknowledge the contributions of many people. To our colleagues and friends at the Hannah Research Institute and the University of Leeds for their help and useful discussions over many years; to Sir Kenneth Blaxter, W. A. Coward, P. R. Murgatroyd, T. G. Parker, J. W. Snellen, G. Spinnler and B. A. Young who have read parts of the manuscript and offered constructive comments; to P. R. Wilson and Mrs Elizabeth Barbour (HRI library) and Mrs Collins and her staff (Univ. Leeds medical library) for their patience and perseverance in acquiring for us the many books and periodicals which we needed to consult; to D. S. W. Cooney and D. J. Johannson for photography; to Mrs Maria Knight for typing the manuscript; to Professor Malcolm Peaker (Director, HRI), Dr Phil Thomas (Head of Department of Animal Nutrition, HRI) and to Professor Brian Jewell (Department of Physiology, Univ. Leeds) for their encouragement and support; and finally to Romaine Hervey (Department of Physiology, Univ. Leeds) and the late James Findlay (Department of Physiology, HRI) who first brought us into the field of calorimetry, gave us endless encouragement and to whom we owe so much. To all these people we are deeply grateful.

J. A. McLean
G. Tobin
1987

To Inez, Unity, Frances and Patrick

1

Introduction

Calorimetry is the measurement of heat. By means of animal calorimetry (the term 'animal' here includes humans) we can estimate the energy costs of living. All life processes including growth, work and agricultural production (milk, eggs, wool, etc.) use energy, the source of the energy being food. The energy content of food is metabolised (i.e. changed) in the body into other energy forms, only some of which are useful in the sense of growth or production. Much of the wasted energy is given off from the body in the form of heat – hence the name calorimetry.

The heat may be measured directly by physical methods (Direct Calorimetry) or it may be inferred from quantitative measurement of some of the chemical by-products of metabolism (Indirect Calorimetry). These alternatives are possible because of the natural constraints imposed on energy transformations by the laws of thermodynamics. Of fundamental importance are the Law of Conservation of Energy (energy cannot be created or destroyed, only changed in form) and the Hess Law of Constant Heat Summation (the heat released by a chain of reactions is independent of the chemical pathways, and dependent only on the end-products). In effect these laws ensure that the heat evolved in the enormously complex cycle of biochemical reactions that occur in the body is exactly the same as that which is measured when the same food is converted into the same end-products by simple combustion on a laboratory bench or in a calorimeter.

The partition of the gross energy of food into its major energy subdivisions is illustrated in Fig. 1.1. Some food is undigested resulting in a loss of energy as faeces. Faeces also contain material originating from the body that has been abraded or secreted into the alimentary tract. The difference between the energy content of the food and the energy content of the faeces is therefore termed the Apparently Digested Energy of the

Food; the adverb (apparently) being included to denote that it is not strictly energy of dietary origin. Cellulose and other vegetable matter which are indigestible to non-ruminants can be converted into digestible end-products by bacterial growth in the rumen or gut of herbivorous animals. This results in loss of energy in the form of the waste product

Fig. 1.1. Partition of the gross energy of food.

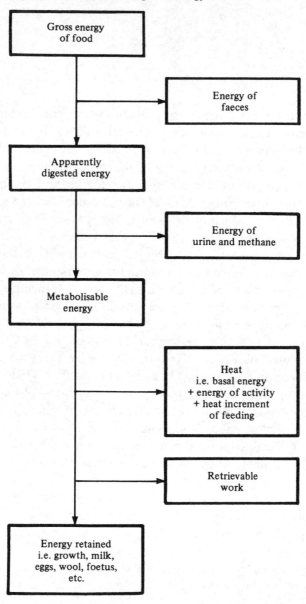

methane. The remaining part of the digestible energy represents nutrients which are absorbed into the blood stream, and waste products from their further metabolism are lost as the energy excreted in urine. The remaining energy is the Metabolisable Energy of the Food. It has to provide for the energetic requirements of the body. First there is the basal energy requirement for maintenance of respiration, blood-circulation and other vital functions. The minimal rate of energy utilisation by a resting subject in a comfortable environment is known as the Basal Metabolic Rate; to this must be added the extra energy cost that occurs after taking a meal (the Heat Increment of Feeding) and any additional energy required for activity, thermoregulation or other muscular work. Finally, if any of the metabolisable energy is left over from meeting these demands, energy may be retained in the body in chemical form as new tissue-growth or as the production of milk or eggs (Retained Energy).

If metabolisable energy of the food is insufficient to meet the demand then energy retention may be negative and the body may consume its own energy reserves. Normally the level of food intake is regulated to meet the demands of maintenance plus growth in the young, or production in the pregnant or feeding mother; but otherwise an excess of food energy intake leads to fattening, or a deficiency to eventual emaciation.

The gross energy of the food, the energy retained in the body and the energy of the faeces, urine and methane are all forms of chemical energy. The energy categories which are concerned with body function and activity are non-chemical and generally this energy is converted into heat. The only exception to this, and then only in very limited circumstances, is the mechanical energy of work. When work energy is stored in some retrievable mechanical form, such as winding up a spring, its eventual loss as heat is delayed until the spring is released. But in most forms of exercise the mechanical energy expended, like the basal energy and the increment due to feeding, is immediately transformed into heat. The rate at which metabolic heat is produced in the body is often termed either Heat Production or Metabolic Rate.

According to the Law of Conservation of Energy a complete balance must exist between the various categories, and this may be expressed as the energy balance equation:

$$
\begin{aligned}
\text{gross energy of food} = \ & \text{energy of faeces} \\
& + \text{energy of urine} \\
& + \text{energy of methane} \\
& + \text{energy retained} \\
& + \text{retrievable mechanical work} \\
& + \text{metabolic heat produced} \qquad (1.1)
\end{aligned}
$$

In nutritional and physiological studies it is possible, with varying degrees of difficulty, to measure each one of these energy categories. The chemical energy content of food, faeces, urine and methane may be estimated by collection, sampling and analysis of these materials. Laboratory analysis of the energy or heat content of food or excreta samples is done in a bomb calorimeter. This is the simplest form of calorimeter to be described in this book, but it is also of fundamental importance for determining the essential constant factors on which animal calorimetry is based. Retained energy is also chemical in form but, except for the energy of milk and eggs, is not readily quantifiable for an intact animal. The amount of energy retained by animals over a period of a feeding trial may be measured by the comparative slaughter technique, in which the energy content of a group of carcases is determined by analysis and compared with that of a control group slaughtered at the start of the trial. This method is reasonably accurate but has obvious limitations and is not at all suitable for humans! Measurement of heat production by means of animal calorimetry combined with measurement of the chemical energy of food and excreta makes it possible to estimate energy retention from the energy balance equation by difference. For this method it is necessary that the animals are maintained on a fixed diet which has been fed at regular intervals for several days prior to measurements (Blaxter, 1956a); also a complete analytical balance of all forms of energy intake and output is essential. Blaxter (1967) has analysed the errors of this procedure and concluded that using indirect calorimetry the energy retention of sheep could be estimated with an accuracy of $\pm 1\%$ of energy intake. From the early work of Atwater & Benedict (1899) it can be concluded that they also achieved similar accuracy in their calorimetric trials on humans. However, these levels of accuracy were only achieved in trials lasting several days, conducted with scrupulous care under laboratory conditions and with full collection and analysis of both intake and excreta.

It is not only energy metabolism that can be studied in this way. The metabolism of individual chemical elements and of the major food categories, protein, fat and carbohydrate, may also be elucidated in calorimetric studies. Bomb calorimetry provides the fundamental information on the relationships between the amounts of oxygen consumed and the amounts of heat and carbon dioxide and water produced during the combustion of protein, fat, and of nitrogenous waste products such as urea. With this information available it is possible to devise a number of possible alternative indirect techniques for the determination of the rates of heat production and the rates of protein and fat deposition. From measurement of expired carbon dioxide alone heat production may be estimated with $\pm 10\%$ accuracy. From measurement of oxygen exchange in respiration heat production may be estimated with $\pm 1.5\%$ accuracy. A full respiratory analysis including oxygen, carbon dioxide and, for ruminants methane also, not only provides highly accurate measurement of heat production but also an estimate of the net contributions of fat and carbohydrate. If urinary nitrogen excretion is also measured the protein metabolism may be elucidated and a complete breakdown is achieved. An alternative method of obtaining the same information is to measure the total retention of the elements carbon and nitrogen in the body.

All these methods are examples of indirect calorimetry which measures the rate of heat generation in the body. Direct calorimetry by contrast measures the rate of heat dissipation from the body. The term Heat Production is often applied rather loosely to both concepts, but they are not the same. Eventually all heat produced must be dissipated, but there are periods when heat is produced faster than it is dissipated and vice-versa; these are periods of body heat storage when the mean temperature of the body is changing.

Calorimetry is used to estimate nutritional requirements of man and farm livestock and to evaluate different foods. It is also a powerful research tool used to study fundamental nutritional and physiological life-processes, and to evaluate stresses imposed by abnormal or severe environments. Finally, it is used as a diagnostic tool in hospitals for the investigation of metabolic disorders.

This book is intended to be useful to those who practise or wish to study or practise calorimetry. Emphasis is placed on details of technique and essential precautions necessary to obtain accurate measurements. The subject is complex and full of potential pitfalls for the practitioner. The authors' main justification for writing the book is that they have at one time or another fallen into many of these pits and they hope to help prevent others from doing the same.

2

Historical

Measurement of the heat output from animals had its beginnings just over two hundred years ago. Adair Crawford in Glasgow measured the rise in temperature of a water jacket surrounding an animal chamber and, at about the same time, Lavoisier in Paris measured the heat output from a guinea-pig by recording the quantity of ice melted inside an insulated chamber which contained the ice and the animal. There is some doubt as to whether Crawford or Lavoisier made the first of these measurements (Mendelsohn, 1964; Blaxter, 1978). From the earlier work by Boyle, Hooke, Mayow and Priestley it was known that 'pure air' was necessary to support both animal life and the flame of a candle, and it was believed that both the animal and the candle 'phlogisticated' the air, thus gradually rendering it unfit for the continuing support of either. Both Crawford and Lavoisier compared the amount of heat given off by their animals with that given off by a lighted candle. Crawford believed that both animal heat and the heat produced by 'combustible bodies' was evolved from the air as phlogiston was formed but it was Lavoisier who made the breakthrough by recognising that the vital process involved was not the production of phlogiston, but the consumption of a new element which he named 'oxygene'. The heat from living animals, like that from the burning candle, was evolved during the oxidation of carbon and the formation of carbon dioxide gas. Lavoisier thus laid the foundation on which both direct and indirect calorimetry is based when he concluded that: 'La respiration est donc une combustion.' Sadly, the world was abruptly deprived of Lavoisier's genius, which had made so many major contributions to the advancement of chemistry and biology, when in 1794 he was executed before the Paris mob. Nearly a century elapsed before any further major breakthrough was made in animal calorimetry.

Lavoisier had demonstrated a relationship between the amount of carbon dioxide which an animal produced and the amount of heat given off by the same animal when placed in his calorimeter. But the measurements were not simultaneous and the cold environment of his ice calorimeter was unnatural and could have affected the animal. The question remained: was the heat formed due solely to the combustion of carbon? In 1822 the Academie de Sciences in Paris nominated 'The Determination of the Source of Animal Heat' as the subject of a prize study. Despretz (1824) and Dulong (1841) competed for this prize and each independently followed almost identical lines of research. They sought to prove that animal heat was derived from the combustion of both carbon and hydrogen. They argued that since each volume of carbon dioxide produced was

Fig. 2.1. Dulong's apparatus for measurement of oxygen consumption and carbon dioxide and heat output of a rabbit.

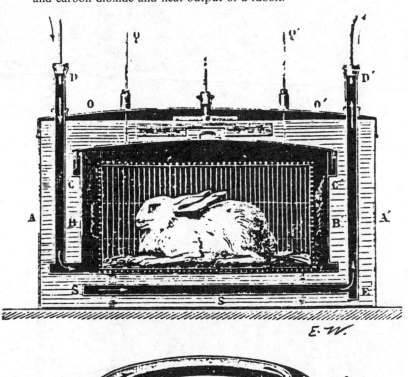

8

Fig. 2.2. Regnault & Reiset's closed-circuit system for measuring respiratory gases.

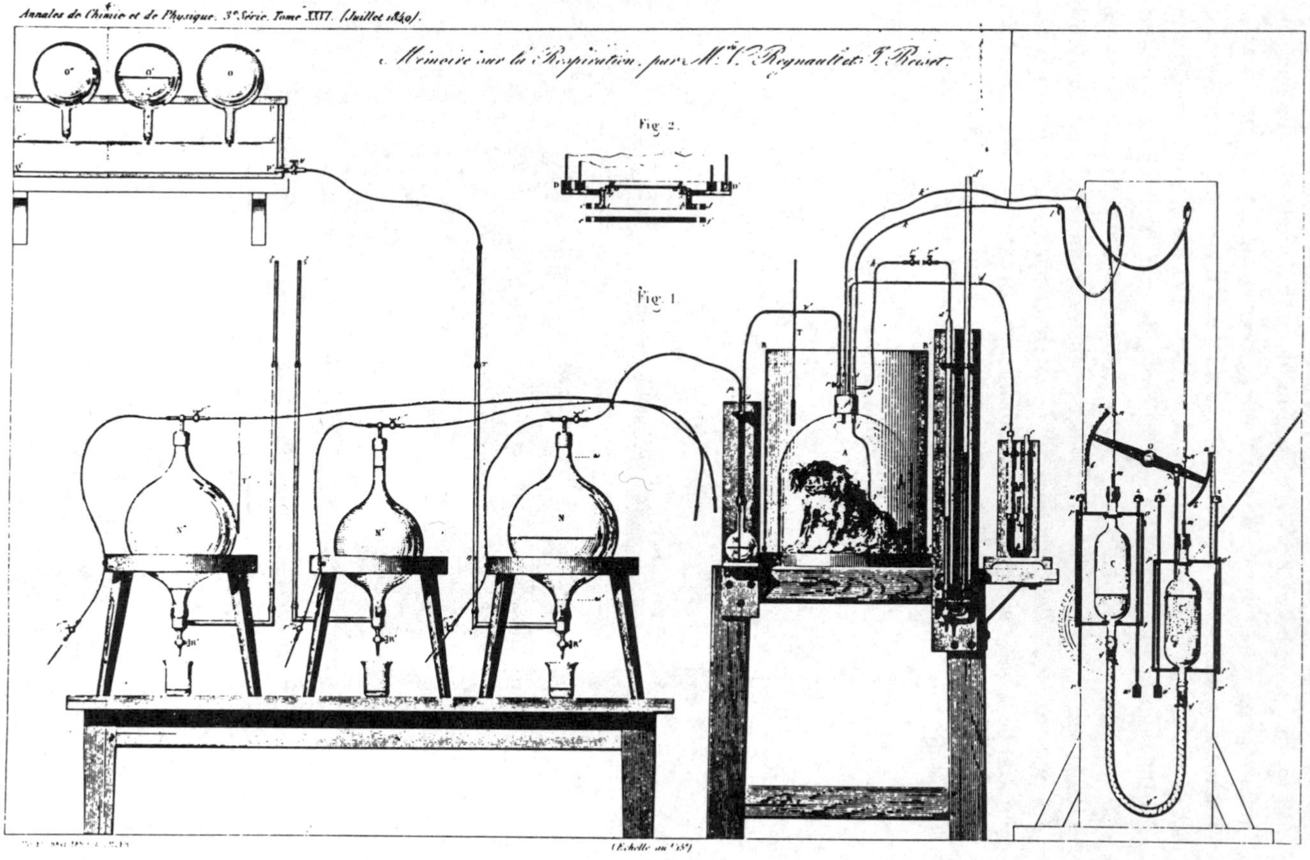

formed from the same volume of oxygen, any oxygen in excess of the carbon dioxide produced must have been used to oxidise hydrogen. By simultaneously measuring oxygen consumption and carbon dioxide production they could calculate the amounts of carbon and hydrogen oxidised, and hence the amount of heat produced in their combustion; this should equate with the physical output of heat from the animals. Dulong and Despretz each built very similar calorimeters in which the heat of their animals (rabbits) was transferred to a surrounding water bath, whilst at the same time the respired air was collected for analysis. These were the first true respiration calorimeters, although that name was not yet in use. Dulong's apparatus (Lefèvre, 1911) is illustrated in Fig. 2.1. Both Dulong and Despretz found that the observed heat was some 20 % higher than that calculated from the gas volumes. They concluded that oxidation could not be the sole source of animal heat.

A further fundamental scientific advance, vital to the subsequent development of calorimetry, was made in 1842 when the German Robert Meyer, a ship's doctor, formulated the Principle of Conservation of Energy, now more generally called the First Law of Thermodynamics. This generalised law may be expressed in many specific ways including the Energy Balance Equation (equation (1.1) and Fig. 1.1) which relates heat output from an animal to the chemical energy of food and excreta; this is the underlying principle of Indirect Calorimetry. Another expression of the Principle of Conservation of Energy is the Heat Balance Equation:

$$
\begin{aligned}
\text{heat elimination} = {} & \text{heat lost by radiation} \\
& +\text{heat lost by convection} \\
& +\text{heat lost by conduction} \\
& +\text{heat lost by evaporation}
\end{aligned}
\tag{2.1}
$$

This equation provides the basis for Direct Calorimetry. The Law of Constant Heat Summation, although postulated before the Principle of Conservation of Energy, is in fact yet another expression of the latter (Kleiber, 1961, p. 123) and a particularly useful one in aiding the understanding of calorimetric principles.

Regnault & Reiset (1849) were the first to describe a closed-circuit system for measuring respiratory gases. Fig. 2.2, reproduced from their original paper, shows their apparatus in delightfully meticulous detail. The animal was introduced into the bell-shaped chamber through an aperture in the base, as detailed in the inset. The reciprocating pipettes on the right, containing potassium hydroxide solution, pulled air in and out of the chamber and absorbed carbon dioxide from it. The oxygen used up by the animal was replenished from one of the three volumetric flasks on the left. Regnault & Reiset studied the respiration of various small animals and birds in this apparatus and reported both their successes and disasters in full and frank detail. A scientific account written today would surely omit reporting on the experiment in which they fed 'une pâtée abondonte de pain et d'eau grasse' to a hungry dog immediately before placing it in the chamber and: '...Lorsqu'il fut renfermé dans la cloche, le chien eut une indigestion et vomit toute la nourriture qu'il venait de prendre, mais il l'a ravala immédiatement. Un second vomissement suivit bientôt le premier; les matières vomies furent de nouveau mangées, avec avidité, et le chien ne manifesta plus aucun signe de malaise pendant le reste de l'expérience.' There was also a pigeon that laid an egg in the chamber and then ate it! Despite these tribulations Regnault & Reiset found that the ratio of carbon dioxide produced to oxygen consumed varied according to the type of food eaten rather than according to the species of animal. They also reported a very small volume of nitrogen given off by most animals. Presumably some systematic error, perhaps associated with lack of temperature control of the oxygen reservoir, contributed to this erroneous conclusion. They made no attempt to estimate heat output from their measurements. On the contrary, they thought this impossible, and they refuted the hypothesis of Lavoisier, Dulong and Despretz that in the combustion of carbon the portion of the oxygen consumed but not recovered as part of the exhaled carbon dioxide had served to transform hydrogen into water. In some of their experiments, Regnault & Reiset had actually found more oxygen in the exhaled carbon dioxide than had been supplied by respiration. 'This fact alone' they claimed 'demonstrates the inexactitude of these hypotheses, and frees us from discussing them further.' Although this abrupt dismissal and the interpretation of their results are both now seen to be incorrect, there can be no doubt as to the value of their work in the development of animal calorimetry. The principle of the Regnault–Reiset apparatus is essentially the same as that used in closed-circuit respiration chambers of the present day. Its main defect was its dependence on temperature and although Regnault & Reiset attempted to use a larger version for sheep, and one for horses was also built by Zuntz (1905), these were found to be inaccurate and the method was not applied again to large animals until after developments in direct calorimetry had led to large chambers with very precise control of temperature.

Regnault & Reiset's results were reinterpreted by Bischof & Voit (1860) who had analysed the chemical composition of food and excreta. They noted that when meat was fed to dogs virtually all the nitrogen from the meat was excreted in the urine; but starved dogs also excreted a high level of nitrogen in the urine with no nitrogenous intake. They concluded that during starvation the dog lived by 'consuming its own flesh'. To study this problem further Voit felt it necessary to measure the total turnover in the body of the elements carbon and nitrogen. He encouraged Pettenkofer to build an apparatus for this purpose in their laboratory at Munich. Pettenkofer's (1862) was the first open-circuit respiration chamber. The quantity of air drawn through the chamber was measured by a gasmeter, and a sample of that air through another gasmeter and then through bottles of sulphuric acid and barium hydroxide solutions for absorption of water vapour and carbon dioxide respectively. An extremely sensitive balance was used to measure the weight changes over a given time of the subject (a man) and of the absorbers; the weights of food and water consumed and of urine and faeces excreted were also measured. The weight of oxygen consumed was then calculated from the weight changes (subject before plus food and water, less subject after plus urine, faeces and carbon dioxide and water vapour absorbed). The total amounts of carbon and nitrogen were calculated from the weights and chemical compositions of the food, urine and faeces. In this way Pettenkofer & Voit (1866) were able to determine the proportions of carbon, nitrogen and oxygen involved in the metabolism of their subjects under different dietary situations. They were able to demonstrate that metabolism could be fully interpreted in terms of the oxidation of three types of food substance, protein, fat and carbohydrate. An example of how the results of one of their experiments was calculated in full may be found by referring to Lusk (1928) pp. 26–30. This is the so-called carbon–nitrogen balance method of calculating respiratory metabolism. It will be noted that Pettenkofer & Voit *calculated* the rate of oxygen consumption; they were unable to measure it directly with their apparatus. This refinement was added in the calorimeter of Atwater & Benedict (1905).

Rubner, also working in Voit's laboratory at Munich, studied the quantitative relationships between different types of food consumed and the heat evolved. The calorific value of different foods and of urea, when burned in a bomb calorimeter, were by this time known, and Rubner (1902) demonstrated his 'Isodynamic Law' – that protein, fat and carbohydrate may replace each other in heat production in the body in accordance with the heat-producing value of each food stuff in a bomb calorimeter. He went on to estimate the exact amount of heat per unit weight of each substance metabolised (4.1, 9.3 and 4.1 cal/g for protein, fat and carbohydrate). The values for metabolism of fat and carbohydrate were the same as those found in the bomb calorimeter, because the end products were the same in both instances; however the value for protein was less than that determined on the laboratory bench because in the body the protein was only partially oxidised and some of its chemical energy was converted into the chemical energy of products excreted in urine and faeces.

Later in his own laboratory at Marburg, Rubner built a direct calorimeter in which he actually measured the heat given off by a dog (Rubner, 1894). The animal chamber was a double-walled structure of copper sheet with an air gap in between; the whole being surrounded by a constant temperature water jacket. Heat from the dog passing through the surfaces slightly raised the temperature of the inner wall and therefore the mean temperature of the air in the gap, whose change in volume was recorded by a 'spirometer'. The effects of water temperature fluctuations were compensated by a second 'correction-calorimeter' consisting of a series of interconnected copper boxes contained within the same water jacket and having the same total volume as that of the air gap surrounding the chamber proper. The animal chamber was also connected to a Pettenkofer-type system for respiratory analysis.

In a series of experiments on dogs given meat and fat diets and starved, Rubner found complete agreement (within $\pm 1\%$) between the heat calculated indirectly from the Pettenkofer system and the heat actually recorded by his calorimeter. He thus proved (Rubner, 1894) what he called 'the law of maintenance of energy in the animal body'. Atwater & Benedict (1903) confirmed this finding in a series of experiments on men. Nowadays we see such proof as unnecessary, but, prior to the days of Rubner and Atwater, it was not appreciated that conservation of energy is a fundamental principle of nature applicable to all things whether animate or inanimate.

Rubner's calorimeter was an isothermal one, i.e. the temperature inside remained steady and the heat generated within was estimated from the very small temperature rise it generated as it flowed through the surfaces. This system was combined with an open-circuit Pettenkofer-type system for respiratory analysis. Atwater's calorimeter for man was an adiabatic or heat-sink calorimeter. The principle had first been introduced by d'Arsonval (1886) who had built a small calorimeter for rabbits. A liquid-cooled coil inside the chamber, together with the subject, was so regulated that the heat absorbed by it exactly balanced out the heat produced by the subject. Thus the net heat produced inside the chamber was always zero and there was no heat flow through the walls. The rate of heat extraction via the coil was estimated from the temperature increase and quantity of the cooling liquid. The original Atwater calorimeter (Atwater & Rosa, 1899), built at Middletown, Con., also employed the open-circuit Pettenkofer system for indirect estimation of heat production using the carbon–nitrogen method for calculation. It was Atwater & Rosa who conceived the term 'respiration calorimeter' to indicate that their apparatus consisted of a respiration chamber which did not of itself measure heat, and a calorimeter which did. After it had been established that the estimate of heat production possible from respiratory analysis did in fact agree with the direct measurement of heat, the term indirect calorimeter came to be applied to respiration chambers. The name has stuck, although purists will continue to claim that it is bad terminology.

The Atwater calorimeter was rebuilt several times and successive improvements are described in a series of bulletins issued by the United States Department of Agriculture Office of Experimental Stations, and later by the Carnegie Institution of Washington. By 1905 they had started to recirculate the exhaust air after first removing the carbon dioxide and water vapour in absorbers, and replacing the oxygen which had been consumed from a gas cylinder. They thus converted the measurement of respiratory gases to the closed-circuit system of Regnault and Reiset, so enabling heat production to be calculated by the respiratory quotient (RQ) method instead of from carbon–nitrogen balance. The meticulously detailed description of this apparatus (Atwater & Benedict, 1905) makes fascinating reading and demonstrates the profound appreciation that these early workers had of the potential errors and precautions necessary for accurate direct and indirect calorimetry; an appreciation that is manifestly lacking in many much later expositions on the subject. It is interesting to note that the two main forms of direct animal calorimetry practised today, isothermal and adiabatic, or temperature gradient and heat sink, both originated at about the same time and both were perfected through the ingenuity of men, Rubner and Atwater, whose early scientific training had been under the direction of Carl Voit in Munich.

Mention should also be made of two other types of direct calorimeter built in the nineteenth century. In one, the heat of the animals was transferred to a fixed mass of water whose temperature was recorded. Despretz (1824) and Dulong (1841) had used this method and it was employed in the bath calorimeter of Liebermeister which was criticised and improved by Lefèvre (1911). Lefèvre also originated the convection or 'air' calorimeter (Fig. 2.3), in which the subject was confined in a ventilated tunnel. Heat output was estimated from the flowrate and temperature increase of the airstream. Lefèvre (1911) wrote a very comprehensive review of calorimetry systems used up to that time.

The ratios between the volume of oxygen consumed, the volume of carbon dioxide produced and the quantity of heat produced when different foods were burned were known from chemical analysis and bomb calorimetry. The ratio of carbon dioxide to oxygen, which Pflüger named

Respiratory Quotient (RQ), was known to be 0.7 for fat, 1.0 for carbohydrate and about 0.83 for protein. The quantity of protein metabolised
could be estimated from the nitrogen excreted in the urine and hence also
the volumes of oxygen and carbon dioxide involved in nitrogen metabolism. The ratio between the remaining carbon dioxide and oxygen, the so-
called non-protein RQ, could be used to quantify the metabolism of heat
produced by fat and carbohydrate. This type of calculation (the RQ
method) was much easier to do than the carbon–nitrogen calculation, but
depended on knowing the oxygen consumption; but only the Regnault–
Reiset type of respiratory apparatus recorded the rate of oxygen consumption, and this was at that time not very accurate. A simple but
accurate method of measuring both oxygen and carbon dioxide for small
animals was the open-circuit gravimetric apparatus of Haldane (1892)

Fig. 2.3. The 'air calorimeter' of Lefèvre.

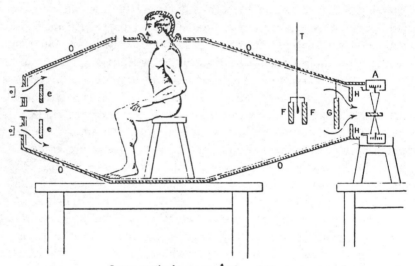

Coupe verticale, montrant le sujet avec son capuchon.

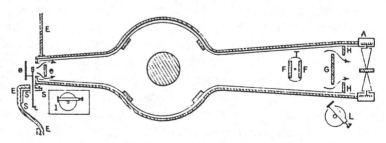

Projection horizontale de l'appareil.

which is illustrated in Fig. 2.4. Room air was continuously drawn through the system of absorption bottles and the animal chamber by a pump. Carbon dioxide was removed in absorber 1 and water vapour in absorber 2. Any water vapour given off by the animal in the chamber (*Ch*) was removed by absorber 3, and any carbon dioxide by absorbers 4 and 5. Water vapour produced by the animal was found from the weight increase of absorber 3, and carbon dioxide from that of 4 and 5 combined. Since neither hydrogen nor carbon had been lost or gained by the combined bottles Ch, 3, 4 and 5 the net weight gain of these represented the oxygen consumed by the animal. This system was simple and effective; it relied only on an accurate balance and whether the animal could be induced to remain still during weighings; and it lent itself ideally to calculations by the RQ method. It was, however, not an automatic system and it is not much used now.

One other indirect chamber system originating in the nineteenth century is the confinement method of Laulanie (1894). The animal was confined in a sealed chamber and the changes in the composition of the air inside were estimated from analysis of samples taken at intervals. Obviously the chamber could only be operated for a limited period. This method was not greatly used, but was revived by Charlet-Lery (1958) for pigs and more recently has been developed in the form of large animal systems (Blaxter, Brockway & Boyne, 1972; Aulic *et al.*, 1983) in which measurement is made nearly continuous by periodically flushing out the chamber and recharging it with fresh air.

At the start of the twentieth century emphasis had been placed on proving the agreement between direct and indirect calorimetry. Once this had been established by the work of Rubner on dogs and Atwater on men,

Fig. 2.4. Haldane's gravimetric apparatus for measuring respiratory gaseous exchange of laboratory animals.

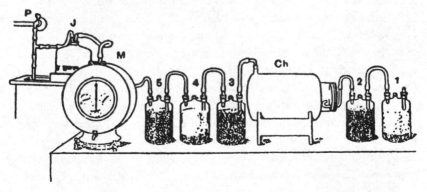

the main thrust was placed on the development of indirect rather than direct systems. A number of large respiration chambers were built to study the food energy requirements of farm animals; some of the more notable were those of Armsby (1904), Møllgard & Anderson (1917), Ritzman & Benedict (1929), Mitchell & Hamilton (1932), Kleiber (1935) & Marston (1948). These all worked on the open-circuit principle of Pettenkofer but with the important addition which had been first introduced by Sonden & Tigerstedt (1895) that aliquot samples of outlet air were collected, with subsequent analysis for the concentration of oxygen as well as carbon dioxide and methane. Gas analysis was entirely by chemical means. There were numerous types of apparatus (Lunge & Ambler, 1934) including that of Haldane (1898) and later modifications of it which are still in use today.

For indirect measurement of heat production of humans the main development was of portable systems which would eliminate the restriction of a respiration chamber. Various workers had attempted, during the nineteenth century, to collect and analyse respiratory gases of subjects breathing into mouthpieces or facemasks; an interesting review of these early methods was written by Carpenter (1933a). In the Tissot (1904) spirometer, all the air expired by a subject was collected in an inverted metal bell suspended and counterbalanced over a water seal. The subject wore a facemask equipped with lightweight valves to separate inspired from expired air. A logical development from this was the Douglas (1911) bag, in which the collection vessel was a rubberised bag carried on the back of the subject. Such total collection methods were of course limited by the size of the collection vessel required to contain all the air breathed out. To overcome this, mask methods were developed using the Regnault–Reiset principle of recirculation and absorption of carbon dioxide and water vapour. A spirometer connected to the system was used by Krogh (1916) to estimate oxygen consumption from the rate of diminution of total gas volume. This method was further developed by Benedict, Roth and Collins (Benedict & Collins, 1920; Roth, 1922) and a version of their apparatus was even adapted for use on animals up to the size of cows at the University of Missouri (Brody, 1945, pp. 313–19). These closed-circuit mask methods still involved connecting the subject to some static device for measuring the change in volume. An open-circuit apparatus which could be carried on the back and which allowed long-term measurement of energy expenditure during a range of activities, had been developed in Germany by Zuntz *et al.* (1906). The apparatus (Fig. 2.5) comprised a mask with manually operated taps, a dry gasmeter and a mercury tonometer for collection of expired air samples. It was the forerunner of the later instrument developed by Kofrayni and Michaelis which later became known as the Max Planck respirometer (Müller & Franz, 1952). This is an open-circuit mask system with a lightweight dry gasmeter and an aliquoting device which collects a sample from each expiration into a small sampling bag. The whole apparatus is designed to be carried in a

Fig. 2.5. The first portable respirometer. (Designed by Zuntz *et al.*, 1906.)

small back-pack. An excellent account of portable systems from the Tissot spirometer up to the Max Planck respirometer and including much practical advice on their use is included in the laboratory manual written by Consolazio, Johnson & Pecora (1963, pp. 12–50).

During the first half of the twentieth century direct calorimetry was generally held to be too complex and expensive for use where heat output could alternatively be estimated by indirect methods. Not many direct calorimeters were in use but there were some notable exceptions. The Atwater–Benedict calorimeter at Middletown, Con., in which the early verification of agreement between direct and indirect calorimetry was made has already been mentioned. Benedict & Carpenter (1910) later built a series of man-sized calorimeters in their new laboratory at Boston including a large group chamber capable of holding 12 individuals in bed; and Williams, Lusk & Dubois built similar calorimeters for investigation of metabolic disorders at Cornell University and the New York Medical Center (Lusk, 1928, pp. 69–74). All of these were adiabatic (heat-sink) direct calorimeters and incorporated facilities for simultaneous measurement of respiratory exchange by either closed- or open-circuit principles. These machines were built before the days of automatic control and analysis. They were operated by observers (usually three or four) who continuously switched indicating meters to observe different temperatures, pressures, etc., and made manual adjustments to valves and rheostats to maintain constant conditions in and around the chamber. Gas samples had to be collected manually over mercury and transferred to volumetric apparatus for chemical analysis; in fact nearly all measurements and analyses were laborious operations. The wealth of detailed measurements recorded, of accurate deductions made and new knowledge accumulated by these pioneers provides a legacy which we today with our autoanalysers, data-loggers and computers find hard to live up to.

Research into the physiology of body temperature regulation during the period between the two world wars led to the development of two other direct calorimeters of special interest. Murlin & Burton's (1935) calorimeter embodied both heat-sink and gradient principles. The chamber was a glass cylinder 1.93 m long by 0.58 m in diameter and much of the (human) subject's non-evaporative heat was removed by a water cooled heat exchanger also made of glass. Heat escaping through the walls was estimated, using a nickel differential thermometer, from the temperature difference across a 3.75 cm-thick layer of 'kapok-like' insulating material surrounding the chamber. This chamber which can be regarded as the immediate forerunner to the modern gradient-layer calorimeter was also operated as a closed-circuit respiration chamber. There was good agreement between direct and indirect measurements. The partitional calorimeter (Winslow, Herrington & Gagge 1936*a*, *b*; Gagge, 1936) was an ingenious device for estimating each parameter of the heat balance equation (equation (2.1)) without actually measuring any of them directly. The subject sat on a balance which recorded his weight loss, and was surrounded by reflecting walls whose radiant temperature could be altered. Respiratory measurements using a mask system allowed calculation of heat production, metabolic water and metabolic weight losses. Temperature measurements of the deep-body, skin and surroundings allowed calculation of body-heat storage and heat losses by radiation and convection. Evaporative heat loss was calculated from those weight losses attributable to evaporation of moisture. An ingenious series of experiments was devised to calibrate the instrument after which it was found to be remarkably accurate. Its advantage was a very high speed of response; its disadvantage, stressed by the authors, was that the heat flow coefficients so painstakingly determined were applicable only to their special conditions (a semi-reclining subject confined in the slow-moving airstream of their calorimeter and with no drip-off of sweat from the body).

Direct calorimeters for working with large animals were at that time even more rare than those for humans, but adiabatic chambers were built at Cambridge for work on sheep and dogs (Hill & Hill, 1914) and on pigs and young cattle (Capstick, 1921; Deighton, 1926). A new principle, differential calorimetry, was also tried. Two identical cylindrical chambers were mounted side by side in a large duct through which preconditioned air was passed by a fairly elaborate ventilation system. One chamber contained the animal and the other an electric heating element. The current in the heating element was continuously adjusted so that the rise in temperature of both chambers was the same. The rate of delivery of heat to the element in one chamber was then taken to be equal to the non-evaporative heat output of the animal in the other. The advantage of this system was its potentially high speed of response to rapid changes in heat output, because the chambers could be constructed of very thin metal. This method was successfully used for various small animals and geese by Benedict & Lee (1937) and for poultry by Deighton (1939).

The major advances in electronics and instrumentation made during the Second World War have transformed the practice of animal calorimetry. Electronic methods of gas analysis, automatic control systems and computer controlled data logging and processing have combined vastly to increase the ease, accuracy and speed of collecting information, although the basic calorimetric principles used are still those invented about 100 years ago. The biggest advance in calorimetry itself has been the introduction by Benzinger & Kitzinger (1949) of thermoelectric measurement of the mean temperature gradient across a thin plastic layer. In their gradient-layer calorimeter for man (Benzinger *et al.*, 1958) the temperature gradient was measured not across a still-air gap or kapok lining, but across a thin plastic (plexiglass) sheet. A series of very many interlaced thermocouples all connected in series measured the mean temperature gradient across this thin sheet which formed a complete internal lining over all surfaces of the chamber. The output voltage from the thermocouple network gave an accurate and rapidly responding measurement of the non-evaporative heat output from the subject. Using a series of heat exchangers lined with similar gradient layers, which he called platemeters, Benzinger was also able to measure the evaporative heat losses from his subjects. Close control of temperature of both the chamber and platemeters was essential. A number of large gradient-layer calorimeters have since been built for measurements on cats, dogs and monkeys (Prouty, Barret & Hardy, 1949; Lawton, Prouty & Hardy, 1954; Hammell & Hardy, 1963), man (Tiemann, 1969; Young & Carlson, 1952; Spinnler *et al.*, 1973), sheep (Pullar, 1969) and cattle (McLean, 1971). Most of these calorimeters were painstakingly constructed by the investigators themselves. Nowadays prefabricated gradient-layer panels and even whole calorimeter chambers are available commercially.

The heat-sink principle has also been combined with modern instrumentation; amongst calorimeters of this type that have been described in detail are those for pigs (Mount *et al.*, 1967) and man (Dauncey, Murgatroyd & Cole, 1978; Jacobsen, Johansen & Garby, 1982; Murgatroyd, 1984). The air flow–temperature rise or convection principle for direct calorimetry has also been revived for man (Visser & Hodgson, 1960; Carlson *et al.*, 1964; Tschegg *et al.*, 1979; Snellen, Chang & Smith, 1983) and pigs (Kelly, Bond & Heitman, 1963). This method relies on very accurate measurement of the temperature increase of air ventilating a tunnel in which the subject is placed.

Large respiration chambers for studying the efficiency of food conversion by farm animals have been constructed in many countries and many of these were described at the first Symposium on Energy Metabolism of the European Association of Animal Production held at Copenhagen in 1958 (EAAP, 1958). Further calorimetric developments have been described at subsequent symposia held since at three-yearly intervals (EAAP, 1961–82). At the Fifth Symposium a list of laboratories with facilities for measuring energy metabolism was issued which contained over 70 entries world wide. The Regnault–Reiset closed-circuit principle has been employed in chambers for sheep (Blaxter, Graham & Rook,

1954; Alexander, 1961) and cattle (Wainman & Blaxter, 1958*a*) and the confinement method has been used in a chamber for cattle (Blaxter *et al.*, 1972). But most of the more recently described chambers for farm animals have made use of the Pettenkofer open-circuit system modified to include on-line analysis of inlet and outlet airstreams (Thorbek & Neergaard, 1970; Verstegen *et al.*, 1971; Vermorel *et al.*, 1973; Young, Kerrigan & Christopherson, 1975; Thomson, 1979; Cammell *et al.*, 1981). Lundy, McLeod & Jewitt (1977) have described a multicage system for poultry in which a common gas analysis module is switched sequentially between exhaust gases from each chamber in turn. A multicage system for rats has been described by Tobin & Hervey (1984).

Open-circuit chambers for long-term measurements on humans have been described by Dauncey, Murgatroyd & Cole (1978) and by Jequier & Schutz (1983). Facemasks or head hoods (Kinney *et al.*, 1964) have sometimes been used to avoid the slow response of open-circuit respiration chambers, which results from the long time taken for gas concentration levels to equilibrate in the necessarily large volume of the apparatus. But with the use of computer controlled data-logging systems it is now possible to process the data so as to effectively reduce the response time, even of a large chamber, to only a few minutes (McLean & Watts, 1976).

Electronic advances have also made possible the development of improved lightweight portable instruments for measurement of respiratory exchange. In the Integrating Motor Pneumotachograph or IMP (Wolff, 1958*a*) a pneumatic signal proportional to the volume flow of expired air is amplified sufficiently to drive a lightweight motor and so to activate an aliquoting device. The integrated volume of expired air is obtained from the number of revolutions of the motor, but the gas sample still has to be analysed at a later date. The Metabolic Rate Monitor (Webb & Troutman, 1970) is an open-circuit mask system ventilated by a lightweight blower which is servo-controlled so as to ensure a constant decrement in the oxygen concentration of the ventilating air. Airflow is estimated from the speed of the blower to which it is linearly related. This instrument eliminates the need for collection of gas samples, and because its basic measurement is the product of flowrate and oxygen concentration difference (see Chapter 4) it is potentially a very accurate means of measuring heat production.

The ultimate level of portable calorimetry is perhaps the suit calorimeter of Webb, Annis & Troutman (1972). The (human) subject wears an undergarment made of water-cooled plastic tubing next to the skin, and over this he wears insulated outer clothing. The temperature of the cooling water is controlled so as to maintain the subject in comfortable thermal balance, and may be adjusted to accommodate quite heavy work rates by the subject without sweating being induced. It is possible for the subject to wear the suit calorimeter and at the same time have his heat production recorded by the metabolic rate monitor (Webb & Troutman, 1970). Combined use of the two systems makes it possible to analyse all heat balance parameters simultaneously over 24 h long periods whilst the subject remains free from the restraints and confinement of a calorimetric chamber. Previously a suit calorimeter embodying the gradient-layer principle had been described by Young, Carlson & Burns (1955). The advantage of their suit was that it could distinguish between heat loss rates from different body regions, but its use was limited in time by the accumulation of sweat inside the garment.

The difficulty of getting animals to accept facemasks and the total unsuitability of mask methods for studies on animals whilst grazing has led to the development of methods involving tracheostomy, in which normal breathing through the nasopharynx is bypassed. Measurements have been made in this way on sheep (Young & Webster, 1963), pigs (Charlet-Lery & Laplace, 1976) and cattle (Flatt *et al.*, 1958).

A quite different approach to the estimation of heat production of grazing animals was introduced by Corbett *et al.* (1971). They continuously infused radiocarbon in the form of $NaH^{14}CO_3$ into sheep. After approximately three hours the level of ^{14}C in the body fluids had equilibrated and it was possible to estimate the rate of carbon dioxide production by the animal from the ratio of the rate of infused activity to the specific activity of the urine. The method has also been used for grazing cattle (Young, 1970). It gives an estimate of heat production of about 4–8 % accuracy (Whitelaw, Brockway & Reid, 1972).

A new technique for estimating carbon dioxide production involves the use of doubly labelled water in the form of $^2H_2^{18}O_2$ (Lifson, Gordon & McClintock, 1955). When taken orally this material quickly equilibrates throughout the pool of body water. Since both isotopes are stable they remain in the body until washed out by normal metabolic processes. But whereas deuterium is only lost as water the ^{18}O is also lost as carbon dioxide and therefore becomes diluted more rapidly. The rates of dilution of both isotopes may be calculated from the results of mass-spectrometric analysis of two samples of urine passed at different times; from these the rate of carbon dioxide elimination may be calculated. The method was originally very expensive due to the cost of the isotopes and its use was confined to small animals. Recently, however, increasing sensitivity of mass spectrometers have reduced the amount of isotope required and the method is being used for measurements on humans (Schoeller & Webb,

1984; Schoeller & van Santen, 1982; Coward *et al.*, 1984). Agreement with mask methods for heat production is of the order of 10%. The doubly labelled water technique, like the radiocarbon method, measures only the carbon dioxide production which is not as accurate an indication of heat production as is oxygen consumption. However, the doubly labelled water method has the great advantage that it is totally non-invasive and allows the subject to be completely free of all masks, chambers or measuring instruments. It has great potential for providing reasonably accurate estimates of heat production of human subjects going about their normal daily tasks or of animals in the field.

Calorimetry has thus developed in many forms during the two hundred year period since Crawford and Lavoisier made their first measurements of animal heat. However the main advances have been made in the last one hundred years. An interesting account with photographs or drawings of many of the early calorimeters has been published very recently (Webb, 1985).

3

Calorific equivalents

Measurement of the rate of metabolic heat production by indirect calorimetry depends on two assumptions:

(1) It is assumed that the end result of all the biochemical reactions which occur in the body amounts effectively to the combustion or synthesis of three substances – carbohydrate, fat and protein.

(2) It is assumed that for each of these substances, when it is oxidised in the body, there are fixed ratios between the quantities of oxygen consumed, carbon dioxide produced and heat produced.

By any standards these are gross oversimplifications. The first completely ignores the metabolism of minerals, which represent nearly 7% of the total bodyweight. Although the mineral status of the body may be nearly stable in unproductive adults, major changes occur during phases of skeletal growth, pregnancy and lactation. The second assumption implies a uniformity in the properties of fat and protein that would seem improbable in view of their extremely varied chemical compositions. The only real justification for the assumptions is that over many years of practical application it has been found that indirect calorimetry is remarkably consistent and in close agreement with direct calorimetry.

The first step in evaluating a theoretical basis for indirect calorimetry must be to establish the values of the calorific factors for carbohydrate, fat and protein. These are determined from the results of combustion of the materials in a bomb calorimeter.

3.1   Bomb calorimetry
3.1.1   *Adiabatic bomb calorimetry*

As the name of the calorimeter implies, the energy content of a sample is determined by combustion in a closed system which can neither gain heat from, nor lose heat to the outside environment. The adiabatic property of the calorimeter is achieved by surrounding the system with a water jacket whose temperature is controlled to match precisely that of the system. The temperature rise of the system is thus a true measure of the energy liberated by the combustion of the sample.

The first proposal for an adiabatic bomb calorimeter was probably made by Holman in 1895; however much of the real progress in its early development can be attributed to Richards (1905, 1909). Although he recognised that heating coils could be used to control the temperature of the jacket to match that of the system, he preferred to use a closely controlled exothermic chemical reaction within the jacket. It was left to Benedict & Higgins (1910) to introduce the first serious use of a heating coil to control the temperature within the water jacket. However, none of these methods could attain the rapid and close matching of temperatures in the system and jacket necessary for truly accurate adiabatic bomb calorimetry. In 1916 Daniels finally achieved this by placing heating elements in the water jacket and making the whole jacket become a conductor. Although the jacket temperature was still manually adjusted to that of the system, it could be changed rapidly and matching to within ± 0.01 °C achieved. In 1957 Raymond, Canaway & Harris automated this temperature control system and produced an adiabatic bomb calorimeter which became the basis for the current commercially available systems.

Fig. 3.1 shows the principal features of the calorimeter. It is composed of three main components: the stainless steel bomb itself; a can (the calorimeter vessel) containing a known mass of water; and an outer water jacket whose temperature can be controlled by a heating and cooling system.

The first requirement is to determine the combined heat capacity of the water, the bomb, the calorimeter can and its support, i.e. how much energy must be released to raise the temperature of the system by 1 °C (its water equivalent). For this purpose thermochemical grade benzoic acid, containing 26.442 kJ of energy per gram, is used. Dry benzoic acid is placed in a crucible (typically of nickel chromium) and compressed to form a pellet; the weight of the pellet is determined to an accuracy of 0.1 mg. The crucible is then mounted in a holder fixed to the lid of the bomb. The holder additionally acts as an electrode; a second electrode also projects from the bomb lid. The electrodes are connected by a length of nickel chromium or platinum wire to which is tied a piece of cotton thread; the other end of the thread is placed firmly in contact with the sample (both lengths are constant from determination to determination). The lid is placed on the body of the bomb and held securely in place by the closure ring. Oxygen is introduced via a valve in the lid and the pressure in the bomb raised to 3 MN m^{-2} (30 atmospheres). The bomb is placed in the calorimeter vessel which contains a known weight of water (the weight is usually kept constant, to a precision of ± 0.5 g, from determination to determination). The temperature of the water (typically 22–24 °C) is several degrees above that of the cold tap water or room temperature.

The bomb and calorimeter vessel are mounted on an insulating tripod inside the calorimeter and the lid closed. Attached to the lid and projecting into the water in the vessel are: a Beckman thermometer, accurate to 0.001 °C; a water stirrer; and a thermistor. The lid also contains the firing circuit contact which connects to one of the electrodes in the bomb to initiate combustion. An air space separates the calorimeter vessel from the surrounding water jacket (of which the lid is part); immersed in the water jacket is a powerful heating element and a second thermistor which has been carefully matched to the first. Water is circulated through the jacket and an external cooling circuit. The thermistors, which sense the temperatures of water in the vessel and jacket, form two arms of a Wheatstone bridge circuit; the two other arms consist of two high-stability resistors and a potentiometer used to balance the bridge when the calorimeter is initially set up. Thereafter if there is a change in the temperature of either

Fig. 3.1. The adiabatic bomb calorimeter.

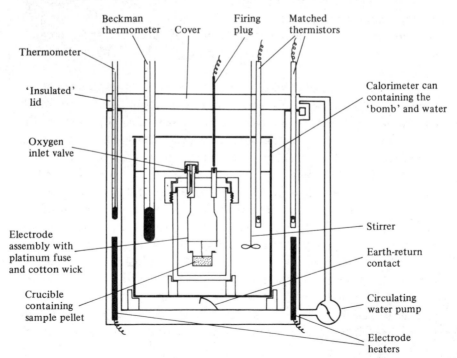

thermistor, its electrical resistance alters and the bridge is unbalanced. In practice the change is usually brought about by the lower temperature of the water in the jacket with respect to that in the calorimeter vessel. The signal from the bridge is then amplified and used to drive a relay which switches current to the heating elements. The circuit ensures that the temperature of the water in the jacket remains equal to that in the calorimeter vessel. This provides dynamic insulation and gives the calorimeter its adiabatic properties.

After loading the bomb and vessel into the calorimeter the water temperature in the vessel is allowed to stabilise (typically 22–24 °C). The temperature is measured by the Beckman thermometer. The pellet of benzoic acid is then ignited in the bomb by triggering the electrical firing circuit. Heat is liberated as the sample is completely burned in the high oxygen atmosphere and the temperature of the bomb and the surrounding vessel of water increases. The temperature of the water jacket increases in parallel as the heating elements are activated by the Wheatstone bridge circuit. Once a new steady state temperature has been achieved it is measured by means of the Beckman thermometer. The system is then disassembled and the water jacket cooled. The procedure takes 20–30 min to complete.

The heat capacity can be determined by carrying out replicate analyses in which samples of benzoic acid containing *c.* 30 kJ are burned; the capacity is obtained by dividing the energy content of the benzoic acid sample by the temperature rise produced by its combustion. A preferable alternative is to burn a series of samples containing amounts of energy in

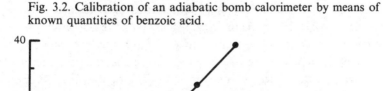

Fig. 3.2. Calibration of an adiabatic bomb calorimeter by means of known quantities of benzoic acid.

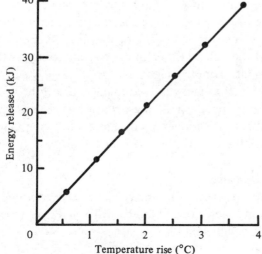

the range 0–40 kJ; a regression of the energy liberated on the temperature rise of the system gives the heat capacity (Fig. 3.2).

The heat capacity of each combination of bomb and calorimeter vessel must be determined and if any component of the bomb is changed (e.g. the quantity of water or the length of fuse) the measurement must be repeated. Once the heat capacity is known, samples of other materials can be burned and from the weight of sample and the temperature rise the energy density of the sample can be calculated.

A small correction may be necessary to correct for extra heat released in the bomb when sulphur in the sample is converted to sulphuric acid (under normal conditions it would be converted to sulphur dioxide). In addition nitrogen in the air in the bomb is converted to nitric acid whereas under normal combustion conditions it would not undergo any chemical reaction. Any acids formed in the combustion are determined after washing out the bomb and titrating the washings for total acidity and nitric acid; sulphuric acid is determined by difference. The total heat released is then adjusted for the small contribution of nitric and sulphuric acid formation; in most situations the error involved is sufficiently small to ignore the need for a correction.

Although the method is capable of very high precision and accuracy, it is slow and laborious. However, it is now possible to add micro-processor control to the calorimeter to remove the routine manual monitoring of temperatures and to carry out the calculations. Furthermore several calorimeters can be controlled by a bench-top computer; this can provide automatic triggering of the individual firing circuits once the computer detects that an initial steady state temperature has been achieved. The computer can determine the increased steady state temperature after combustion of the sample. From the sample weight, the heat capacity of the bomb calorimeter and the rise in temperature the energy content of the sample can be calculated by the computer. The operator is left simply to prepare the bombs, to load and unload the calorimeters, and to supply the computer with the appropriate sample weights.

Although adiabatic bomb calorimetry is used mainly for dry or near-dry samples, methods are available for handling wet samples such as milk or urine (Nijkamp, 1969, 1971) and samples which cannot be dried because of their high content of volatile materials (Owens *et al.*, 1969).

3.1.2   *Ballistic bomb calorimetry*

This was a technique developed originally by Fox, Miller & Payne (1959). An improved version is described in full by Miller & Payne (1959) and the system is available commercially.

It was meant to be considerably simpler than the adiabatic calorimeter with a faster throughput of samples. Fig. 3.3 shows the basic design which in many respects resembles an inverted bomb from an adiabatic calorimeter. The sample is ignited in an oxygen rich atmosphere and the energy liberated increases the temperature of the bomb; the temperature rise over the surface of the bomb is, however, uneven. There is no attempt to prevent dissipation of this heat and thus after reaching a peak the temperature of the bomb decreases. The peak temperature is measured by a thermocouple mounted directly in the bomb. The thermocouple is connected to a galvanometer via a series resistance, the value of which is selected to give full-scale deflection of the galvanometer for a known rise in temperature. The bomb calorimeter can be calibrated as before with benzoic acid.

Whilst the apparatus does achieve its goals of simplicity and high throughput there are limitations to its accuracy and precision. Miller & Payne reported that for several materials the mean standard error for six determinations was $\pm 1.1\%$ (range 0.6–1.7). Blaxter (1956a) has argued that a standard error of less than $\pm 0.5\%$ of the energy value is essential in accurate energy balance studies; on the basis of Miller & Payne's data this would not usually be attained with a ballistic bomb calorimeter for most samples unless about 15 replicate analyses were carried out. This has

Fig. 3.3. The ballistic bomb calorimeter.

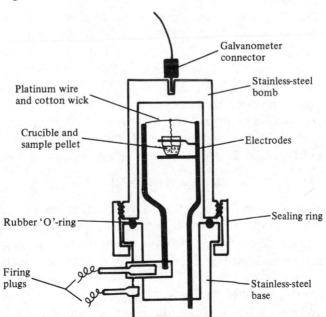

been confirmed by Sibbald (1963). He compared the two calorimeters and reported that to obtain the same precision as an adiabatic bomb calorimeter, six times as many replicate samples would have to be analysed in a ballistic bomb calorimeter. Since the ballistic bomb calorimeter is only 2–3 times faster in determining the energy content of a sample than the adiabatic bomb calorimeter it does not provide a serious alternative to the adiabatic system for accurate energy balance studies.

3.2   The oxidation of foods and excreta

In the following sections a brief description is given of the results obtainable by bomb calorimetry of individual chemical substances representative of carbohydrate, fat and protein, and of the inferences that may be drawn from applying such information to indirect calorimetric measurements. This approach to the subject has been described in many textbooks on energy metabolism and nutrition (e.g. Peters & van Slyke, 1946; Brody, 1945); it is included here for completeness and because it helps to provide an understanding both of the fundamental assumptions of indirect calorimetry and of the significance of respiratory quotient. But the calculation procedure is cumbersome and is now largely superseded. The same conclusions may be drawn from a shorter approach which is described in Section 3.2.5.

3.2.1   *Carbohydrate*

The name carbohydrate implies that its elementary composition consists of carbon and water, $C_x(H_2O)_y$. On oxidation each atom of carbon combines with one molecule of oxygen to form one molecule of carbon dioxide; hence the Respiratory Quotient (r), that is the volumetric (and molar) ratio of carbon dioxide produced to oxygen consumed, is equal to one. This may be illustrated by the chemical formula for the oxidation of glucose (molecular weight 180.2), which on combustion in a bomb calorimeter is found to release 2817 kJ/mole (673 kcal/mole) or $Q = 15.64$ kJ/g (3.736 kcal/g) of energy, which is lost as heat:

$$C_6H_{12}O_6 + 6O_2 \rightarrow 6CO_2 + 6H_2O + 2817 \text{ kJ}$$

(We prefer to use a positive sign here to indicate that heat is a product of the combustion, rather than the conventional negative sign for enthalpy changes.) Since the six moles of oxygen required to oxidise 1 mole of glucose occupy 6×22.41 l (at s.t.p.) the ratio of heat to oxygen consumed (q) is $2817/(6 \times 22.41) = 20.95$ kJ/l (5.01 kcal/l). The quantity of oxygen consumed (a) is $6 \times 22.41/180.2 = 0.746$ l/g of glucose oxidised.

Most carbohydrates in the diet of man and animals consist of saccharides or polysaccharides; the results of combustion of the most common forms are shown in Table 3.1. Although the heat per gram of substance consumed (Q) is variable, the uniformity of the values of the heat per litre of oxygen consumed (q) and of the respiratory quotient (r) suggests that if it were possible to estimate the quantity of oxygen formed or of carbon dioxide produced by carbohydrate metabolism, this would provide an accurate means of estimating the quantity of heat evolved regardless of the type or mix of carbohydrate.

3.2.2 Fat

Fats are triglycerides (triacylglycerols), i.e. compounds of glycerol and three fatty-acids. The fatty-acid composition is extremely variable but the majority of animal fats include high proportions of the saturated acids palmitic (C16), stearic (C18) and the monounsaturated oleic acid (C18, 1). Vegetable fats are rich in unsaturated fatty-acids, particularly linoleic acid (C18, 2). Palmitic acid, for example, has a heat of combustion of 10040 kJ/mol (2398 kcal/mole) or $Q = 39.16$ kJ/g (9.35 kcal/g).

$$C_{15}H_{31}COOH + 23O_2 \rightarrow 16CO_2 + 16H_2O + 10040 \text{ kJ}$$

The respiratory quotient (r) is $\frac{16}{23} = 0.696$ and the ratio of heat to oxygen consumed (q) is 19.48 kJ/l (4.65 kcal/l).

The variation between the calorific factors Q, a, r and q of different fatty-acids is considerably greater than that of different carbohydrates. Nevertheless the value of the factor q is always within limits of 19.4–19.6 kJ/l. The variation is further reduced in practice due to the fact that animal and vegetable fats contain a mixture of different fatty-acids, and the mean value is slightly increased by the contribution of glycerol ($q = 21.2$ kJ/l) which is a minor constituent of all fats. Palmitic acid has been used in the example because its heat output per litre of oxygen consumed

Table 3.1. *Calorific factors for carbohydrates*

	Q Heat of combustion (kJ/g)	a Oxygen consumed (l/g)	q $q = Q/a$ (kJ/l)	r Respiratory quotient
Monosaccharides ($C_6H_{12}O_6$) glucose, galactose, fructose	15.64	0.746	20.95	1.00
Disaccharides ($C_{12}H_{22}O_{11}$) sucrose, lactose, maltose	16.51	0.786	21.02	1.00
Polysaccharides ($C_6H_{10}O_5$)n starch, glycogen, cellulose	17.50	0.830	21.11	1.00

and respiratory quotient are close to the average values found for most animal fats ($q = 19.75$ kJ/l or 4.718 kcal/l and $r = 0.71$).

When a mixture of fat and carbohydrate is oxidised, given the values $q_F = 19.75$ and $r_F = 0.71$ for fat and $q_K = 21.0$ and $r_K = 1$ for carbohydrate, it is possible from measurements of both oxygen consumed and carbon dioxide produced to estimate both the heat output and the composition of the mixture. Suppose, for example, that the quantities of oxygen and carbon dioxide are 100 and 90 l respectively. If x l of the oxygen are used for fat oxidation, then the carbon dioxide formed from fat is $0.71x$ l and that from carbohydrate is $(100-x)$ l. Hence the total quantity of carbon dioxide is $0.71x + 100 - x = 90$, therefore $x = 34.5$ l. The remaining 65.5 l of oxygen must have been used to oxidise carbohydrate. The total heat output is therefore $(65.5 \times 21.0) + (34.5 \times 19.75) = 2063$ kJ (491 kcal). This type of calculation forms the basis for the well known table of values for analysis of mixtures of fat and carbohydrate which was modified by Lusk (1928) from original data of Zuntz and Schumburg. The original table gives values of the calorific factors for respiratory quotients from 0.707 to 1.00 in steps of 0.01, an abbreviated version is given in Table 3.2.

3.2.3   *Protein*

Animal and vegetable proteins consist of a variable mixture of some 19 different amino-acids. Unlike fat and carbohydrate they cannot be completely oxidised to carbon dioxide and water. In land mammals the nitrogen from the protein is mostly converted into urea which is excreted in the urine. Measurement of urinary nitrogen excretion can thus provide an estimate of the quantity of protein metabolised.

Table 3.2. *Analysis of the oxidation of mixtures of carbohydrate and fat.* (*Converted from the table of Lusk, 1928*)

Respiratory quotient	Percentage of total oxygen consumed by Carbo-hydrate	Fat	Percentage of total heat produced by Carbo-hydrate	Fat	Heat per litre of oxygen (kJ)
0.707	0	100.0	0	100.0	19.62
0.75	14.7	85.3	15.6	84.4	19.84
0.80	31.7	68.3	33.4	66.6	20.10
0.85	48.8	51.2	50.7	49.3	20.35
0.90	65.9	34.1	67.5	32.5	20.61
0.95	82.9	17.1	84.0	16.0	20.87
1.00	100.0	0	100.0	0	21.13

Synthesis of urea occurs through a complex cycle of biochemical reactions; but according to the Law of Constant Heat Summation (Law of Hess) the heat evolved is independent of the chemical pathway. We may therefore estimate the heat evolved in converting an amino-acid into urea from the individual heats of combustion. For example, the oxidation of the amino-acid alanine (molecular weight 89.094) is represented by

$$4CH_3CH(NH_2)COOH + 15O_2$$
$$\rightarrow 12CO_2 + 14H_2O + 2N_2 + 4 \times 1623 \text{ kJ}$$

and the oxidation of urea (molecular weight 60.056) by

$$2(NH_2)_2CO + 3O_2 \rightarrow 2CO_2 + 4H_2O + 2N_2 + 2 \times 634.6 \text{ kJ}$$

Subtracting the two equations gives

$$4CH_3CH(NH_2)COOH + 12O_2$$
$$\rightarrow 2(NH_2)_2CO + 10CO_2 + 10H_2O + 5223 \text{ kJ}$$

This equation shows that the oxidation of 4 moles (4×89.094 g) of alanine consumes 12 moles (12×22.41 l) of oxygen and produces 2 moles (2×60.065 g) of urea and 10 moles (10×22.41 l) of carbon dioxide; the heat evolved is 5223 kJ or $q = 5223/(12 \times 22.41) = 19.42$ kJ/l of oxygen (4.64 kcal/l). The respiratory quotient (r) is 0.833.

There are considerable variations between the values of these factors for different amino-acids; also some of the nitrogen is converted into other compounds such as uric acid and creatinine. As with fats, the averaging

Table 3.3. *An example of the calculation of heat production by the traditional method*

During 4 h a subject consumed 57.25 l O_2	
produced 48.30 l CO_2	
produced 1.28 g urinary N_2	
Assume (from Lusk, 1928) that for each g of N_2 from protein:	
O_2 consumption	$= 5.94$ l
CO_2 production	$= 4.76$ l
heat evolved	$= 18.68$ kJ/l O_2
$\therefore$ O_2 for protein oxidation	$= 1.28 \times 5.94 = 7.60$ l
CO_2 from protein oxidation	$= 1.28 \times 4.76 = 6.09$ l
Heat from protein oxidation	$= 7.60 \times 18.68 = 142$ kJ
Non-protein O_2 consumption	$= 57.25 - 7.60 = 49.65$ l
Non-protein CO_2 production	$= 48.30 - 6.09 = 42.21$ l
Non-protein respiratory quotient	$= 49.65/42.21 = 0.850$
kJ per l O_2 (from Table 3.2)	$= 20.35$
Non-protein heat production	$= 49.65 \times 20.35 = 1010$ kJ
Total heat production	$= 1010 + 142 = 1152$ kJ
Rate of heat production	$= 1152 \times 1000/(4 \times 60 \times 60)$
	$= 80.0$ W

effect of the mixture of amino-acids in protein reduces the variation and a set of values can be selected which is reasonably representative of the metabolism of protein to urine.

It is therefore possible, from measurement of the quantity of urinary nitrogen, to estimate not only the amount of protein metabolised but also the amounts of oxygen consumed and of carbon dioxide produced in association with the protein. The remaining (non-protein) parts of the total oxygen and carbon dioxide are associated with fat and carbohydrate metabolism, and their contributions to the total heat may be calculated as already described. An example showing the steps taken in the full calculation is given in Table 3.3.

3.2.4   *Incomplete oxidation*

In herbivorous animals methane gas is produced by bacterial fermentation of cellulose and expelled through the mouth. Formation of methane uses energy which would otherwise be lost by ruminant animals as heat. Consequently, methane production must also be taken into account in assessing the overall heat balance. The combustion of methane (molecular weight 16.015) may be represented as:

$$CH_4 + 2O_2 \rightarrow CO_2 + 2H_2O + 882 \text{ kJ}$$

The energy which, but for the production of methane, would have been dissipated as heat is $q = 882/22.41 = 39.4$ kJ (9.41 kcal) per litre of methane, or 19.7 kJ (4.70 kcal) per litre of oxygen. Adjustments for the volumes of oxygen and carbon dioxide associated with methane must also be included in the calculation of heat production of ruminant animals.

Hydrogen formed by fermentation in the gut or rumen may be included in the calculation in the same way as methane. However, the production of hydrogen and methane in non-ruminants and of hydrogen in ruminants is so small that it can normally be neglected.

3.2.5   *A shorter approach*

The methods just described both for estimating the calorific factors and for calculating heat production were gradually developed during the late eighteenth and early nineteenth centuries and are now mainly of historical and educational interest. Newer methods based on algebraic analysis of the calculation procedures are shorter and more versatile. The end-result is that heat production may be expressed by a simple linear equation

$$M = \alpha V_{O_2} + \beta V_{CO_2} + \gamma N + \delta V_{CH_4}$$

where V_{O_2} is oxygen consumption, V_{CO_2} is carbon dioxide, N urinary nitrogen and V_{CH_4} methane production. It is desirable to understand how this equation is derived so that values may be chosen for the 'constants' α, β, γ and δ that are appropriate to the measurement conditions.

In this approach the chemical nature of the metabolised substances is not considered; the only factors affecting the end-result are the heats of combustion and the elemental composition of these materials. The breakdown of the metabolic mixture into the major categories of carbohydrate, fat and protein is retained because the metabolism of these individual categories is of interest to dieticians and clinicians; but the analysis may be extended to other substances if they are produced and measureable as end-products of the overall metabolism. Although parts of this unified approach have been described by a number of authors (Weir, 1949; Brouwer, 1958; Kleiber, 1961; Consolazio, Johnson & Pecora, 1963; Bursztein *et al.*, 1977), it has not, to our knowledge, been described in its entirety; although a paper by Abramson (1943), which is not usually accorded the credit it deserves, gave a nearly complete description.

The first step, originally described by Lehmann *et al.* (1893) is to derive the respiratory quotient and the oxygen consumption per gram of substance burned from its elemental composition. This may be illustrated by a very simple example: suppose that 1 g of a substance consists of 0.75 g carbon, 0.1 g hydrogen and 0.15 g oxygen with atomic weights 12, 1 and 16 respectively. Then for complete combustion to carbon dioxide and water:

$$O_2 \text{ for converting C to } CO_2 = 0.75 \times 32/12 = 2.000 \text{ g}$$
$$O_2 \text{ for converting } H_2 \text{ to } H_2O = 0.10 \times 16/2 = 0.80 \text{ g}$$
$$\text{net oxygen consumption} = 2.00 + 0.80 - 0.15 = 2.65 \text{ g}$$
$$= 2.65 \times 22.4/32 = 1.855 \text{ l}$$
$$\text{carbon dioxide consumed} = (0.75 \times 44/12) \times (22.4/44) = 1.40 \text{ l}$$
$$\text{respiratory quotient} = 1.40/1.855 = 0.755$$

These simple calculations can be generalised in the form of two equations which are derived in Appendix I:

$$\text{oxygen consumed } (a) = (1.8658 f_C + 5.558 f_H - 0.7004 f_O) \text{ l/g} \quad (3.1)$$
$$\text{respiratory quotient} = 1/(1 + 2.9789 f_H/f_C - 0.3754 f_O/f_C) \quad (3.2)$$

where f_C, f_H and f_O are the fractions (by weight) of carbon, hydrogen and oxygen in the substance burned. The numerical constants in these equations are derived from correct atomic weights rather than the approximations 12, 1 and 16 used in the example.

Equations (3.1) and (3.2) are applicable to all substances including carbohydrate, fat, protein, methane and any other metabolite. Protein contains an additional fraction (f_N) of nitrogen; this means that $f_C + f_H + f_O$ does not add up to one; but it does not affect the validity of the equations provided the combustion is complete (i.e. to carbon dioxide, water and nitrogen). Some proteins also contain small quantities of sulphur.

If the heat of combustion of the substance is Q kJ/g, then the heat produced per unit of oxygen consumed is

$$q = Q/a \text{ kJ}/1.0_2 \tag{3.3}$$

Equations (3.1), (3.2) and (3.3) allow calculation of the factors a, r and q for any substance of known elemental composition. It will, however, generally be simpler for pure substances to derive the constants from the chemical formula of the oxidation reaction as in the examples described for glucose, palmitic acid, alanine and urea.

The next step is to derive equations for the quantities of oxygen consumed (V_{O_2}), carbon dioxide produced (V_{CO_2}) and heat produced (M).

Suppose that the quantities of metabolites and the values of their calorific factors Q, a, r and q are as defined by Table 3.4. Then from columns 1 and 3 we may form the equation

$$V_{O_2} = Ka_K + Fa_F + Pa_P - V_G a_G - Ua_U \tag{3.4a}$$

It should be noted that the terms in K, F and P are positive indicating combustion and those in G and U are negative indicating formation of the

Table 3.4. *Symbols used to represent the quantities and calorific factors of substances oxidised or synthesised*

Column	1	2	3	4	5
		Q	a	r	$q = Q/a$
	Weight oxidised	Heat of combustion	Oxygen consumed	Respiratory	
Substance	(g)	(kJ/g)	l/g	quotient	(kJ/l)
Carbohydrate	K	Q_K	a_K	r_K	q_K
Fat	F	Q_F	a_F	r_F	q_F
Protein	P	Q_P	a_P	r_P	q_P
Methane	V_G†	Q_G‡	a_G‡	r_G	q_G
Urinary nitrogenous substances	U	Q_U	a_U	r_U	q_U

† volume of CH_4 (l), ‡ per 1 CH_4

substance. By incorporating the values from columns 4 and 5 we may also write

$$V_{CO_2} = Ka_K r_K + Fa_F r_F + Pa_P r_P - V_G a_G r_G - Ua_U r_U \qquad (3.4b)$$

$$M = Ka_K q_K + Fa_F q_F + Pa_P q_P - V_G a_G q_G - Ua_U q_U \qquad (3.4c)$$

Also, if the quantity of urinary nitrogen is N g and if f_{NP} and f_{NU} represent the fractions (by weight) of nitrogen in protein and in urinary–nitrogenous compounds, then assuming that all nitrogen from metabolised protein appears in the urine

$$P = N/f_{NP} \text{ and } U = N/f_{NU} \qquad (3.5)$$

Substituting these values of P and U into equations (3.4a, b and c) gives

$$V_{O_2} = Ka_K + Fa_F + Na_N - V_G a_G \qquad (3.6a)$$

$$V_{CO_2} = Ka_K r_K + Fa_F r_F + Na_N r_N - V_G a_G r_G \qquad (3.6b)$$

$$M = Ka_K q_K + Fa_F q_F + Na_N q_N - V_G a_G q_G \qquad (3.6c)$$

where

$$a_N = (a_P/f_{NP}) - (a_U/f_{NU}) \qquad (3.7a)$$

$$r_N = [(a_P r_P/f_{NP}) - (a_U r_U/f_{NU})]/a_N \qquad (3.7b)$$

$$q_N = [(a_P q_P/f_{NP}) - (a_U q_U/f_{NU})]/a_N \qquad (3.7c)$$

The factors a_N, r_N and q_N are the calorific factors for the oxidation of protein to urine, carbon dioxide and water, with a_N expressed per gram of urinary nitrogen excreted. It can be seen that the calorific factors in equations (3.4)–(3.7) are all fixed quantities and may be evaluated solely from the elemental composition and heats of combustion of the dietary and waste substances. For most purposes they may be regarded as constants, although the exact values used will vary slightly according to species and dietary situation. This will be discussed further in Section 3.3.

The final step in the unified approach is to solve equations (3.6a, b and c) for the three unknown quantities K, F and M in terms of the measurable quantities V_{O_2}, V_{CO_2}, N and V_G and the fixed calorific factors (see Appendix II). The solution for heat production is

$$M = \alpha V_{O_2} + \beta V_{CO_2} + \gamma N + \delta V_G \qquad (3.8)$$

where

$$\alpha = (q_F r_K - q_K r_F)/(r_K - r_F) \qquad (3.9a)$$

$$\beta = (q_K - q_F)/(r_K - r_F) \qquad (3.9b)$$

$$\gamma = a_N[q_N(r_K - r_F) - q_F(r_K - r_N) - q_K(r_N - r_F)]/(r_K - r_F) \qquad (3.9c)$$

$$\delta = -a_G[q_G(r_K - r_F) - q_F(r_K - r_G) - q_K(r_G - r_F)]/(r_K - r_F) \qquad (3.9d)$$

Solutions for K and F, in the same form as equation (3.8) are also given in Appendix II. These parameters refer to the net quantities of carbohydrate and fat metabolised (i.e. total metabolised less total synthesised) rather than to the quantities of dietary carbohydrate and fat metabolised.

The calculation of heat production from indirect calorimetry is thus reduced to a simple formula (equation (3.8)) involving four measured quantities and four fixed coefficients. It only remains to evaluate the coefficients.

Equation (3.8) may be extended to accommodate other substances which may be metabolised in special circumstances. For example ketosis gives rise to the formation of ketone bodies, and diabetes to the formation of sugars in the urine. Provided that the quantities of these or any other compounds containing only carbon, hydrogen and oxygen are measurable, and that their calorific factors are known, they may be treated in exactly the same way as methane. It is only necessary to insert additional terms $- G'a'_G$, $- G'a'_G r'_G$ and $- G'a'_G q'_G$ into equations (3.4a, b and c) for each substance synthesised. This will result in an additional term $\delta'G$ in the final equation (3.8) with δ' being defined by an equation exactly the same as equation (3.9d), but with a_G, r_G and q_G replaced by a'_G, r'_G and q'_G.

The only commonly metabolised foodstuff to which the equations are not readily adaptable is alcohol. Ethyl alcohol ($r = 0.667$, $q = 20.36$ kJ/l) on combustion releases approximately 5% more heat per litre of oxygen consumed than would be predicted from its respiratory quotient by extrapolation of Table 3.2. Equation (3.8) would therefore underestimate that part of the metabolic heat derived from alcohol by 5%. The appropriate amendment to the equations could theoretically be made by incorporating terms such as $G a'_G$ (with positive signs to denote combustion). The new variable G' so introduced would represent the quantity of alcohol metabolised during the period of measurement; but this would only be known if the measuring period were long enough for the complete oxidation of a measured amount of alcohol consumed; an unlikely situation in practice.

A further possible refinement of equation (3.8) allows for the sulphur compounds of protein. Urinary sulphur may be treated in the same way as urinary nitrogen; this will result in slight modifications to the values of Q_N, a_N, r_N and q_N brought about by the inclusion of additional terms ($- a_s r_s/f_{SU}$ and $- a_s q_s/f_{SU}$) in equations (3.7a, b and c).

3.3   Review of published values
3.3.1   *Calorific factors*

What then are the appropriate values to choose for the calorific factors to be used in indirect calorimetry? It is helpful in answering this question to consider an extended form of the energy balance diagram previously described. Fig. 3.4 illustrates the net result of the food/energy transformations that occur in the body. Energy categories are indicated by rectangular boxes. In reality, retained energy (*RE*), e.g. tissue growth, is built up from the breakdown products of the food consumed; the same is true of some of the faecal energy (*FE*). *RE* and *FE* are not shown in this position in Fig. 3.4 because calorimetric measurements do not distinguish between consumption and synthesis, and the quantities P, K and F appearing in equations (3.4)–(3.7) are seen to be the net quantities (*GE-FE-RE*) of carbohydrate, fat and protein that are oxidised to carbon dioxide, methane, water, heat and the nitrogenous products of protein metabolism. The breakdown of energy shown by Fig. 3.4 is merely depicted as taking an alternative pathway which, in accordance with the Law of Hess, achieves the same end-result.

The calorific factors used in indirect calorimetry must therefore be based on the transformations enclosed within the dotted lines of Fig. 3.4. It is important to realise that these are not the same factors as those used in compiling tables of the physiologic fuel value of foods (e.g. Paul &

Fig. 3.4. Diagram illustrating the *net result* of the food/energy transformations which occur in the body. Rectangular boxes indicate forms of energy. This diagram *does not* indicate the actual pathways of metabolism, but only pathways which are equivalent according to the Law of Hess.

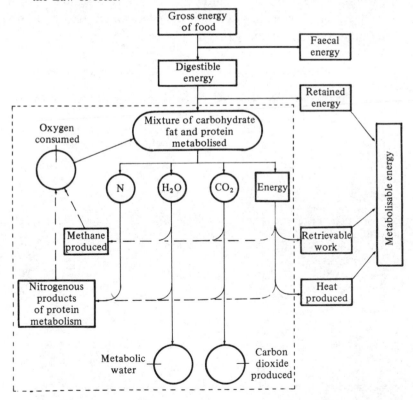

Southgate, 1978) employed by nutritionists and dieticians. The latter factors include allowance for energy lost as faeces and are therefore smaller in terms of energy output per gram of substance metabolised. In passing it should be noted that it is possible and indeed generally practised by dieticians to estimate metabolic rate (in this context it is usually referred to as energy intake) from an exact knowledge of the quantity and type of food eaten using the food tables. This method (of 'counting one's calories') is of great value in assessing dietary requirements of individuals and populations, but as a means of measuring metabolic rate it is of limited accuracy (Passmore, 1967). It cannot rival indirect calorimetry for accuracy because it assumes fixed levels of digestibility for foods.

The calorific factors for use in calorimetry quoted in most textbooks published prior to 1960 (see Tables 3.5, 3.6 and 3.7) are nearly all derived from original analytical data determined nearly a century ago by various pioneering scientists including Rubner, Atwater, Zuntz and Benedict. Most of the early data is expressed in heat units of calories, and oxygen is often measured by weight. In compiling Tables 3.5, 3.6 and 3.7 the conversion factor 4.184 kJ/kcal and oxygen density 1.429 g/l have been used. Values are quoted with numbers of significant figures comparable with the original data; brackets denote values which have been derived from other original data, e.g. $(q) = Q/a$.

Table 3.5. *Calorific factors for carbohydrate* (*starch*) *and some other compounds*

Material Reference	Q_F Heat of combustion (kJ/g)	a_F Oxygen consumption (l/g)	r_F Respiratory quotient –	q_F $= Q_K/a_K$ (kJ/l)
Starch				
Magnus-Levy (1907) (after Rubner)	17.2	0.829	1.00	21.1
Lusk (1928) (after Zuntz & Loewy)	17.51	0.8288	1.00	21.13
Abramson (1943) (after Benedict)	17.6	0.829	1.00	21.18
Kleiber (1961)	16.7	(0.800)	1.00	20.9
Brouwer (1965)	17.6	0.829	1.00	21.2
Elliot & Davison (1975)	–	–	–	21.11
	Q	a	r	q
Methane (*per l*)	39.36	2.00	0.50	19.68
Ethanol	29.77	1.462	0.667	20.36
Glucose	15.64	0.746	1.00	20.95

The work of Rubner and of Atwater was reviewed by Morey (1936) and Widdowson (1955). It is clear that their objective was to determine the physiologic fuel value to the body (as now used in food tables), rather than factors for use in respiration calorimetry which had not yet been developed. This is the reason why Rubner's factor for protein ($Q = 17.2$ kJ/g or 4.1 kcal/g) used by Magnus-Levy and later writers in estimating metabolic rate is lower than those published subsequently. Lusk's (1928) values for carbohydrate and fat are those which he used for compiling the table frequently to be found in textbooks and given in short form in Table 3.2. Cathcart & Cuthbertson (1931) made careful analyses of fat from human cadavers. They confirmed (approximately) the values previously reported for adipose tissue but found different values for liver and muscle fat. They argued that since it is the liver and muscle fat that is most readily mobilised, the values for fat from these regions are more appropriate for use in estimating heat production. Whilst this may be true during exercise, the overall conversion in the long term must be of dietary fat. Kleiber (1961) gave no reason for his unusually low values of heat output from carbohydrate. Elliot's & Davison's (1975) recommended values are means of previously published values, which they reviewed. Abramson (1943), quoting values of Benedict, distinguished between values for animal fat

Table 3.6. *Calorific factors for fat*

Reference	Q_F Heat of combustion (kJ/g)	a_F Oxygen consumption (l/g)	r_F Respiratory quotient –	q_F = Q_F/a_F (kJ/l)
Magnus-Levy (1907) (after Rubner)	38.9	2.019	0.71	19.6
Lusk (1928) (after Zuntz & Loewy)	39.60	2.0193	0.707	19.62
Cathcart & Cuthbertson (1931)				
Liver and muscle fat	38.4	1.937	0.718	19.82
Adipose tissue	39.8	2.001	0.711	19.88
Abramson (1943) (after Benedict)				
Animal fat	39.8	2.013	0.711	19.82
Human fat	39.9	1.992	0.713	20.1
Brouwer (1965)	39.8	2.013	0.711	19.8
Dargol'tz (1973)				
Avian fat	38.9	2.03	0.71	19.3
Elliot & Davison (1975)				19.61
Ben-Porat *et al.* (1983)	39.74	2.028	0.705	19.60

and human fat. Brouwer's (1965) values refer to farm animals, and those of Dargol'tz (1973) to birds.

The choice of appropriate calorific factors for protein is more complex than for carbohydrate and fat. Fig. 3.4 illustrates the fact that the relevant reaction is the oxidation of the net protein consumed into carbon dioxide, water and nitrogenous products of protein metabolism. The latter term is used rather than 'urine' to allow for the small quantity of metabolised protein nitrogen which, along with undigested dietary protein, appears in the faeces. The net protein consumed may, in different circumstances, include either straightforward oxidation of dietary protein, or anabolism of body protein from dietary protein or katabolism of body protein. Also the urine will have different compositions in each circumstance as was shown by the work of Rubner (Lusk, 1928). The composition of urine also varies between species. The primary nitrogenous waste product is urea in land mammals, uric acid in birds and ammonia in fishes. It is obvious that any calorific values chosen for protein must be approximations, and perhaps the best compromise that has been made is to base calculations

Table 3.7. *Calorific factors for the partial oxidation of protein to urinary waste products*

Species Reference	$1/f_{NP}$ Protein– nitrogen ratio	a_N Oxygen consumption (l/g N)	r_N Respiratory quotient –	q_N (see equation (3.7c)) kJ/l
Land mammals				
Magnus-Levy (1907) (after Rubner)	(6.24)	6.030	0.81	(17.8)
Peters & van Slyke (1931) (after Rubner)	(6.25)	5.939	0.801	18.77
Lusk (1928) (after Zuntz & Loewy)	(6.14)	(5.940)	0.802	18.68
Abramson (1943) (after Benedict)	–	5.741	0.809	19.3
Kleiber (1961)	6.25	6.7	0.808	18.8
Brouwer (1965)	6.25	5.98	0.809	19.2
Dargol'tz (1973)	6.25	6.45	0.85	19.3
Elliot & Davison (1975)	–	–	–	19.25
Birds				
Dargol'tz (1973)	6.25	5.97	0.74	19.17
Braefield & Llewellyn (1982)	6.27	5.85	0.72	19.43

on 'standard protein', i.e. 53% carbon, 7% hydrogen, 23% oxygen, 16% nitrogen and 1% sulphur (Kleiber, 1961) with heat of combustion $Q = 23.7$ kJ/g.

Table 3.7 shows that prior to 1960 there was good agreement between the published calorific values for protein, which were mostly calculated on the basis that the nitrogenous waste products were those actually found in urine. Since then higher values have been published, particularly for the amount of oxygen consumed per gram of urinary nitrogen, these are calculated assuming the waste product is urea (Kleiber, 1961) or urea and ammonium sulphate (Dargol'tz, 1973; Braefield & Llewellyn, 1982). It is questionable whether they are preferable to the earlier estimates. The values given by Ben-Porat *et al.* (1983) are derived from a thermodynamic consideration of the heats of combustion and formation of the compounds involved in oxidising protein to urea. Dargol'tz (1973) and Braefield and Llewellyn (1982) also gave recommended values for use in metabolic measurements on birds; these are based on uric acid as the primary waste product and are characterised by lower values for the respiratory quotient and for the oxygen consumed per gram of urinary nitrogen. One value in Table 3.7 that is agreed on by nearly all observers is the value 6.25 for the conversion from urinary nitrogen to protein.

In 1958 the First Symposium on Energy Metabolism sponsored by the European Association for Animal Production appointed a committee to consider and recommend constants and factors for use in metabolic calculations. The committee's recommendations were endorsed at a subsequent symposium and their report (Brouwer, 1965) is reproduced in Appendix III. Their recommended factors, which are included in Tables 3.5, 3.6 and 3.7, have gained widespread acceptance as being applicable to metabolic measurements on farm animals. A similar committee set up in 1985 by a Workshop on Human Energy Metabolism sponsored by the European Economic Community is presently considering factors for use in human calorimetry.

3.3.2 *Tables and equations for respiratory exchange measurements*

Lusk's (1928) table of calorific values for different carbohydrate/fat mixtures (Table 3.2) has been frequently reproduced; the only challenge to it was made by Cathcart & Cuthbertson (1931). They suggested slight modification of the table based on their analyses of human fats. Use of tables is, however, now largely superseded by versions of equation (3.8).

$$M = \alpha V_{O_2} + \beta V_{CO_2} + \gamma N + \delta V_{CH_4} \qquad (3.10)$$

The definitions of the constants α, β, γ and δ (equations (3.9)) may be rewritten in slightly simplified form by inserting the value for the respiratory quotient of carbohydrate ($r = 1$) about which there is no dispute.

$$\alpha = (q_F - q_K r_F)/(1 - r_F) \tag{3.11a}$$

$$\beta = (q_K - q_F)/(1 - r_F) \tag{3.11b}$$

$$\gamma = a_N[q_N - q_F(1 - r_N)/(1 - r_F) - q_K(r_N - r_F)/(1 - r_F)] \tag{3.11c}$$

$$\delta = -a_G[q_G - q_F(1 - r_G)/(1 - r_F) - q_K(r_G - r_F)/(1 - r_F)] \tag{3.11d}$$

It is clear that the calorific coefficients α and β, for oxygen consumption and carbon dioxide production, depend only on the values of q (kJ/l) and respiratory quotient for carbohydrate and fat. The coefficients for urinary nitrogen (γ) and for methane (δ) are, however, more complex functions and are particularly dependent on the relatively small differences between the respiratory quotients of the various food categories. Fortunately, the

Table 3.8. *Published values of the coefficients in the equation*
$$M = \alpha V_{O_2} + \beta V_{CO_2} + \gamma N + \delta V_{CH_4}$$

Coefficient for:	α Oxygen (kJ/l)	β Carbon dioxide (kJ/l)	γ Urinary nitrogen (kJ/g N)	δ Methane (kJ/l)
Species References (and source of calorific factors employed)				
Humans				
(1) Abramson (1943) (Lusk)	16.0	5.15	−7.8	−
(2) Abramson (1943) (Benedict[a])	16.2	4.94	−5.8	−
(3) Abramson (1943) (Benedict[b])	17.2	3.93	−6.7	−
(4) Consolazio, Johnson & Pecora (1963) (Rubner)	15.8	4.86	−12.5	−
(5) Weir (1949) (Loewy, Zuntz, Cathcart & Cuthbertson)	16.50	4.63	−9.08	−
(6) Ben-Porat *et al.* (1983)	16.37	4.57	−13.98	−
Farm animals Brouwer (1965)	16.18	5.02	−5.99	−2.17
Birds Farrell (1974)	16.20	5.00	−1.20	−
Fishes Braefield & Llewellyn (1982)	15.94	5.15	−9.14 (per g ammonia)	

[a] based on calorific factor for animal fat, [b] human fat.

consequent uncertainty as to the exact values of γ and δ is of little practical importance because these terms make only a minor contribution to the total heat production.

Various published values of the coefficients α, β, γ and δ are summarised in Table 3.8. The differences between the various sets of coefficients are best illustrated by an example. Taking the data previously analysed in Table 3.3, the individual contributions of the terms in V_{O_2}, V_{CO_2} and N for each of the six equations listed as applicable to humans as well as for the Brouwer equation are shown in Table 3.9. In all cases the oxygen term contributes approximately 80% and the carbon dioxide term approximately 20% to the total value of M. The nitrogen term is very variable, reflecting the uncertainty as regards the calorific factors for protein and nitrogenous waste products, but it contributes only about -1% to the total in this example; for a high protein diet its contribution could be twice as great. The value of M calculated from the equation of Consolazio *et al.* is noticeably lower than that from all the others; this is due to the former authors basing their equations on Rubner's calorific factors which, as discussed earlier, are not the correct ones for analysis in indirect calorimetry. Unfortunately Consolazio *et al.*'s equation has been much used and has been repeated in tables and textbooks (e.g. Diem & Lentner, 1970; Passmore & Draper, 1965; Durnin & Passmore, 1967). Ben-Porat *et al.* (1983) also used the same constants as Consolazio *et al.* (1963) for protein, together with Merrill's & Watt's (1955) constants for fat and the constants of glucose (see Table 3.5) for carbohydrate. In the remaining equations it is notable that the sum of the coefficients for V_{O_2} and V_{CO_2} are, in nearly every case, close to 21.1. This uniformity reflects the fact that when respiratory quotient is 1 the two coefficients have the same multiplier (i.e. $V_{O_2} = V_{CO_2}$), and their combined value is representative of the heat of

Table 3.9. *Contributions of the individual terms of the equations for calculating heat production proposed by different authors.*

Equation used (see Table 3.8)	αV_{O_2} (kJ/l)	$+ \ \beta V_{CO_2}$ (kJ/l)	$+ \ \gamma N$ (kJ/g)	$= M$ (kJ)
(1) Abramson–Lusk	916	249	-10	1155
(2) Abramson–Benedict	927	239	-7	1159
(3) Abramson–Benedict	985	190	-9	1166
(4) Consolazio *et al.*	905	235	-16	1124
(5) Weir	945	224	-12	1157
(6) Ben-Porat *et al.*	937	221	-18	1140
(7) Brouwer	926	242	-8	1160

(Data taken from the example of Table 3.3. For the equations see Table 3.8.)

combustion of carbohydrate. When respiratory quotient is less than 1 the uncertainty as to the appropriate calorific factors for fat is reflected in the relative importance accorded by different authors to the V_{O_2} and V_{CO_2} terms. Both Weir's equation and Abramson's third equation are based on calorific factors for human fat and may be especially appropriate to conditions such as exercise which involve catabolism of body reserves. Equations based on Lusk's calorific factors are likely to be more appropriate to most normal conditions. The coefficients of the Abramson–Lusk equation differ only very slightly from those of Brouwer.

Any distinction between different constants to suit special conditions is however bound to be a fine one. It must be appreciated that the oxidation of a particular substrate at a given time represents but a step in the chain of reactions whose overall effect is the combustion of a protein–fat–carbohydrate mixture; also that both catabolic and anabolic processes are nearly always occurring simultaneously. Hunt (1969) suggested that in the postabsorptive state, because oxygen has already been used to change the form of fuel before metabolic measurements are taken, the calorific constants must be appreciably reduced to account for the specific dynamic action of the fuel. If this were correct there would have to be an enormous difference between the calorific values of the substrates formed during digestion and those of the dietary constituents themselves. Ben-Porat *et al.* (1983) have used thermodynamic reasoning also to dismiss Hunt's (1969) calorific factors, which are clearly incorrect. Hunt's factors and the coefficients of the metabolic rate equation for the postabsorptive state which have been derived from them (Bursztein *et al.*, 1977) are therefore not included here in Tables 3.8 and 3.9 respectively.

Brouwer's (1965) equation was endorsed at an international meeting and is widely accepted as appropriate for farm animals. Endorsement of a similar equation for human metabolic measurements will hopefully be given by the EEC committee currently considering the matter.

Lusk's (1928) table of calories per litre of oxygen consumed covered the respiratory quotient range 0.7–1. This left the interpretation of results when respiratory quotient was greater than 1 open to doubt. The method of calculation using an equation by-passes the difficulty. Levels of respiratory quotient of up to 1.2 are not uncommon in productive animals which are synthesising fat and protein from largely carbohydrate diets; they are also encountered in undernourished humans during the recovery stage following starvation. These situations present no exception to the arguments used in deriving the equations provided that the two basic assumptions stated at the beginning of this chapter still hold. Ben-Porat *et al.*'s (1983) thermodynamic analysis confirms that the same constants are indeed applicable to situations involving catabolism and anabolism of both fat and protein. If for example an animal consumes an all-carbo-hydrate diet and at the same time synthesises body-fat ($r = 0.7$), then the synthesis will involve a greater depletion of oxygen consumption than of carbon dioxide production. The net effect will be that the animal will produce more carbon dioxide than the volume of oxygen it consumes. The value of F (see Appendix II) will be found to be negative in accordance with the fact that fat is being synthesised rather than consumed. The value of P (equation (3.5)) can never be negative. When body protein is syn-thesised, provided its nitrogen is of dietary origin, the net protein met-abolised (i.e. metabolisable minus retained) must still be positive.

The parameters P and F are therefore somewhat vague concepts and their actual values derived from the equations (equation (3.5) and Ap-pendix II) have limited practical significance. Although Ben-Porat *et al.* (1983) have suggested that the value of F may be used to distinguish between the postabsorptive state in critically ill patients during hyper-alimentation ($F > 0$) and the fasting state ($F < 0$). The parameter K, however, is a reasonably good estimate of the quantity of carbohydrate metabolised and M (equation (3.8)) is an accurate estimate of heat pro-duction.

The relative dominance of the oxygen term in equation (3.8) and the fact that oxygen consumption and carbon dioxide production are both of the same order of magnitude means that an approximate estimate of heat production is obtainable from oxygen consumption alone by assuming some value for respiratory quotient (Blaxter, 1962, p. 45). The errors involved in this approximation and a possible way of avoiding them in open-circuit calorimetry are discussed in Chapter 5.

3.3.3 Calculation from carbon–nitrogen balance

If the amounts of carbon and nitrogen retained in the body are measured, the amount of energy retained may be calculated by assuming that all body tissue consists of proteins and fat of known compositions and heats of combustion.

Suppose that the fractions by weight of nitrogen in protein, carbon in protein and carbon in fat are f_{NP}, f_{CP} and f_{CF}, respectively, and the heats of combustion of protein and fat are Q_P and Q_F. Then the quantity of protein retained is N/f_{NP}, and its energy content is $Q_P N/f_{NP}$ where N is the quantity of nitrogen retained. The quantity of carbon incorporated into protein is $(f_{CP}/f_{NP}) \times N$, and the remainder of the total carbon re-tained (C) must have been incorporated into fat, i.e. $[C - (f_{CP}/f_{NP}) \times N]$. Its energy is $(Q_F/f_{CF}) \cdot [C - (f_{CP}/f_{NP}) \times N]$.

The total energy retained (RE) is then the sum of the protein energy and the fat energy:

$$RE = \frac{Q_F \cdot C}{f_{CF}} - \left(\frac{Q_F f_{CP}}{f_{CF}} - Q_P\right)\frac{N}{f_{NP}} \tag{3.12}$$

This type of calculation was done by Pettenkofer & Voit (1866), who by precise weighing and analysis of food intake and waste products were able to demonstrate a very accurate balance between all forms of energy intake and output. It has been extensively used to estimate RE from carbon and nitrogen balance measurements alone. However the estimation depends on assuming fixed values for the various constants in Equation (3.12); it also ignores the presence of body-tissue constituents other than protein and fat.

The constants used are based on chemical analyses and bomb calorimetry of fat-free muscle and of depot fat; those recommended by the European Association of Animal Production's sub-committee (Brouwer, 1965, see Appendix III) are $f_{NP} = 0.16$, $f_{CF} = 0.52$, $f_{CF} = 0.767$, $Q_P = 23.8$ kJ/g and $Q_F = 39.7$ kJ/g, and this leads to the formula

$$RE = 51.83 \times C - 19.40 \times N \tag{3.13}$$

The factors used in deriving this equation were, for consistency, the same as those used by the committee in deriving the equation which relates heat production to respiratory gaseous exchange. Blaxter (1962, p. 49), however, used a statistical approach to derive a best-fit equation based on chemical analysis of actual tissues (muscle, brain and livers) of cattle (Blaxter & Rook, 1953)

$$RE = 52.51 \times C - 28.87 \times N$$

This equation seems, for cattle, to be more firmly based than the theoretical equation of the Brouwer committee.

3.3.4   *Metabolic water*

In water balance studies it is necessary to determine the quantity of water evolved in metabolic oxidative processes. This, too, can be calculated from the respiratory gaseous exchange. From consideration of either the elemental composition or the chemical formula of the substance values may be derived for the weight of water produced per litre of oxygen consumed (w) in the oxidation of protein, fat, carbohydrate and methane. The quantities w_N, w_F, w_K and w_G are then exactly analogous to q_N, q_F, q_K and q_G and may replace them in Equations (3.8) and (3.9) to provide an estimate of metabolic water.

Van Es (1967) calculated the constants for water on the basis of the

Brouwer (1965) recommendations. The values ($w_N = 0.439$, $w_F = 0.529$, $w_K = 0.671$ and $w_G = 0.804$ grams of water per litre of oxygen) do not exhibit the same uniformity as do the values of q and the resulting formula for the metabolic water (W g) can therefore be only approximate:

$$W = 0.181 \times O_2 + 0.490 \times CO_2 - 0.826 \times N - 0.756 \times CH_4$$

3.4   Summary and recommendations

Estimation of the amount of heat produced (M) from analysis of respiratory gases employs factors which are based on the heats of combustion and elementary composition of carbohydrate, fat, protein and urinary nitrogenous substances.

Calculations may be reduced to an equation of the form:

$$M = \alpha V_{O_2} + \beta V_{CO_2} + \gamma N + \delta V_{CH_4} \text{ kJ}$$

where V_{O_2} = oxygen consumption (l at stp); V_{CO_2} = carbon dioxide production (l at stp); V_{CH_4} = methane production (l at stp); N = nitrogen excreted in the urine (g).

The values of the factors α, β, γ and δ may vary slightly according to the nutrition, activity and growth pattern of the subject but for most normal practical purposes the following equations may be used:

For man (Abramson, 1943):

$$M = 16.0 \times V_{O_2} + 5.15 \times V_{CO_2} - 7.8 \times N \text{ kJ}$$

For farm animals (Brouwer, 1965):

$$M = 16.18 \times V_{O_2} + 5.02 \times V_{CO_2} - 5.99 \times N - 2.17 \times V_{CH_4} \text{ kJ.}$$

For estimation of retained energy (RE) from carbon and nitrogen balance (g) the following equation is used:

$$RE = 51.83 \times C - 19.40 \times N \text{ (kJ)}$$

The equations in this chapter have all been expressed in terms of quantities of heat (kJ), gases (l) and urinary nitrogen (g). They may also be expressed in terms of metabolic rate (kW), gas consumption (l/s) and urinary nitrogen excretion (g/s) e.g.

$$\dot{M} = \alpha \dot{V}_{O_2} + \beta \dot{V}_{CO_2} + \gamma \dot{N} + \delta \dot{V}_{CH_4}$$

4

Indirect calorimeters

Whilst direct calorimeters measure the rate of heat *dissipation* of a subject, indirect calorimeters measure the rate of heat *generation*; averaged over a long period of time the two rates will be equal or very nearly so. Thus it is a mistake to believe that because of its name, indirect calorimetry is a second-rate means of measuring heat production.

Indirect calorimetry estimates heat production from quantitative measurements of materials consumed and produced during metabolism. Most methods involve estimation of respiratory gas exchange and these may be classified according to their operating principles as confinement, closed-circuit, total collection, and open-circuit systems.

In confinement systems the subject is held in a sealed chamber and the rates of change of gas concentration in the chamber are recorded.

In closed-circuit systems the subject is again held within or breathes into a sealed apparatus; the carbon dioxide and water vapour produced by the subject are measured as the weight gain of appropriate absorbers, and the amount of oxygen consumed is measured by metering the amount required to replenish the system. In total collection systems all the air expired by the subject is accumulated in order to measure subsequently its volume and chemical composition.

There are two major forms of open-circuit calorimetry. In one the subject breathes directly from atmosphere and by means of a non-return valve system expires into a separate outlet line. In the second form, the subject inspires from, and expires to, a stream of air passing, by means of a pump or fan, across the face. In both cases the flow of air is measured either on the inlet or outlet side of the subject. Air from the outlet is either collected continuously or periodically for later analysis, or is sampled continuously for on-line analysis.

Most of these respiratory exchange methods depend mainly on the measurement of oxygen consumption, either alone or combined with measurements of carbon dioxide and methane production, and sometimes urinary nitrogen excretion. But it is also possible to estimate heat production approximately from carbon dioxide production alone. Alternatively heat production can be calculated accurately from the rate of turnover in the body of carbon and nitrogen (C—N balance). Whilst this eliminates the need for measurement of oxygen, it does involve measurement of the total quantities and chemical composition of all food consumed and excreta produced. Details of the C—N balance, which is not much practised now, are not described here, but the methods of analysis will be found in a chemical textbook (for example Peters & van Slyke, 1932) and the relevant equation for calculation of heat production is given in Chapter 3.

Other indirect methods are based on measurement of carbon dioxide turnover by isotopic tracer techniques, or by recording heart rate as a correlate of heat production.

Fig. 4.1. Schematic diagram of the confinement system for cattle at the Rowett Institute. (Reproduced from Blaxter, Brockway & Boyne, 1972.)

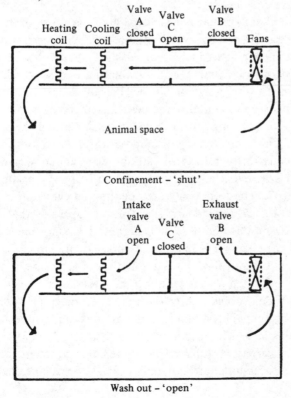

4.1   Confinement systems

The confinement principle for estimating oxygen consumption is said to have been first used by Lavoisier for small animals (Kleiber, 1961). It has also been used for guinea pigs, rabbits and dogs (Laulanie, 1894), pigs (Charlet Lery, 1958), and cattle (Turner & Thornton, 1966). In all these applications the animal was confined for a period in a hermetically sealed chamber and the amount of oxygen consumed was estimated from the change in its concentration within the fixed volume of the system. Measurement had to be confined to the relatively short period before the air in the chamber became seriously depleted in oxygen. Blaxter, Brockway & Boyne (1972) overcame this difficulty in their cattle chambers at the Rowett Research Institute in Aberdeen by periodically flushing out the chambers for a few minutes with a stream of fresh air. They named this device 'the open and shut case', and they operated it with a repeated 53 min cycle of 50 min shut and 3 min open. The principle is illustrated in Fig. 4.1 and the description that follows is modified from that originally given by Blaxter *et al.* (1972).

4.1.1   *The Rowett Research Institute confinement respiration chambers*
4.1.1a   *Chambers and ventilation*

The two chambers each measure $2.4 \times 1.8 \times 3.6$ m overall. The surfaces of the second chamber were made of 4.75 mm sheet steel, after problems had been encountered with leaks in the first chamber which was built of plywood. In each chamber the ventilating and air-conditioning system is contained within a 0.45 m slice of the full length and height of the chamber. This section contains the heat exchanger coils, valves and three continuously operating 30 cm diameter fans. During the washout periods the flow of air is diverted by the internal valve, C, so that air enters by port A, circulates through the air-conditioning system, mixes with the chamber air and is vented at port B. Valves A and B are 45 cm square and seat into oil wells 7.5 cm deep. They are operated by cables from a low-speed electric motor. A second low-speed motor operates the internal valve C. Total air flow is about 15 000 l/min and air movement in the animal space of the chamber is maximally about 15 m/min.

The chamber is entered through a 1.2 m wide door made of plastic-faced chipboard and fitted with a window. The door can be sealed onto a 2.5 cm thick expanded rubber gasket. There is a second small door in the side of the chamber, capable of admitting a man; it is similarly sealed. The overall dimensions in which an animal can be held are $1.3 \times 3.2 \times 2.4$ m. Within this area a variety of stalls can be erected, and the chamber has in fact been used to accommodate not only a single beast but also small flocks or individual sheep or a human being. One hundred electrical conductors are led through the wall of the chamber for servicing of equipment and for physiological measurement purposes. A drinking water supply to the chamber is also provided.

4.1.1*b*   Pressure control

The system has a fixed physical volume and to avoid leaks pressure inside the chamber has to be maintained equal to that outside. There are two factors which may lead to disturbance of the balance: the outside (atmospheric) pressure may alter during the course of a confinement period; and the animal does not necessarily replace the volume of oxygen it consumes with the carbon dioxide and methane it produces and hence the pressure of the chamber alters. The solution adopted to ensure zero pressure differential between the chamber and outside is to adjust the temperature of the circulating air.

A heat exchanger is supplied with glycol at a preset temperature, below the operating temperature within the chamber. Provision is made for the collection of water vapour condensed on this heat exchanger. Water is circulated through another heat exchanger at a temperature close to the operating temperature within the chamber (12–30 °C). An electric heating grid rated at 400 W is placed after this heat exchanger. A water manometer senses the difference between chamber and atmospheric pressures; electrodes at the water surface in the manometer detect the sign of the pressure differential and operate relays to reduce the flow of water to the second heat exchanger and hence cool the chamber when its pressure rises or to switch on the electric heating grid in the event that chamber pressure falls. The change in temperature required to maintain zero pressure differential in the course of a 50 min confinement period is less than 1 °C.

Air temperature in the animal space shows a vertical distribution of less than 2 °C and no detectable horizontal gradients. Wet- and dry-bulb temperatures are measured immediately before the ventilating fans in a position that gives a representative value for the temperature distribution in the animal space.

4.1.1*c*   Gas sampling, analysis and calculation

Shortly after the start and before the end of each confinement period the airstream is sampled through a 5 cm bore pipe situated after the fans. Composition of the samples is measured using infra-red gas analysers for carbon dioxide and methane, and a paramagnetic analyser for oxygen. The change in gas concentrations produced by a heavily milking cow during the confinement period is approximately 0.8 % for both carbon dioxide and oxygen. The internal volume of the chamber less the space occupied by the animal is adjusted to STPD. The adjusted volume multiplied by the concentrations of oxygen, carbon dioxide and methane then gives the volumes of these gases present at each measurement and, by difference, the amounts consumed and produced.

4.1.2   Pressure control by volume adjustment

Aulic *et al.* (1983) have constructed a man-sized respiration chamber of similar design to that just described, but using variable volume rather than variable temperature as the means of maintaining zero pressure differential between the chamber and outside. The chamber air supply is connected to a 10 l low-resistance spirometer whose piston moves whenever the pressure differential exceeds 0.5 mm Hg, and to a larger motorised spirometer of 120 l capacity through which part of the airstream is circulated. Whenever the piston of the small spirometer reaches one of its limits the large spirometer is driven in a compensatory direction until the small spirometer returns to the mid-point of its travel. This system provides improved control of the chamber temperature.

4.1.3   Design theory

The relationships are now considered between the confinement-chamber volume (V), the ventilation rate during the washout ($\dot{V}_0$) and the carbon dioxide concentration (F) in the chamber for an animal producing the gas at a rate $\dot{V}_G$. The same arguments, but with sign reversal, apply for oxygen.

When shut: during the confinement period the rate of increase of carbon dioxide concentration is given by the ratio of the production rate to the total volume,

$$\frac{\mathrm{d}F}{\mathrm{d}t} = \dot{V}_G / V \tag{4.1}$$

If the duration of the confinement is t_s min and the initial carbon dioxide concentration is F_{min} then the final concentration (F_{max}) is obtained by integration of equation (4.1)

$$F_{max} = F_{min} + (\dot{V}_G / V) t_s \tag{4.2}$$

When open: during washout the rate of carbon dioxide removal from the chamber consists of that exhausted less that produced by the animal and that entering with the fresh-air stream (concentration F_I)

$$-V \frac{\mathrm{d}F}{\mathrm{d}t} = \dot{V}_0 F - \dot{V}_G - \dot{V}_0 F_I \tag{4.3}$$

rearranging and integrating gives

$$(\dot{V}_G + \dot{V}_o F_I - \dot{V}_o F) = A e^{-\dot{V}_o t/V} \tag{4.4}$$

The integration constant A may be evaluated from the initial conditions when $t = 0$ and $F = F_{max}$.

$$A = \dot{V}_G + \dot{V}_o F_I - \dot{V}_o F_{max}$$

and substituting for A in equation (4.4) gives

$$F = (\dot{V}_G/\dot{V}_o + F_I)(1 - e^{-\dot{V}_o t/V}) + F_{max} e^{-\dot{V}_o t/V} \tag{4.5}$$

At the end of the open period (duration t_o) the carbon dioxide concentration has dropped to F_{min} and after the first few cycles F_{min} and F_{max} will attain fixed levels

$$F_{min} = (\dot{V}_G/\dot{V}_o + F_I)(1 - e^{-n}) + F_{max} e^{-n} \tag{4.6}$$

where $n = \dot{V}_o t_o/V$ is the number of complete air changes that occur during the open period.

Solving equations (4.2) and (4.6) for F_{min} and F_{max} gives

$$F_{min} = \dot{V}_G/\dot{V}_o + F_I + (\dot{V}_G/V) t_s \cdot \frac{e^{-n}}{(1 - e^{-n})} \tag{4.7}$$

$$F_{max} = \dot{V}_G/\dot{V}_o + F_I + \frac{(\dot{V}_G/V) t_s}{(1 - e^{-n})} \tag{4.8}$$

Also

$$\frac{F_{min} - \dot{V}_G/\dot{V}_o + F_I}{F_{max} - \dot{V}_G/\dot{V}_o + F_I} = e^{-n} \tag{4.9}$$

Fig. 4.2. Relationships between the limiting gas concentrations, F_{min} and F_{max}, and the durations of the open and shut phases (t_o and t_s) of a confinement respiration chamber.

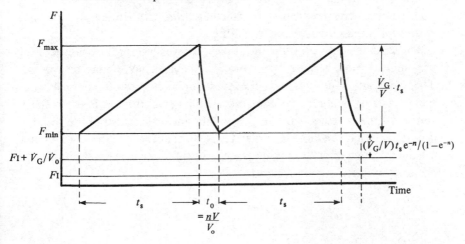

These relationships are illustrated in Fig. 4.2. F_{min} and F_{max} are each made up of three components:

(1) F_I, the concentration of carbon dioxide in fresh air ($= 0.0003$)

(2) $\dot{V}_G/\dot{V}_o$, the increase in concentration that would occur across the chamber if it were left running continuously in the open condition.

These two components together represent the minimal level from which concentration increments can conceivably start.

(3) The third component consists, for F_{min} of a fraction of the previous concentration increment not washed out during the open period, plus for F_{max}, the concentration increment of the present open period.

The factor e^{-n} in equation (4.9) represents the fraction of the total available concentration increment 'wasted' by incomplete washout.

Maximum sensitivity is obtained by making the measured concentration difference $(F_{max} - F_{min})$ as large as possible. However, F_{max} must be kept below about 0.01 to avoid stress on the subject; F_{max} may be set at this level by choosing the duration of confinement t_s according to the size of the animal. The value of F_{min} must then be kept as small as possible. The first component is small and not alterable; the other two components are both kept small by making the ventilation rate large. The value of $\dot{V}_o$ must be at least $100 \times \dot{V}_G$ and must provide at least 2.5 air changes per minute to keep components 2 and 3 both below 0.001.

For large animal chambers a practical limit to $\dot{V}_o$ may be set by the capacity of the air-conditioning plant needed to bring the large quantity of air to the required temperature. The time taken for a given number of air changes is also reduced by making the chamber volume small. This shortens the length of both confinement and washout periods but has no affect on the proportion of the total time occupied by measurements.

The quantities of the respiratory gases exchanged by the animal during each confinement period may be calculated directly from equation (4.1) with appropriate adjustment to STPD.

The confinement method without periodic washout of the chamber is of very limited value because of the brief time over which it may be applied. The 'open and shut' system is however almost continuous, the measurements occupying almost 90% or more of the total time. It is particularly suited for measurements on large animals since it avoids the use of large quantities of carbon dioxide and water absorbents needed in closed-circuit systems. It also avoids flowrate measurements which are a limitation to accuracy in open-circuit systems.

4.2   Closed-circuit

4.2.1   Small animal respirometers

The closed-circuit principle for respiratory analysis, which was first introduced by Regnault & Reiset (1849), is still widely used today in respirometers of various forms. The method is particularly suited to metabolic measurements on small animals. The animal is placed, together with carbon dioxide and moisture absorbers, in a closed space and the quantity of oxygen needed to replace that used up by the animal is measured as it is admitted to the system. In most versions, only oxygen consumption is measured. Respirometers usually operate either on a constant pressure-variable volume or constant volume-variable pressure principle.

4.2.1a   Constant pressure systems

The type of chamber in which pressure is kept constant has been widely used. The simplest design (Lilienthal, Zierler & Folk, 1949) used a chamber with a long, coiled, graduated tube inserted into it. A rat was placed in the chamber, together with containers of soda lime and anhydrous calcium chloride to absorb carbon dioxide and water respectively, and the apparatus was put into a waterbath. A drop of coloured water was introduced into the end of the tube, and was drawn along it as the animal used oxygen and any carbon dioxide produced was absorbed. The passage of the drop between two volumetric graduations was timed. Similar systems using soap bubbles rather than a drop of water have also been used (Watts & Gourley, 1953; MacLaury & Johnson, 1968).

A more satisfactory low cost system, used with chicks, has been developed by Strite & Yacowitz (1956) and is shown in Fig. 4.3; a modified version has been used by Charkey & Thornton (1959). The animal chamber was connected through a three-way stopcock to the top of a graduated burette; another limb of the stopcock was connected to an oxygen reservoir. The chamber also had an outlet through a tap to atmosphere and contained a solution of potassium hydroxide to absorb any carbon dioxide produced. A thermometer was mounted in the chamber. A levelling bulb, which contained water, was connected by means of rubber tubing to the bottom of the burette so that in effect the burette formed one limb of a water manometer. The burette was connected to the oxygen reservoir after turning the stopcock, then by lowering the bulb, it was filled with oxygen. The animal chamber was isolated from atmosphere and the stopcock turned to connect the animal chamber with the burette. When the water level was the same in the burette and levelling bulb, the reading in the burette was noted and a stopwatch or clock started. As the animal used up oxygen and any carbon dioxide produced was absorbed, the water level rose in the burette. During this time the levelling burette was adjusted to keep pace with the water level in the burette. When the burette was almost full of water the levelling bulb was placed in a position so that its water level was slightly above that of the burette. When the water levels were equal the reading and time were noted. The movement of the levelling bulb ensured that when the readings are made the pressure in the chamber was atmospheric. Finally, the animal chamber was opened to atmosphere.

The volumes were adjusted to STPD using measured barometric pressure and the chamber temperature. The method has the advantage that, providing there is no change in temperature and the readings are made at constant pressure, it is not necessary to know the volume of the system. It does continue to have, however, the disadvantage of a short duration of measurement.

The duration can be increased by connecting to the chamber, a reservoir of greater capacity than the burette described above. The reservoir usually employed is the spirometer which is composed of an inverted bell dipping into water and supported by a balancing counterweight (Fig. 4.9(a)). The amount of oxygen consumed is estimated from the linear movement of the

Fig. 4.3. The closed circuit system used by Strite & Yacowitz (1956).

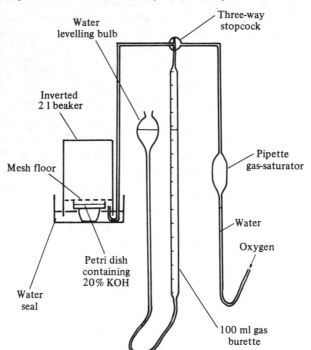

bell. Further extension to the duration of measurement can be achieved by automated refilling of the spirometer by placing, at the lower and upper limits of the bell's travel, microswitches which initiate and terminate recharging with a fixed amount of oxygen (Heusner, Roberts & Smith, 1971).

One must not neglect the fact that if animals are confined for long periods of time, there will be a build up of materials in the chamber, for example ammonia, which will degrade the environmental conditions to which the animal is exposed. Commercial supplies of oxygen often contain impurities, such as the inert gases, which may cumulate during long periods of measurement. Errors can also arise due to variations in barometric pressure and the temperature in both the reservoir and the chamber.

Miller *et al.* (1981) have described a system which comprises an animal chamber, a dummy chamber and an oxygen reservoir and eliminates the effect of fluctuations in temperature and pressure. All the components are immersed in a cooling tank. The reservoir takes the form of a rigid tank from which, by means of liquid injection, oxygen is displaced at a rate that maintains a zero pressure differential between the animal and dummy chambers.

4.2.1*b* Constant volume systems

Errors due to temperature and pressure variations are more easily prevented or compensated in respirometers operating on the constant volume principle, which is now more widely employed than the constant pressure systems.

An early apparatus of the constant volume type is that of MacLagan & Sheahan (1950) used for mice. The equipment comprised a 6 l capacity vacuum desiccator attached to a mercury manometer through a tap in the desiccator lid. Inside the desiccator was a cylindrical plastic animal chamber with, above and below, two trays of soda lime. The volume of the desiccator was determined by filling it completely with water and then siphoning off successive portions for weighing. An allowance was made for the additional volume of the tubing to the manometer, and the reduction in the volume caused by the presence of the animals, the cage and soda lime. Once the animal was placed in the desiccator, oxygen was introduced through the tap to achieve approximately 50% enrichment of the air, after which the tap was closed. The manometer was then quickly connected to the desiccator and the tap reopened. As oxygen was used up and the expired carbon dioxide absorbed, the pressure in the system, measured by means of the manometer, decreased. The total pressure fall could be as much as 300 mm.Hg. After an initial 30 minute period, it was assumed that temperature in the desiccator remained constant. The oxygen consumption was then calculated using the ideal gas law.

The system, although simple, had several disadvantages:

(1) Saturated atmosphere: there was no attempt to remove respiratory moisture and the atmosphere rapidly became saturated, possibly causing distress to the animal.

(2) Pressure drop: there are two main objections to the extent of the fall in pressure in MacLagan & Sheahan's equipment: firstly, it is difficult to eliminate leaks at such negative pressures and, secondly, the effects on the animals may be stressful under such unphysiological environmental conditions.

Grad (1952) eliminated the need for a substantial pressure drop in constant volume systems by the use of an automated delivery of oxygen. The animal chamber was connected to an oxygen reservoir. Both chamber and reservoir were placed in a water bath. A water-filled tube led from the water bath and through a stopper into the oxygen reservoir; the heights of the ends of the tube were such that water would not syphon across into the reservoir. As oxygen was used up and pressure decreased in the system, water was drawn through the tube into the reservoir. The amount of water transferred was measured volumetrically to give oxygen consumption. Waring & Brown (1965) used a similar water displacement method for poultry.

Morrison (1951) introduced an automated system which also gave a record of oxygen consumption. A primitive mercury pressure transducer in line with the animal chamber operated a solenoid valve which allowed oxygen to enter to restore the original chamber pressure. Operation of the valve was recorded on a kymograph. Oxygen consumption was determined from the amount of oxygen required to restore the original pressure and the number of occasions during the measurement that the valve was activated. Dowsett, Kekwick & Pawan (1963) described an almost identical system. An advanced form of these systems using a pressure transducer was described by Cline & Watts (1966). Gentz *et al.* (1970) described a similar system for measuring the oxygen consumption of piglets or human neonates. A float of fixed volume was used to activate electromagnetic valves. As oxygen was consumed and the float chamber completely depleted, the valve connecting it to the chamber was closed and that to the oxygen reservoir was opened. When the float chamber was refilled the status of the valves was reversed. However, these systems were still subject to errors introduced by fluctuations in temperature or pressure or both.

The equipment described by Smothers (1966), and illustrated in Fig. 4.4

eliminated these errors, but it relied on a manual delivery of oxygen. The method was similar to that described by Bailey, Kitts & Wood (1957). It consisted of two sealed chambers containing soda lime and each connected through a stopcock to a common capillary manometer. The animal chamber was connected to an oxygen reservoir made from a 20 or 40 ml test tube with an overflow tube joined to the bottom. The reservoir was filled with oxygen from a low-pressure source. Oxygen was displaced from the reservoir by injecting water from a syringe pipette which could in turn be refilled from a water reservoir. Initially both chambers were open to the atmosphere using the three-way stopcocks, and the manometer fluid adjusted to the index line. The oxygen reservoir was then filled with water from the syringe and its connection to the animal chamber clamped at A; it was then filled with oxygen. Sufficient oxygen was admitted to fill the reservoir forcing water out through the overflow which was then clamped at B.

The animal in a cage was placed in the animal chamber and the lids on both chambers secured. The stopper in the animal chamber lid was temporarily removed whilst time was allowed for temperature equilibrium. Once equilibrium was achieved the clamp at A was removed, the stopper replaced and the stopcocks turned to connect both chambers to the manometer. The timing was then commenced and the plunger of the syringe depressed to introduce a fixed quantity of water (1 or 2 ml) into the oxygen reservoir. This increased the pressure in the animal chamber and this was indicated by the manometer. The syringe was recharged with water ready to bring about a displacement of oxygen from the reservoir

Fig. 4.4. The closed-circuit system used by Smothers (1966).

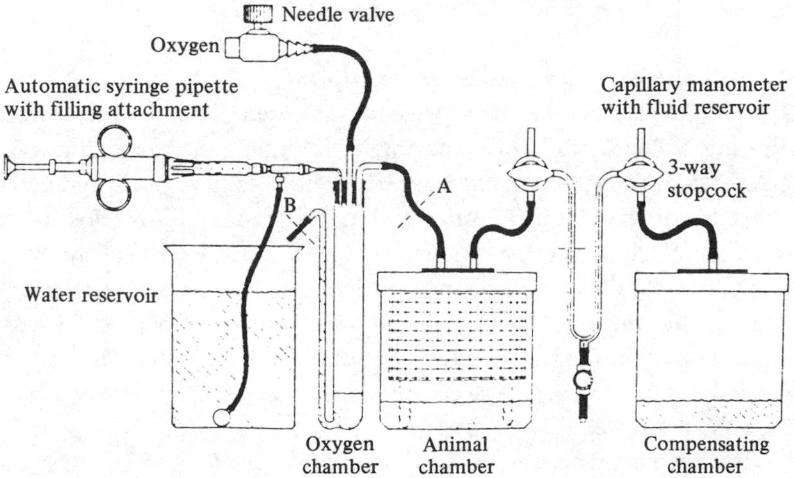

the instant that the manometer indicated a zero pressure differential between the chambers. The time taken for pressure to equalise and the amount of water used to displace oxygen from the reservoir gave the rate of oxygen consumption. The procedure was repeated to give the desired number of replicate estimates.

The system is simple and uses only standard laboratory equipment. The effects of barometric pressure and temperature variations are compensated by the second chamber (effectively this is a thermobarometer), although the volume of oxygen displaced must still be adjusted to STPD. The size of all the chambers and reservoirs can be selected according to the size of the animal, but it is desirable to keep the chamber volumes small for maximum sensitivity.

Improvements have been added to this basic system to provide automatic operation. The earliest was that of Clayton (1970), who used a thermo-barometer system and combined it with automated delivery of oxygen from a pipette driven by a pressure-sensitive tambour placed in the thermo-barometer. Stock (1975) used a micro-differential pressure switch to trigger operation of an automatic syringe dispenser. Recently two systems have been described (Arundel, Holloway & Mellor, 1984; Howland & Newman, 1985) in which differential pressure transducers control the operation of peristaltic pumps which introduce oxygen into the animal chamber; the amount is determined from the number of revolutions of the pump. Temperature stabilisation is achieved in these three systems by means of a temperature-regulated water jacket; the pressure-differential switches and transducers maintain the pressure in the chambers fixed with respect to barometric pressure. Again the volumes need to be adjusted to STPD. These systems are easily calibrated by withdrawing a measured volume of gas from the apparatus and recording the response.

4.2.2   *Large animal respiration chambers*

Regnault & Reiset's closed-circuit principle was greatly refined in Atwater & Benedict's (1905) calorimeter for man, but the method was not greatly used for large animal chambers until Blaxter and his co-workers revived it at the Hannah Institute. The first of a series of closed-circuit respiration chambers for studies on sheep was described by Blaxter, Graham & Rook (1954), and successive improvements on the basic design and modifications to suit special experimental requirements have been reported by Wainman & Blaxter (1958b), Kelly (1977) and Stokes (1977). A version large enough to accommodate an adult steer or cow was also built (Wainman & Blaxter, 1958a).

The sheep chambers are open-topped boxes made of 0.3 cm thick mild steel and measuring approximately $180 \times 65 \times 125$ cm. The lid, another shallow box constructed of similar material, is sealed either by dipping into a water-filled trough around the periphery of the chamber and can be completely lifted off; or by screw clamps compressing a rubber gasket, and hinged at one end. One chamber has been fitted with a door at the front to ease the entry of animals, which otherwise must be hoisted in using a sling. This chamber, which is illustrated in Fig. 4.5, also has an integral cooling jacket welded onto the outside; ethylene glycol, from a tank with both a refrigerator and heater, is pumped through the cooling jacket to provide temperature control of the chamber over the range -10–$40\,°C$.

The animal is confined in an inner cage constructed of sheet steel and weldmesh, which also carries additional equipment for excreta collection, air distribution, lighting, a drinking bowl, trough and food hoppers. The hoppers may be operated by solenoid valves from outside the chamber to control the entry of food into the trough. Water is also fed to the drinking bowl from a supply outside the chamber.

Air is drawn from the chamber by a rotary compressor and returned to the chamber via three circuits A, B, and C (see Fig. 4.6) each controlled by a diaphragm valve. Circuit A is a simple by-pass and includes a gas sampling loop with a three-way valve. Moisture is absorbed in mild steel cylinders, 112 cm long by 10 cm wide, packed with silica gel. Carbon dioxide is absorbed in 20 l polythene bottles each containing 5 l of potash

Fig. 4.5. One of the respiration chambers for sheep at the Hannah Institute. (Reproduced from Stokes, 1977.)

solution (50 % w/w KOH). Circuit B, the main absorption circuit, consists
of three silica gel cylinders, two potash bottles and a further three silica gel
cylinders to remove moisture evaporated from the potash bottles. Circuit
C consists of two silica gel cylinders. Adjustment of the proportion of the
total circulation between circuits B and C provides some measure of
humidity control in the chamber. All the absorbers are mounted together
on a wheeled trolley (see Fig. 4.5) which can be quickly replaced by another
standby trolley prepared with fresh absorbers. After weighing, the used
silica gel cylinders are dried out with a stream of heated air; they are
reused repeatedly.

Oxygen is automatically replenished as it is used up in the chamber from
a counterbalanced spirometer. The spirometer, which varies little in design
from the original described by Blaxter & Howells (1951), contains a
supply of oxygen over water at atmospheric pressure; it is isolated from
the air in the chamber only by a water bubbler. When the supply of oxygen
in the bell reaches a predetermined level, a micro-switch causes the oxygen
supply line to the chamber to be temporarily closed whilst the spirometer
is refilled to another predetermined level with oxygen from the cylinder.
Another micro-switch then causes reconnection of the spirometer to the
chamber. The number of refills is registered on a counter.

Methane produced by the animal accumulates in the chamber. Its rate
of production is estimated from analysis of samples from the chamber
taken at the beginning and end of the experimental period. Corrections to
the volumes of oxygen consumed and carbon dioxide produced are also

Fig. 4.6. Diagram of the ventilatory circuit of the Hannah Institute
respiration chamber for sheep.

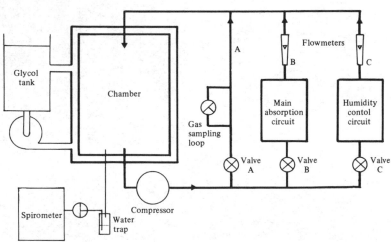

calculated from these analyses. Other measuring devices include wet and dry bulb thermometers which monitor spirometer temperature and chamber temperature and humidity, and also a water manometer for measuring chamber pressure relative to the atmosphere.

Upon starting an experiment the animal is harnessed, placed in the open chamber and the equipment for the collection of excreta connected up. A charged and weighed absorption unit is connected into circuits *B* and *C* which remain closed-off at first. The spirometer is filled to the upper level and the oxygen supply tap closed. The rotary compressor is then started and the chamber lid closed with circulation initially round circuit *A* only. A gas sampling bag is connected to the three-way tap and rinsed out twice with chamber gas before collection of the 'begin' sample. During rinsing of the bag, the barometric pressure, spirometer temperature, counter and scale reading are noted. Whilst the sample is being collected the chamber wet- and dry-bulb temperature are measured. The compressor is momentarily switched off, the manometer pressure read, and the compressor switched on again. Valve *A* is then closed, and valves *B* and *C* opened sufficiently to give the required flow through the circuits. The oxygen tap is then opened and the chamber is left running, usually for 24 h. During this time the spirometer reading is noted from time to time.

At the end of the 24 h period, valve *A* is opened and valves *B* and *C* and the oxygen supply tap are closed. An 'end' gas sample is then collected and barometric pressure, temperatures and the spirometer reading measured as before for the 'begin' sample. The rotary compressor is switched off, the manometer read and the chamber opened. If the procedure is to be repeated, the absorption train is replaced with a freshly charged and weighed one and the spirometer recharged and read again. Urine and faecal samples are removed, fresh containers inserted and the harness checked. The lid is closed, the compressor switched on and the procedure repeated as before. The whole procedure takes about 10 min.

Calculation of results is best explained by reference to the specimen calculation sheet shown in Fig. 4.7 taken from Kelly (1977). The primary data from gas analysis are the percentages listed under O_2, CO_2 and CH_4 for the 'begin' and 'end' air samples. The percentage of N_2 is calculated by difference and the nitrogen ratio (begin/end) is the volume-correction factor used to adjust the 'end' values to the same volume reference as the 'begin' values. This is analogous to the Haldane transformation made in open-circuit calorimetry (equation (4.18)). The concentration differences (Δ) are then obtained by subtraction.

The chamber gas volumes at STPD are calculated from the chamber pressure and temperature measurements noted in the mid-left hand box.

Fig. 4.7. A specimen sheet for the calculation of gas exchanges in a closed-circuit respiration chamber. (Reproduced from Kelly, 1977.)

EXPT No.	$N47(2)$	BEGIN	14/12/75	END	15/12/75	DAY No.		1
ANIMAL	209	PERIOD	III	RATION		$5+B$		
CHAMBER	B	SCALE FACTOR		51·65				

O_2 CONSUMPTION				GAS ANALYSIS						
TIME	TEMP	COUNT	SCALE		CO_2	O_2	CH_4	N_2	RATIO B/E
0854	10·0	290	68·75	BEGIN	0·22	20·22	0·17	79·39	
1610	11·6	290	45·10	END	0·40	17·38	2·06	80·16	0·9904
21·10	11·2	290	25·30	CORR	0·40	17·21	2·04		
0715	11·0	291	37·70						
08 37	11·0	291	33·10	Δ	+0·18	−3·01	+1·87		
Av.	11·0	1	35·65						

TIME FACTOR = 1·0098
MEAN BAR = 770·37

	0857	0843		
TIME	0857	0843	OXYGEN	
DRY BULB (°C)	16·0	17·8	UNITS	87·30
WET BULB (°C)	14·4	14·7	FACTOR	x 3·133
PRESSURE (p)	+7	+1	LITRES	273·48
BAR	771·75	769·00	g.a. DAY	50·09
			TOTAL	323·57
0.0735 p	+0·51	+0·07	TOTAL (24 hr)	
VAPOUR PRESS.	−11·52	−11·01		326·75
CORR. BAR	760·74	758·06	CARBON DIOXIDE	
TEMP. FACTOR	0·9446	0·9388	WEIGHT	729
JOINT FACTOR	0·9455	0·9364	FACTOR	x 0·5058
VOLUME	1762	1762	LITRES	368·73
GAS VOLUME NTP	1666·01	1649·95	g.a. DAY	2·93
			TOTAL	371·66
N_2 VOLUME NTP	1322·65	1322·60	TOTAL (24 hr)	375·31
			METHANE	
			g.a. DAY	31·16
			TOTAL	31·16
			TOTAL (24 hr)	31·46

The temperature factor is $273/(273+T_{Ch})$ and the pressure factor is $(P+\Delta P-P_w)/760$ where P is barometric pressure, ΔP the water mano-meter reading (converted by the factor 0.0735 to mm Hg), and P_w is the vapour pressure of the chamber obtained from tables of wet- and dry-bulb temperatures. The value 1762 is the physical volume of the chamber in litres. The STPD adjusted volume multiplied by the nitrogen correction then gives the volume of nitrogen in the system, which should be the same for both 'begin' and 'end'; this comparison provides a routine con-firmation that the chamber is free from leaks. The STPD adjusted 'begin' gas volume multiplied by the gas concentration differences $(\Delta/100)$ then gives the quantities accumulated in the chamber during the day (g.a. DAY): note that for oxygen the value listed is negative accumulation i.e. deficit.

The spirometer UNITS refers to the overall drop in the level (cm) and consists of the scale reading difference ('begin' minus 'end') plus a number of refills each of 51.65 cm (scale factor). This has to be multiplied by a factor representing the STPD adjusted volume of oxygen per cm change in height of the spirometer bell; it is obtained from specially prepared tables based on the equation

$$\text{Factor} = V_s \times (P-P_s)/760 \times (273+T_s)$$

where V_s is the physical volume of per cm of travel of the spirometer bell (3.257 l), T_s is the mean spirometer temperature and P_s is the saturation vapour pressure at T_s. The amount of carbon dioxide absorbed is cal-culated from the weight gain of the absorber divided by the density of carbon dioxide at STPD ($= 1/0.5058$). The combined quantities of gas accumulated and absorbed are then multiplied by a time factor which adjusts the values to a full 24 h collection period. Heat production may then be calculated using the Brouwer equation (see Chapter 3, Table 3.8).

The estimation of oxygen consumption could be improved if spirometer and barometric pressure readings were taken more frequently and the STPD adjusted volume computed between each set of readings. This would involve readings during the night period and the improvement in accuracy would be only marginal. The apparatus in its present form has the merit of simplicity; the measurements being confined to gas con-centrations, temperatures, pressures, and weights. Gas concentrations were determined by chemical analysis when the apparatus was first built.

The collection of urine and faeces allows the determination of a com-plete carbon–nitrogen balance. Using these chambers Blaxter (1962) demonstrated excellent agreement between energy retention estimated from respiratory exchange and by the carbon–nitrogen balance technique.

Further refinements of the technique have been included in four sheep-sized respiration chambers built at the Rowett Research Institute, Aberdeen (Brockway, McDonald & Pullar, 1977). These include cooling of the air before passage through circuits *B* and *C*, as this improves the efficiency of silica gel drying. Also compensating water towers connected to each chamber allow the volume of the system to be altered by ± 30 l; this manual adjustment of volume is used to assist in maintaining a steady level of oxygen despite the accumulation of methane.

The closed-circuit respiration chamber for large animals is thus simple to operate and very accurate. Its main disadvantages are the need for it to be airtight and that it is suited only to long-term (*c.* 24 h) measurements.

4.2.3   *Human respiration chambers*

Closed-circuit systems now are of little importance in the measurement of gas exchange in man, however no coverage of calorimetry would be complete without a discussion of them.

4.2.3a   *The Atwater and Benedict respiration chamber*

This was a landmark in human calorimetry combining direct and indirect calorimetry; we have described its historical background in Chapter 2. It provided considerable evidence in support of the use of gas exchange as a measure of metabolic rate; and largely on the basis of this system, it would be surprising to find anyone who would now challenge the view that oxygen consumption provides an accurate measure of energy expenditure.

In its first form, the Atwater & Rosa calorimeter (Atwater & Rosa, 1899), the measurement of gas exchange was by means of the Pettenkofer–Voit open-circuit method. It was only in the later design (Atwater & Benedict, 1902, 1905) that the open-circuit was replaced by a closed-circuit system similar to that of Regnault & Reiset.

The design of the chamber is shown in Fig. 4.8. Air was drawn from the chamber by a blower pump and passed through an external circuit. Moisture in the air was absorbed by passage through a container of sulphuric acid, and this was followed by absorption of carbon dioxide in soda lime; the moisture given off from the soda lime was then absorbed by a second container of sulphuric acid.

The fall in pressure in the system, caused by the uptake of oxygen by the subject and the absorption of any expired carbon dioxide, was detected by the movement of a rubber membrane in a device called the tension equaliser. Oxygen was introduced into the circuit from a previously weighed cylinder in order to counteract movement of the rubber membrane; the amount introduced into the circuit was measured by weighing the oxygen cylinder at the beginning and end of the study, and at regular intervals throughout. For the weighing procedure, the circuit was closed-off from the chamber, and the cylinder sealed, disconnected and then replaced by another weighed cylinder. After opening the new cylinder, the circuit was reconnected to the chamber; the replaced cylinder was reweighed and the volume of oxygen dispensed over the period calculated as described below.

In a later version of the closed circuit (Benedict, 1912) the tension equaliser was replaced by a spirometer through which the circulating air passed. As pressure in the chamber decreased, the spirometer bell fell; when the bell reached its lowest point, a micro-switch completed an electrical circuit and released a valve on the oxygen inlet line. As oxygen flowed from the cylinder into the chamber, the spirometer bell rose, the electrical circuit was once more broken and the valve closed. Each hour

Fig. 4.8. The Atwater & Benedict respiration chamber. The walls of the chamber are insulated but the chamber has a viewing window (7), and a porthole (6) that gives access to the chamber. Direct calorimetry measurements are made by the system represented by (1)–(5). Water enters through (1) and leaves through (4) and the amount of water flowing through the system is determined by collecting it in a reservoir (5). The temperature change of the water is determined by measuring inlet and outlet water temperatures by means of thermometers at (2) and (3). The indirect measurements are made using the circuit in which air is drawn from the chamber at (8) and returned with replacement oxygen at (9). The details of the means of operation of the chamber are given in the text. (After Bell, Davidson & Scarborough, 1965.)

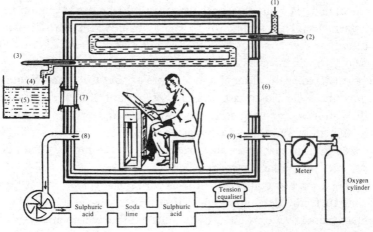

the closed circuit was sealed off from the main chamber and the spirometer filled with oxygen from the cylinder to restore the level of the bell to that from which it had started one hour previously. The cylinder was then replaced and reweighed as described above and the circuit reconnected to the chamber. The delivery of oxygen from the cylinder could also be measured by a meter.

The sulphuric acid and soda-lime containers were regularly replaced in the circuit; the change in the weight of the first sulphuric acid container gave the amount of respiratory water produced by the subject. The amount of carbon dioxide produced was obtained from the change in weight of the soda-lime container plus the second sulphuric acid container. (Although carbon dioxide is taken up by the soda lime, weight is lost as moisture is evolved and this must be collected in a second sulphuric acid container.)

The volume of oxygen can be determined from its weight by using the Avogadro's law (one gram is equivalent to 700 ml); and the volume of carbon dioxide from its density at the ambient temperature (it is not an ideal gas and does not obey Avogadro's law).

Because the chamber was so large (5100 l), any slight change in the gas concentration within it from one measurement to another would produce a considerable error in the amount of oxygen consumed and carbon dioxide produced; for example, a change in gas concentration of only 0.1 % would lead to an error of 0.5 l. To correct for this the chamber air was analysed just before the gravimetric measurements described above. From the chamber 10 l of air were removed and passed through three U-tubes, containing in turn sulphuric acid, soda lime, and sulphuric acid; and the amount of carbon dioxide determined. Using an assumed respiratory quotient, the gas concentrations in the chamber could be calculated. The sample of gas was then returned to the chamber through a gasmeter. The amounts of oxygen consumed and carbon dioxide produced, as measured in the closed circuit, could then be adjusted for any change in chamber gas concentration. An allowance could also be made for any change of chamber temperature or pressure which would affect the pressure in the closed circuit and chamber and thus the measurements of the volumes of oxygen consumed and carbon dioxide produced. Atwater & Benedict were able to make accurate measurements of the oxygen consumption, carbon dioxide production and respiratory quotient in their subjects, sometimes over long periods of time.

There were several disadvantages with their system. It was expensive to construct, and demanding and laborious to run; in 12 years of development from 1892, only 22, usually quite short, experiments were completed involving a total of five subjects. However, it is probably true to say that most of the difficulties arose from the direct calorimetry side of the system. One of the chambers greatest limitations was that only a modest range of activities were possible within it; and the need for measurement of the energy cost of everyday activities, particularly those of work and leisure, led to the need for more portable systems than respiration chambers. The absorbers in the circuit, particularly those for water, had a large resistance and created a high pressure in this part of the system making leaks more likely. The absorbers were very heavy, yet their weight change caused by absorption of water or carbon dioxide was small, and this meant that large balances, accurate to 0.1 g, were necessary.

Despite these difficulties, the accuracy of the calorimeter was impressive, even by current standards: calorimetric measurements showed that 99.8 % of the carbon dioxide and 100.6 % of the water produced, and 101.0 % of the oxygen consumed were recovered when a known quantity of alcohol was combusted in the chamber. Overall, however, this type of calorimeter has proved unsuitable for routine use, and it has been largely superseded by open circuit designs.

4.2.4　Spirometers

As well as acting as oxygen reservoirs in closed-circuit chambers as described above, spirometers have been used as closed-circuit systems in their own right for measurement of oxygen consumption. When they are used for the direct measurement of oxygen consumption, the subject is connected to the spirometer, through the inlet and outlet tubes, by means of either a face mask or a mouthpiece and nose clip.

Spirometers are composed of an inverted chamber (usually referred to as the bell) dipping into a container which is filled with water to provide an effective airtight seal. The bell is balanced by a counterweight and this provides the bell with neutral buoyancy in the water. The gas in the bell is assumed, therefore, to be at atmospheric pressure. In the container there is usually a central core, free of water, in which soda lime can be placed. The expired air passes from the inlet tube through this core en route to the bell; by this means, carbon dioxide can be absorbed from the expired air before it enters the bell. In some designs the carbon dioxide absorber is placed externally in the inlet line. Air leaves the bell through a separate outlet tube. The inlet and outlet tubes must have valves which allow the air to flow in one direction only; these valves may be built into the spirometer or may form a separate unit that is attached to the external ends of the inlet and outlet tubes.

4.2.4a   Types of spirometer

There are two main types of spirometer: the Krogh spirometer; and the Benedict–Roth spirometer. Several other distinct types of spirometers, now rarely used, are described by Bartels *et al.* (1963).

Krogh's spirometer (Krogh, 1913, 1916) is shown in Fig. 4.9a. The bell and water container are both rectangular in cross-section; the bell pivots along one edge and is balanced by a counterweight attached to this edge. A pen is fixed to the opposite side of the bell and writes directly onto a kymograph. Although it has a large cross-sectional area for its volume, it maintains high sensitivity by its pivotal arrangement and by mounting the pen where there is most vertical travel when the volume changes.

Fig. 4.9. The (*a*) Krogh and (*b*) Benedict-Roth spirometers and (*c*) a typical recording. a, carbon dioxide absorber; b, bell; c, counterweight block or chain; d, Kymograph recorder; e, water seal; f, pivot; g, pulley wheel; h, thermometer; i, oxygen inlet tap; j, one-way valve. (Fig. 4.9(*b*) modified from Bell *et al.* 1965.)

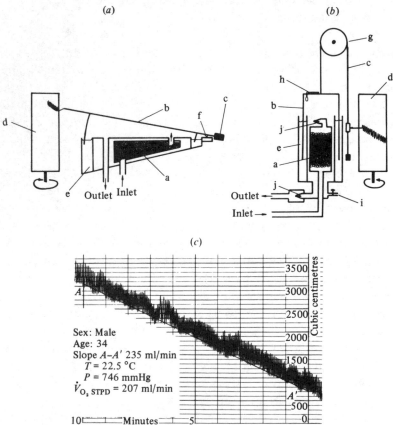

The Benedict–Roth spirometer is the most common type of spirometer currently in use. The basic design was an adaptation by Roth (1922) of an earlier, more complex design due to Benedict (1909, 1912). The bell and water container are cylindrical. The bell typically has a capacity in the range of 6–14 l, and is connected to a chain which passes over a pulley to a counterweight. The pulley arrangement ensures that the change in the length of the chain across it, and hence the distribution of chain's weight, compensates for the effect of buoyancy on the weight of the bell as it moves vertically in the water. A writing arm is attached to the pulley and writes onto a kymograph (Fig. 4.9(*b*)).

The change in volume of the bell is a product of its cross-sectional area and the change in its height above the water level of the container. The greatest reading accuracy and sensitivity can be achieved with bells having a very small cross-sectional area which, for any change in volume of the bell, gives a large vertical movement. Inevitably, if the spirometer is to have a worthwhile volume, there are practical limitations to its height and a compromise is necessary in determining its ideal dimensions. The original Benedict–Roth spirometer, and others designed for measurement of energy expenditure, were of such a size that a 1 mm change in height was equal to a change in volume of 20.73 ml; this change was equivalent to 0.1 kcal at a respiratory quotient of 0.82 (a value typical of a subject eating a mixed diet, and at which 1 l of oxygen is equivalent to 4.825 kcal).

Both of these spirometers, when filled with oxygen, are typically used for measurements lasting 10–20 min. However the duration of measurement can be increased by periodic or continuous addition of known amounts of oxygen to the spirometer; Bartels *et al.* (1963) have reviewed several manual and automated methods by which this may be achieved.

4.2.4*b* *The use of a spirometer*

Before beginning a measurement, the spirometer is usually filled with oxygen. The mouthpiece and nose clip are put in place and the subject is allowed to breathe normally to atmosphere through a bypass in mouthpiece, or, if a facemask is used, through a three-way tap in the inlet and outlet lines. Once the subject has become accustomed to the system, the operator closes the bypass and the subject is connected to the bell. Ideally the subject is unaware of the switch. The cycles of inspiration and expiration appear as vertical oscillations of the bell; which, as it is depleted of oxygen, settles deeper into the water. The height, and thus the volume, of the bell can be either read off a scale attached to the spirometer, or can be displayed on a kymograph recorder by means of a writing arm linked to the bell. If the volume change is read directly off a scale, it is important that there must be no change in the volume of the lungs from the beginning to the end of the period of measurement, since this change would be included in the recorded volume (Willard & Wolf, 1951). To eliminate this source of error, the readings at the beginning and end of the measurement are made at the end of a normal expiration. A more satisfactory way is to calculate the rate of oxygen consumption from the slope of the recording on the kymograph chart paper (Fig. 4.9(c)).

4.2.4c Spirometers as oxygen reservoirs

Spirometers have also been used for this purpose in closed circuit systems for man and animals (described elsewhere in this chapter) to increase the duration of the measurement of energy expenditure. Krogh's spirometer has been used in systems for measurement of energy expenditure in man (Krogh, 1913) and in rats (Christensen, 1946; Moses, 1947; Holtkamp *et al.*, 1955). The Benedict–Roth type of spirometer has been more widely used as a reservoir, for example: for man (Benedict, 1912; Riche & Soderstrom, 1915); for large animals (Blaxter, Graham & Rook, 1954; Wainman & Blaxter, 1958*a,b*; Alexander, 1961) and for small animals (Scharpey–Shafer, 1932; Morrison, 1947; Dewar & Newton, 1948; Smith, 1955; Haldi, Wynn & Breding, 1961; Moors, 1977).

4.2.4d Spirometers for large animals

Although spirometers have acted primarily as oxygen reservoirs in the measurement of oxygen consumption in large animals, Brody (1945) has used them, in varying sizes, as closed circuit systems.

Their use with ruminants presents special problems. The first problem is attaching the animal to the spirometer, though this can be achieved by the use of a mask sealed around the face with petroleum jelly or some such substance. The second problem is more serious and impossible to overcome because the ruminant expires methane in addition to carbon dioxide. Whilst carbon dioxide can be absorbed by soda lime, methane is not absorbed, and it accumulation in the bell causes oxygen consumption to be underestimated. Brody (1945) considers that for estimates made 8–12 h after feeding, the error would be of the order of 3 %; though it might be considerably greater if the measurements were made earlier after feeding (Brody's Fig. 12.8(*b*)). Belching of methane through the mouth and nose is less of a problem since this would be detected on the kymograph recording as a sudden blip on the trace but the slope, from which oxygen consumption is calculated, would be unaffected. The method has not been widely used.

4.2.4e   Calibration and errors

Spirometers may be calibrated by measuring a large number of internal diameters evenly distributed over the whole length and circumference of the bell (Krogh, 1920; Nelson, 1971). They can also be accurately calibrated by introducing into them an amount of air whose precise volume has been determined by passage through an accurately calibrated wet gasmeter (see Chapter 6). Periodic checks by means of burning ethyl alcohol in the system have been recommended (Benedict, 1909, 1925; Carpenter & Fox, 1923; Jones, 1929; Carpenter, 1933b; Barrett & Robertson, 1937; Lewis, Kinsman & Iliff, 1937; Lewis, Iliff & Duval, 1943); however we are not convinced that alcohol burning is a reliable technique (Chapter 7).

There are several sources of error in spirometry:

(1) Leaks: these may occur from the body of the spirometer, the inspiratory and expiratory lines, and from the mouthpiece, nose clip or face mask. The presence of leaks from the body of the spirometer can be checked by filling the bell, closing all the inlets and outlets and detecting a gradual change in the level of the bell when a 30–50 g weight is placed on it. The usual check of the overall system is made by adding a similar weight to the bell during the course of a measurement; a sudden increase in the apparent rate of oxygen consumption indicates a leak. Fowler, Blackburn & Helmholz (1957) have expressed strong reservations concerning the reliability of this overall check to detect anything other than major leaks. Such leaks usually lead to an overestimation of oxygen consumption. Leaks around the mouth and nose of the subject are very difficult to detect and are perhaps the most likely to occur; they cause oxygen consumption to be underestimated.

(2) Temperature changes: gas volumes are influenced by changes in temperature. Although environmental temperatures are unlikely to alter much in the short time over which most observations are made, the absorption of carbon dioxide is exothermic and may alter the temperature of the gas in the bell. The inclusion of a thermometer in the bell or in the outlet line immediately outside the spirometer provides a simple solution and allows the gas volumes to be adjusted appropriately.

(3) Incomplete absorption of carbon dioxide: it is difficult to routinely check whether carbon dioxide is being completely removed from expired air. Regular replacement of the soda lime is an obvious precaution.

(4) Inaccurate calibration of the spirometer and kymograph recorder: although the spirometer may be initially well-calibrated minor damage to the surface of the bell or to the writing arm can introduce significant errors. Regular checks against a wet gasmeter are essential.

(5) Lack of equilibration of gases: when the subject initially begins to breathe 100 % oxygen, the partial pressure of oxygen in the lungs increases whilst that of nitrogen decreases. Until a new steady state is achieved, the change in the volume of the spirometer will not correctly reflect the uptake of oxygen by body tissues. It has been assumed that the steady state is achieved within 5 min (Krogh & Rasmussen, 1922) but this may be optimistic and 10 min may be a more reasonable estimate (Jones, 1950; Nunn, 1977).

(6) Aqueous saturation of the gas: it is assumed that since the gas in the bell is in contact with water it will become saturated at the prevailing temperature; adjustment of the gas volumes to STPD conditions are made on this assumption. Variation in the saturation of the gas in the bell has been discussed by Lewis *et al.* (1937); in a closed circuit the saturation was close to 100 %.

4.3   Total collection systems
4.3.1   Rigid collecting chambers
Probably the best known of the systems based on the collection of expired air in a rigid container is the Tissot gasometer (Tissot, 1904) which was designed for use with man. Although considerably larger, it resembles the Benedict–Roth spirometer described above, though there are some differences. The capacity of the cylindrical bell differs according to the model of gasometer used, but it is typically in the range 100–1000 l. Because of the large capacity of the bell, a fan inside it circulates and mixes the air. The inlet and outlet tubes have individual taps.

Prior to use the water level in the container is checked, and if necessary further water added. The water temperature is allowed to equilibrate with room temperature. With either one or both of the taps open, the bell is raised and lowered over its range to ensure that its movement is not restricted by contact with the container. Once these checks are complete, the outlet tap is opened and the Tissot emptied of gas either by pressing down on the bell gently, or by placing a weight of about 1 kg on the top of the bell. The subject attaches a nose clip and inserts a mouthpiece which is connected to a two-way respiratory valve.

The subject inspires atmospheric air through the inspiratory valve and expires through the expiratory valve; initially the expired air returns to the atmosphere through a bypass (usually a three-way large-bore aluminium tap placed in the expiratory line). To collect the expired air in the Tissot, the inlet tap is opened and the outlet tap and then the bypass are closed. Once the Tissot is about one-third full, the state of the taps is reversed and once more the subject expires to atmosphere. The Tissot is completely emptied as before. The procedure, which is repeated twice more, is designed to ensure that the composition of the dead space air in the system resembles that of the subject's expired air, thus preventing mixing errors.

Immediately before beginning the collection, the initial volume or height of the bell, gas temperature inside the bell and the barometric pressure are recorded. If the Tissot is supplied with a height scale, the reading is converted to a volume using the Tissot calibration factor. A stopwatch is started when the subject is connected to the bell for the fourth time, and the volume or height scale is read at 1 min intervals thereafter. The collection is continued for an appropriate period, depending on the subject's minute volume and the capacity of the Tissot. After a suitable known time, the bypass is opened and the inlet tap closed. The internal fan is switched on to ensure thorough mixing of the gas in the bell; the gas temperature and barometric pressure are read. Temperatures and pressures are always read to an accuracy of 0.1 °C and 0.5 mm Hg respectively. The final volume is then read from the scale, or calculated from the height; all volumes are adjusted to STPD.

A sample of the expired gas is then withdrawn for analysis; this is usually achieved by attaching a gas-tight syringe or collecting bottle to a narrow outlet (sometimes the thermometer hole is used) or sampling tube from the bell. The syringe and outlet or tube are thoroughly flushed, usually at least three times, to remove any dead space air and to ensure that the sample taken has a gas concentration truly representative of the expired air. The sample can then be analysed chemically or, after drying, by means of electrical analysers (a more detailed account of gas analysis is given in Chapter 6). Alternatively, a continuous stream of the expired air can be drawn out from the bell by a pump, dried and passed through electrical analysers. Calculation of oxygen consumption and carbon dioxide production is covered later in this chapter.

The gasometer can be calibrated like the spirometer. Many of the potential errors in the measurements of volume made by the Tissot are similar to those discussed in the section on spirometers. One additional problem with the Tissot, or any other similar system for collection of expired air, is that the non-return respiratory valves sometimes have a resistance which is sufficiently high to cause the subject to hyperventilate. This produces an increase in the respiratory quotient as carbon dioxide is 'blown off'.

Blaxter & Howells (1951) have described a similar system for use with farm animals, and the bell, which was made of light aluminium, was capable of collecting about 200 l of expired air. The animal (a calf) was linked to the apparatus by means of a facemask of sheet rubber sealed around the face with petroleum jelly. The usual counterpoise system was replaced by a modification of one originally used by Tissot. It comprised a glass tube attached by means of a cord over a pulley to the bell; inside the glass tube was a thinner copper tube that entered from the bottom and was connected to the base of the water tank by means of a length of rubber tubing. As the bell rose, water automatically siphoned through the rubber and copper tubes into the counterweight, exactly compensating for the increase in the weight of the bell. The movement of the water level in the glass tube was also exactly equal to movement of the bell and was used to measure it. The water level in the tank was kept constant by means of a constant overflow device.

The operation of the spirometer was similar to that of the Tissot described above; typically expired air was collected for about 30 min. A sample of the expired air was taken for gas analysis.

The major disadvantages of the methods of collecting expired air in rigid containers is that the apparatus is not easily portable and the duration of measurement is very restricted. Like the closed-circuit spirometers and respiration chambers, their use has been limited and they have not proved as popular as the bag systems described below.

4.3.2   Bag systems
4.3.2a   The Douglas bag

The introduction by Douglas (1911) of a canvas bag, which now bears his name, for the collection of expired air from man was a major event in the development of calorimetry. Cunningham's (1964) obituary of Douglas described its origin: 'During a discussion over laboratory tea on the measurement of the respiratory exchange his thoughts went back to the magic lantern shows of his childhood, given by the Vicar of a Leicester parish.' The oxygen and fuel gases for the limelight were stored in rubberised canvas bags, appropriately weighted to supply the correct mixture. 'The suggestion that the expired air should be collected for a known time in a gas bag was only natural', writes Douglas, '...I can claim no originality...'. A few days later he had obtained two such bags and a few days' testing showed the method was 'really promising'.

The Douglas bag was, by the standards of the time, lightweight and could be carried on the subject's back by means of attached shoulder straps. It allowed measurements of energy expenditure to be made on subjects who were able to carry out a range of activities not possible in respiratory chambers or with systems based on gasometers.

Its design has changed little from its inception: the traditional wedged-shaped Douglas bags are constructed of canvas with a vulcanised rubber lining, and are available in several sizes, although 100 or 150 l capacity bags are perhaps the most common. A large calibre flexible rubber hose emerges from the top of the bag and is connected to a wide-bore heavy three-way tap (Siebe, Gorman & Co.). The tap is connected via a length of corrugated tubing to a respiratory valve system and mouthpiece. A short-side tube, held firmly closed by a clip, is connected to the rubber hose to permit the removal of small gas samples from the bag.

Details of the bag and its use have been given by Cathcart (1918) and by Douglas & Priestley (1924). With the tap open, the bags to be used are carefully rolled up tightly to expel any residual air within; the tap is then turned to close the bag to outside air. The subject inserts the mouthpiece and fixes the nose clip in place; the corrugated tube is then attached to a Douglas bag, which may if necessary be placed on the subject's back. The three-way tap is positioned to close off the bag and permit the subject to expire to atmosphere. Once the subject is settled, the tap is turned to collect the expired air in the Douglas bag, and a stopwatch is started. The collection may be for a fixed-time period or the bag filled to a pre-determined volume and the time taken. At the end, the tap leading to the bag is finally closed and the subject expires to atmosphere. It is possible to have a tap arrangement which permits the expired air to be switched to a second bag immediately the first bag is judged to be full, or on the completion of a fixed period of time. By this means the duration of collection can be extended, although handling large numbers of bags without introducing errors requires careful organisation. At high ventilation rates the method is limited by bag size; with moderate levels of exercise a 100 l bag can be filled in one to two minutes, which is probably faster than its contents can be measured.

The methods used to measure the volume of expired air and its gas concentration are described in detail in Chapter 6; only an outline is given here. Sampling and analysis of gas from the bag are as described in the previous section on rigid collection systems. The volume of gas in the bag is measured by means of a wet or dry gasmeter. The bag is connected to the meter, the tap is opened and the bag slowly squeezed to deliver its contents through the meter; as a final step the empty bag is rolled in an attempt to return it to its original volume. The initial and final readings on the meter are taken to calculate the volume of expired air; the temperature of the gas in the meter and barometric pressure are also taken for adjustment of the volumes to STPD. The volume of gas removed for gas analysis must be added. Alternatively a continuous sample of gas can be taken from the outlet of the gasmeter by means of a pump, and passed through a drier to electrical analysers.

The Douglas bag continues to be widely used for the measurement of energy expenditure in man, even now 75 years after its introduction. It is surprising that such an important method was only ever published by Douglas as a communication to the Physiological Society (although he did later describe it more fully in his textbook of practical physiology written with Priestley).

Blaxter & Joyce (1963) have used Douglas bags to measure the energy expenditure of tracheostomised sheep (see Section 4.4.1e). The volume of the bag allowed collection of expired air to be made for up to 30 min. The expired air volume was measured by pumping it from the bag through a wet gasmeter; a sample of gas was collected from the outlet of the meter, dried and analysed for oxygen, carbon dioxide and methane content. Agreement between 'resting levels' of oxygen consumption measured in a respiration chamber and with Douglas bags was good: 16.15 and 16.23 l/h respectively; however, the Douglas-bag method considerably under-estimated carbon dioxide and methane production but much of this probably originated in the rumen rather than the lungs and was presumably lost by belching. In that sense the Douglas-bag method probably gives a better estimate of the 'true' respiratory quotient.

The cost of canvas bags is now substantial and vinyl bags introduced by Perkins (1954) are a cheaper and lighter alternative, although they have a shorter working life, being more prone to damage in our experience.

The major sources of error in the Douglas-bag technique, besides the purely analytical ones which are dealt with in Chapter 6, are leaks from the bag and diffusion of gas through its walls.

(1) Leaks: although the traditional canvas Douglas bag is robust and lasts many years, it may at some point develop leaks. Before use it should therefore be checked: this is usually achieved by filling an empty bag with a known volume of air (measured by passage through an accurately calibrated gasmeter) closing the tap and then placing a weight on the bag; after a period of a few hours the volume of air in the bag is determined by passing it back through the gasmeter. A recovery of less than 100% indicates leakage and recovery must be better than 99% or the bag rejected. This check does not include the possibility of leaks around the subject's mouthpiece and through the nose clip. Blaxter & Joyce (1963) have used a method, which although used with sheep, could be adapted more generally. A bag was filled with an accurately determined volume of air of known nitrogen content. The sheep were allowed to breathe from

this bag and the expired air collected in a second, originally empty, bag. When the inspiratory bag was almost empty, both bags were sealed and disconnected. After the volume of gas and the gas concentrations in the two bags were measured, the volume of nitrogen inspired was compared with the volume of nitrogen expired. The recovery of nitrogen in five such tests was 99.0 % $\pm$ 1.1 %.

(2) Diffusion of gas: rubber is notoriously permeable to gas and this has caused concern to those investigators using Douglas bags. It seems impossible to prevent such diffusion, but it is slow and with careful technique its effects may be negligible. Thorough studies of the problem have been carried out by Balchum *et al.* (1953), Shephard (1955) and Consolazio *et al.* (1963); a study with vinyl plastic Douglas bags was reported by Perkins (1954).

The outcome of these studies is summarised in Fig. 4.10. In essence, there was a substantial reduction in the concentration of carbon dioxide from both rubber and plastic Douglas bags, though the plastic bags were less permeable than the rubber ones: over a 24 h period carbon dioxide concentration decreased by almost half. As carbon dioxide concentration decreased, the concentration of oxygen increased, partly through the decrease in volume with the loss of carbon dioxide, but also because of the diffusion gradient from air. The nitrogen concentration also changes, the direction depending on the balance between the change in volume in the bag and the diffusion gradient for nitrogen (which depends on the respiratory quotient of the expired air). However, the diffusion of oxygen and

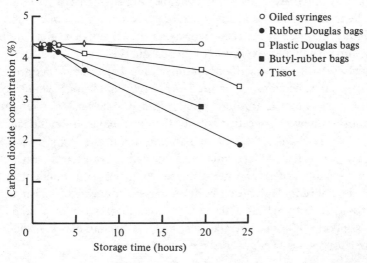

Fig. 4.10. Diffusion of carbon dioxide from several gas storage systems.

o Oiled syringes
● Rubber Douglas bags
□ Plastic Douglas bags
■ Butyl-rubber bags
◊ Tissot

nitrogen was considerably less: the diffusion ratios of carbon dioxide, oxygen and nitrogen have been estimated to be about 11:2:1 (Balchum *et al.*, 1953; Shephard, 1955). The diffusion ratios seemed to be little affected by the age of the bags.

In addition to the permeability of the material of the bag, the other factors effecting the loss of carbon dioxide are those which generally effect the movement of a diffusible substance across a permeable membrane: the diffusion gradient and the surface area of the permeable surface. Thus, for the Douglas bag, there is a positive relationship between the partial pressure of the carbon dioxide inside and its rate of loss from the bag and between the size of the bag and the rate of loss.

The initial volume of gas seems to have little affect on the rate of loss, but the effect of the loss on the gas concentrations is more substantial in bags with a small starting volume than in bags with a large one.

Providing the gas concentrations in the Douglas bags are determined within 10–15 min of collection, the error caused by diffusion is negligible, unless only a small part of the bag's capacity is occupied by gas, in which case an error of up to 1 % of the carbon dioxide production is possible (Shephard, 1955).

4.3.2b   *Other types of bag*

Cooper & Pask (1960) have described a 'home-made' hexagonal PVC bag which they claim is much easier to empty than the Douglas bag and consequently slightly more accurate. The bags will, however, still be permeable to gases. For absolute accuracy, collection in metallised bags, such as those marketed in the United Kingdom by the Signal Instrument Company (Camberley, Surrey), is possible, but at high cost. Metallised bags, described by Johnson *et al.* (1967), are composed of a sheet of aluminium sandwiched between two layers of polyethylene film and are totally impermeable to oxygen and carbon dioxide.

Daniels (1971) has described an ingenious bag system suitable for making serial collections of expired air. Five lightweight 200 l rubber balloons are connected through evenly spaced sidearms to a length of acrylic resin tubing. A collection tube of the same material slides inside the length of tube on which the balloons are mounted; one end is sealed and the other connected to corrugated tubing which connects in turn to a respiratory valve system and mouthpiece. By means of a hole cut in the side of the collection tube each balloon can be connected in turn to the subject and the expired air collected. Gas samples are quickly taken for analysis, taking the usual precautions, and the volume of air in the bags measured by means of a gasometer or flowmeter.

Despite attempts to devise methods for the serial collection of expired air, the main disadvantage of using bags is their low capacity and the short duration of measurement which is possible. Even a 150–200 l bag will be filled by a man working at high rates in two to three minutes, and such large bags rapidly become unwieldy upon filling and limit the subject's activity. This failing has led to the development of several highly portable systems for the measurement of energy expenditure.

4.4 Open-circuit systems

4.4.1 Portable systems (sample collection)

4.4.1a *The Kofranyi–Michaelis (K–M) or Max Planck Respirometer*

Probably the first real portable system was that developed by Zuntz *et al.* (1906) and used in their expedition to study the physiological effects of high altitude; it is illustrated in Fig. 2.5. It was derived from a bulky non-portable system developed by Geppert & Zuntz (1888). The subjects' expired gas volume was measured by a dry gasmeter and a sample of gas manually collected for subsequent analysis. Interestingly, although the apparatus preceded the Douglas bag it was not as widely adopted. The instrument was further improved by Kofranyi & Michaelis (1940) at the Max Planck Institute in Germany; they added a pump that automatically sampled 0.85 ml per l of gas passing through the instrument and this sample was then stored in a rubber bag attached to the instrument

Fig. 4.11. The Kofranyi & Michaelis portable respirometer. (Reproduced from Garry *et al.*, 1955.)

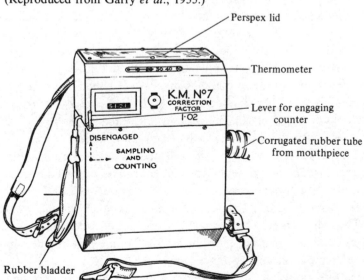

before analysis by the Haldane apparatus. The instrument was made more compact by Muller and Franz (1952) who used a smaller dry gasmeter which drove a counter rather than a pointer around a dial.

This apparatus, called the Kofranyi–Michaelis meter or the Max Planck respirometer, has had a major impact on the measurement of energy expenditure. The K–M meter allowed investigators to extend the measurement of energy expenditure to activities for which the Douglas bag was unsuitable, either because of its bulk or low capacity. The manufacturers have suggested, however, that the instrument is not suitable for flowrates much in excess of 60 l/min.

The K–M meter comprises a metal box (Fig. 4.11) approximately 20 cm wide, 27 cm high and 11 cm deep; its weight, including ancillary equipment such as harness, valves and mouthpiece, is about 3 kg (half that of the Douglas-bag equipment). The instrument can be stood alongside the subject or carried like a rucksack. The metal box contains a dry gasmeter similar to those described in Chapter 6, and comprising two bellows which in turn fill and empty. The temperature inside the meter is measured by a thermometer mounted on the outside of the case.

The subject wears a mouthpiece and nose clip and expired air is delivered by means of a connecting length of corrugated rubber tubing to the gasmeter. Inflation and deflation of the bellows is transmitted by a mechanism which converts their lateral motion into a circular movement to drive a counter that gives a measure of flow. The movement also drives a pump which withdraws a fixed proportion of the expired air, with the remainder being discarded to atmosphere. A three-way valve, operated through an external lever, controls the metering and gas collection which can be turned off and on at either of two sampling rates: nominally 0.3 and 0.6% of the expired air volume. The sample passes to the outside of the meter through a metal pipe and is collected in an attached rubber or plastic bladder of about 500 ml capacity. (The meters are originally supplied with small bladders of about 5–10 ml capacity, but these are best replaced by larger ones.) The gas concentrations of the stored sample can then be analysed chemically or by electrical analysers. At the lowest rate of sampling, the amount of air collected would represent about 150 l of expired air. The duration of measurement can be increased by using several bladders and attaching them in turn to the meter; Consolazio (1971) has used plastic bags of 2 l capacity for measurements of high workrates.

The main sources of error, apart from those associated with the actual measurement of gas concentration, arise from contamination of the collected sample. Primarily this is caused by:

(1) Gas diffusion: the bladders suffer from the same problems of gas

diffusion as the Douglas bag, and if the samples must be stored for prolonged periods, they are best transferred to airtight syringes.

(2) Introduction of dead space air: with the capacity of bladder used, the introduction of even small amounts of dead space air causes substantial errors in the measurement of oxygen consumption. A 3 % contamination of expired air with fresh air would introduce an error of about 1 % in the determined rate of oxygen consumption. To prevent or minimise contamination, the bladders must be tightly rolled up to remove any dead space air before attachment to the meter; and, since there may be such air in the meter, it should be run for several minutes before attaching the bladder.

The other major error is in the estimate of flow through the meter, although this can be reduced by careful calibration. Whilst calibration against a constant stream of air passed through a wet gasmeter or delivered from a Tissot has been used (Montoye *et al.* 1958) a pusatile flow is probably better. Calibration methods are more fully discussed in Chapter 7. Using pusatile calibration Consolazio and colleagues (Riendeau & Consolazio, 1959; Consolazio *et al.*, 1963; and Consolazio, 1971) found that the manufacturer's calibration was somewhat different to their own determined one: of six meters tested, four underestimated, and two overestimated the true volume of air by an average of 4 % and the overall 95 % confidence limits were ±0.9 l/min. Lukin (1963) reported standard deviations of 1–2 % in the accuracy of measurement of minute volume of a pulsatile flow.

Orsini & Passmore (1951) observed that their K–M meter systematically underestimated volume by about 20 % when checked against a known volume of air delivered at different rates from a Douglas bag or a spirometer. Nevertheless over a wide range of work rates, they found that the K–M meter systematically overestimated oxygen consumption by about 4 % compared with the Douglas-bag method and this they considered to be acceptable for the use to which it was to be put. Consolazio *et al.* (1963) have attempted a complete analysis of the errors in measuring energy expenditure by the K–M meter. For accurate reliable use the instrument requires regular careful servicing.

The K–M meter has been widely used for measurement of energy expenditure in man going about his typical everyday activities; and probably most of the estimates of the energy cost of domestic, industrial, sporting and military tasks have been determined with it (for a summary see Durnin & Passmore, 1967; Consolazio, 1971). The additional energy cost of its carriage is negligible.

At minute volumes of about 20 l or less, the resistance to expiration of the meter is similar to that of the Douglas bag and is no more than 8 mm of water. Above this rate of flow, the resistance increases considerably, primarily because of the respiratory effort required to drive the bellows and proportioning pump (Montoye *et al.* 1958).

The instrument has also been used for studies on tracheostomised animals.

4.4.1*b*   *The integrating motor pneumotachograph (IMP)*

The introduction of the IMP by Heinz Wolff in 1958 appeared to offer substantial all-round improvements on the K–M meter, including reduced resistance to flow. In the event it has proved a failure, not through any defects in design, but through the unreliability of several of its components; the design was ahead of the technology of the time.

It was designed to be sufficiently comfortable for prolonged use, if necessary for the whole of a 24 h period, with a sampling capacity that allowed a collection period of not less than one to two hours between changes of sample bag. The apparatus comprised three main units: the combined mouthpiece and flowmeter; the integrating unit and the sampling box; the development of these is described in detail by Wolff (1958*a*).

The flowmeter, which was housed in a clear plastic dome attached to the facemask, comprised a chamber with a side inlet, and an outlet made of a hollow tube mounted in the centre of the base of the chamber, and projecting about two-thirds of the way into it. The top of the chamber was covered with a thin terylene diaphragm whose centre was glued to the upper surface of a nickel-plated disc. The disc was secured to the base of the chamber through three evenly spaced, and slightly stretched, isotonic beryllium springs; at rest these held the disc firmly over the outlet.

Expired air entering the chamber increased pressure slightly and forced the diaphragm and disc up, allowing the air to escape through the outlet. The disc was displaced in proportion to the rate of air flow and this increased the gap between the disc and outlet. The springs provided a restoring force to the disc; the isotonic nature of the springs ensured a near-constant pressure in the chamber despite any increase in displacement of the disc. Thus movement of the disc was linearly proportional to flowrate. In practice, the springs were not perfectly isotonic when stretched and at high flowrates they resisted deformation and produced a non-linear relationship between flowrate and disc displacement. Linearity was restored by an adjustable obstruction placed over the outlet; by empirical adjustment, the obstruction increased pressure in the chamber at high flowrates to cancel out increased tension in the springs. Nevertheless, even at high flowrates, the resistance to flow never exceeded a pressure of about 2.5 cm of water.

Movement of the diaphragm produced a voltage that was measured by a micropotentiometer, and was directly proportional to the rate of flow through the chamber. In use the flow of air through the chamber was pulsatile and, to obtain the minute volume, the voltage output with each expiration had to be integrated with respect to time. The integration was carried out by means of a permanent electric motor with a light, almost inertia-free armature which turned around a stationary iron core. Friction was so low that the motor turned at a voltage as low as 40 mV, and thereafter up to 12 V the speed of the motor was linearly related to the voltage. The rotation of the motor was measured by a revolution counter. The count was the time integral of the voltage applied to the motor from the micro-potentiometer and thus was proportional to the integrated flowrate through the chamber. The output impedance of the micro-potentiometer was too high and variable for direct connection to the motor, so an impedance transformer with two junction transistors was used to provide a low source impedance. This also allowed electrical zeroing. Power for the motor was provided by batteries.

Representative samples of expired air were taken by a single stroke 90 V battery-operated electromagnetic pump periodically driven from the counting mechanism. The sample was passed into a 400 ml laminated PVC or butyl rubber bag. The volume of the stroke of the pump was typically set to 0.3–0.5 ml and a sample was taken from every 1.5 or 4.5 l of expired air by selection through a moveable linkage on the outside of the instrument. To prevent diffusion of gas from the sample bag, it was placed in an aluminium tin filled with expired air, and as the bag filled the surrounding air escaped through a non-return valve in the lid of the tin.

The IMP was demonstrated transmitting, through a radio link, the ventilation rate of soldiers moving across an assault course. Each soldier carried a continuously transmitting lightweight (*c.* 250 g) radio; contacts in the integrating unit switched off the transmitter for a few milliseconds after the passage of a predetermined volume of expired air through the IMP. The transmission could be monitored and recorded by a receiver up to 500 yards away and the breaks in transmission counted (Douglas, 1956; G. R. Hervey, personal communication). Although this application did not achieve any significant use it did demonstrate the potential versatility of the system.

To check the accuracy of the IMP, Wolff (1958*b*) adapted two instruments to allow collection of the total expired air in a Douglas bag after passage through them. Over a range of minute volumes from 6.4 to 81.0 l, the discrepancy in the volume measured by the IMPs, expressed as a percentage of the volume of air in the Douglas bag, ranged from -0.5 to $+0.9\%$ with a negligible difference in the oxygen and carbon dioxide content of the gas.

The failure of the IMP to become widely used, despite its ingenuity and its obvious improvements over both the Douglas bag and the K–M meter, appears at first sight to be surprising. However, in practice the instrument proved unreliable, and in particular the integrating unit was a source of frequent problems. The difficulty seemed to arise mainly from the unstable nature of the transistors: for a fixed voltage output from the micro-potentiometer the input to the motor could change quite rapidly, even within two hours of calibration. Even with no flow through the chamber it was possible to have a voltage input to the motor giving a spurious reading. The batteries also often proved unsatisfactory and this was particularly important since the IMP relied on a steady voltage for stability of calibration. Connections were also prone to damage.

Subsequent portable systems, of which the main ones are described below, have been less adventurous and have used proven components. As a consequence they have been more reliable and successful.

4.4.1c   *The Miser*

The Miser was first briefly described in 1976 (Eley *et al.*, 1976; Goldsmith *et al.*, 1976) and fully reported later by Eley *et al.* (1978). Miser is an acronym for its major design features: a miniature, indicating, sampling, electronic respirometer. It consists of three parts: a gasmeter; a vacuum container and flow regulator for the collection and storage of a gas sample; and a control and display unit.

The facemask has three inspiratory valves and is fitted with a bimetallic thermometer. The flowmeter, which is an electronic Wright respirometer (Chapter 6), is fitted to the single expiratory outlet and generates 320 pulses per litre of flow. A short extension tube with an expiratory valve and a sampling port is attached to the axial outflow of the respirometer. The electrical pulses from the meter and the power for the electro-optical sensor are carried by a flexible lead between the meter and control unit. The sampling port is connected to an electromagnetic valve in the control unit by a narrow-bore, thick-walled flexible PVC tube.

Beyond the valve is an evacuated 110 ml aluminium container with a flow regulator which ensures that flowrate into the container remains constant until its pressure rises to about 700 mm Hg (93 kP). In use the container is clamped to the control unit. The gas sample is analysed chemically or by an electrical analyser; the most difficult problem is extracting the sample from the container. It is removed by connecting an evacuated plastic bag, itself contained in an evacuated chamber, to the container. A valve in the neck of the bag is opened, gas is sucked into the bag and the valve is closed again. A valve in the base of the evacuated chamber is then opened and the chamber raised to atmospheric pressure. The bag is then connected to the analyser, its valve opened and the sample expelled into the analyser aided if necessary by pressurisation of the chamber through the second valve.

The control unit also contains the electrical circuit to record the volume of gas. The output from the meter is divided by 32 by a binary divider, and counted by a decade volume counter and the volume displayed to the nearest 0.1 l; additionally the pulses are divided by 64 and after differentiation are available for storage on a portable cassette tape unit. The frequency of gas sampling through the operation of the electromagnetic valve is also controlled from the control unit. Sample size can be adjusted between 0.1 and 0.5 ml and a sample taken once every 0.4–1.6 l. The internal rechargeable batteries give between 7 and 9 h of continuous use.

Agreement between the Miser and Douglas-bag methods for oxygen consumption was about 2% although this was achieved despite a substantial overestimation of volume (which then needed a correction factor) by the Miser. The instrument has been successfully used for measurement of energy cost of several occupational activities (Fordham *et al.*, 1978; Goldsmith *et al.*, 1978) and in an Antarctic expedition (I. F. G. Hampton – personal communication). Its main weaknesses are the danger of leakage from the facemask and the limited accuracy of the respirometer.

4.4.1*d* *The Oxylog*

The 'Oxylog' was designed by Humphrey & Wolff (1977). It is 18.5 cm wide, 8.2 cm deep and 21.5 cm high; including its carrying case it weighs 2.6 kg. It comprises a close-fitting facemask containing inspiratory and expiratory valves. A thermistor and an electronic Wright respirometer (see Chapter 6) are mounted in the inspiratory valve to measure the temperature and volume respectively of the inspired air. The expired air is conveyed through a flexible corrugated hose, which also supports the electrical cable from the flowmeter, to a gas analyser carried in a case strapped to the waist. The expired air passes through a mixing unit and is continuously sampled by a battery-operated miniature double piston pump. The sample is supplied to a drying tube containing anhydrous calcium sulphate and then to a Beckman oxygen electrode (see Chapter 6);

samples of 'inspired' air are delivered to a second oxygen electrode after drying. Both electrode assemblies also include thermistors. Power is supplied by internal rechargeable batteries with an endurance of up to 24 h.

Suitable electronic circuits calculate the product of the volume and oxygen differential between 'inspired' and expired air. The calculation includes a first-order correction for changes in barometric pressure and temperature, assuming 50 % relative humidity in the inspired air. Values for the total oxygen consumed, total ventilation and rates of oxygen consumption and ventilation are shown on digital displays. There is a recorder output which provides a 5 V positive pulse of 30 ms duration for each 0.1 l of oxygen consumed and each litre of ventilation.

The Oxylog is suitable for ventilation volumes of 6 to 80 l per minute, a total oxygen consumption up to 9999.9 l (with a resolution of 0.1 l) and a rate of consumption within the range 0.25–3.00 l/min (with a resolution of 0.01 l/min).

The respirometer can be calibrated by one of the methods described in Chapter 7; the oxygen electrodes are set up with oxygen-free nitrogen as the zero gas and dry fresh air for the span gas, and the instrument assumed to be linear (Chapter 7).

The instrument has been evaluated by Ballal & Macdonald (1982) and by Harrison, Brown & Belyavin (1982). The Oxylog showed good agreement with the Douglas-bag method for both ventilation rate and oxygen consumption. In general discrepancies were less than about 3 % (although discrepancies were considerably larger over individual one-minute measurements). The main criticisms of the apparatus were the leakiness and discomfort of the facemask and the ease with which it could be dislodged, the difficulty in reading the small digital displays, and the discomfort of carrying the instrument.

Nevertheless the Oxylog is reliable, and it has been successfully used in several field studies (e.g. Golden, Hampton & Smith, 1979, 1980; Dounis, Steventon & Wilson, 1980).

4.4.1e   *Tracheal cannulation techniques*

Tracheal cannulation was used quite extensively nearly 100 years ago as a means of collecting respiratory gases from sheep, cattle and horses. It overcomes the obvious difficulties of making animals wear masks, and makes it possible to conduct measurements on grazing animals. Flatt *et al.* (1958) have summarised the early work. They also devised a portable gas collection method for tracheostomised cattle which used a K–M meter modified slightly to measure greater flowrates; sampling was reduced to 0.1 % of flow and the gas stored in a 100 l plastic Douglas bag.

The trachea was completely transected and the two halves rejoined by insertion of a close-fitting flexible plastic tube. When measurements were required the tube was removed and a catheter of the same flexible material, and with an inflatable cuff, was inserted into the posterior portion of the trachea. Blaxter & Joyce (1963) found the transection method unsuccessful with sheep and used a method similar to that of Cresswell (Webster & Cresswell, 1957; Cresswell, 1961; Cresswell & Harris, 1961). Blaxter & Joyce (1963) formed a fistula by removing a segment from one or two cartilaginous rings on the ventral side of the trachea. After healing a split plug was inserted which formed a T-piece when assembled. For measurements the upper arm of the T-piece was occluded by an inflatable balloon. The methods of Flatt *et al.* (1958) and Blaxter & Joyce (1963) had limited success; problems were encountered with the accumulation of mucus and with the gradual development of a growth or swelling in the trachea. At most the preparations lasted for only one year.

A further disadvantage of these methods was that all the breathing via the normal nasopharyngeal route was abolished during measurements

Fig. 4.12. Re-entrant tracheal cannula. Reproduced from Young & Webster (1963).

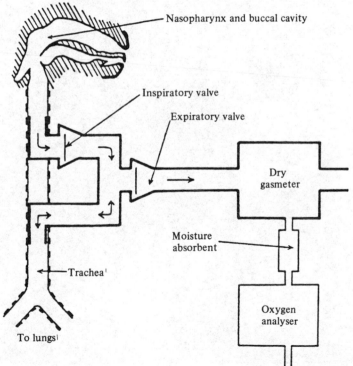

Fig. 4.13. The Mobile Indirect Calorimeter (MIC). (Reproduced from Young & Corbett, 1972.)

and evaporative cooling of the upper respiratory tract prevented; in addition gases eructated from the rumen were not collected. Young & Webster (1963), also working with sheep, overcame these disadvantages. They formed two tracheal fistulae 5 cm apart, each by removal of the ventral half of one cartilaginous ring from the trachea and suture of the tracheal mucosa to the skin. A re-entrant tracheal cannula, shown diagrammatically in Fig. 4.12 could then be inserted into the fistula for measurements. With this system inhalation occurs through the nasopharynx, but exhalation is diverted through the outlet valve of the cannula. Young & Webster (1963) demonstrated that when inspiration through the nasopharynx was prevented during feeding, food intake was depressed and respiratory activity increased.

The same cannulation technique was later combined with a lightweight cart which carried a K–M respirometer (Corbett, Leng & Young, 1967) and was drawn by the sheep (Fig. 4.13). One of the shafts was used as a duct to conduct expired air from the sheep to the meter. The system, which they named Mobile Indirect Calorimetry (MIC) was used to measure energy expenditure of free-ranging sheep. As well as allowing normal inspiration via the nasopharynx, MIC includes full collection of gases eructated from the rumen. This occurs because, as Colvin *et al.* (1957) showed, the normal route first taken by eructated gas is via the trachea into the lungs and not direct expulsion via the nasopharynx. Using this apparatus Young & Corbett (1972) demonstrated a 40–70 % greater level of metabolism in grazing as against penned sheep.

4.4.2 Ventilated flow-through systems

Flow-through calorimetry has become the most popular method for the determination of energy expenditure in man and animals where

Fig. 4.14. The means of collecting expired gas in flow-through systems using: (*a*) a mask or mouthpiece; (*b*) a ventilated hood; or (*c*) a chamber. The dashed circle represents the subject taking up oxygen from, and releasing carbon dioxide to the air.

(*a*)   (*b*)   (*c*)

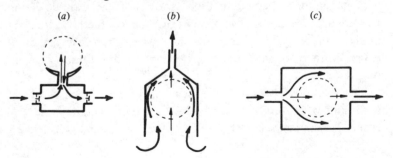

some confinement is acceptable. The principle of the system is that a stream of air is forced to pass across the face of the subject or animal and collects the expired air as it does so. The expired air can be collected in one of three ways (Fig. 4.14): from a mouthpiece or a facemask; a hood or canopy over the head or upper body; or from a chamber, with a single inlet and outlet, that totally encloses the subject. The measurement of energy expenditure requires the determination of the amount of air flowing over the subject and the difference in the gas concentrations of the incoming and outgoing air. The apparatus is highly flexible: by altering only the size of chamber or hood, flowrate and the capacity of the flowmeter, the same equipment can be used for man and animals of greatly differing sizes.

The flow-through method is considerably more demanding of equipment, particularly for gas analysis, than the indirect methods of calorimetry hitherto discussed. The development of highly accurate and precise electrical analysers has been a major stimulus for substantial progress in flow-through calorimetry. Calibration and checks of overall accuracy (covered in detail in Chapter 7) must be of the highest order, but unfortunately this is not always appreciated by those who use such systems.

In addition to continuously measuring gas concentrations on-line, some investigators (for example van Es and his colleagues) also continuously sample the expired gas over the experimental period and then measure its gas concentration (often chemically) at the end of the period. This single estimate provides a check on the value derived from the very many individual measurements made on-line by the physical gas analysers. Schoffelen *et al.* (1984) has described such a system.

An outline of the flow-through method will be described in this section but the principles of the flowmeters and gas analysers often used as part of it are discussed in detail in Chapter 6.

4.4.2a   *Open-circuit chambers*

The open-circuit chamber method was one of the earliest types of calorimeters, having been devised by Pettenkoffer & Voit in 1875, although at that time only the volume of air and its carbon dioxide content could be measured. With the development of chemical methods for the accurate analysis of oxygen, it became widely used for measurement of energy expenditure in animals, particularly those of agricultural importance, and man.

The method fell into disuse because of the difficulty of chemically analysing large numbers of samples to a high accuracy and the relative ease of other methods. The development of accurate electrical gas analysers has reawakened interest in the method. There are a large number of reports in the literature of open-circuit chambers and we do not propose to list them all. We will mention here only a few that the reader might consult. Jequier & Schutz (1983) have recently described an open-circuit chamber for human studies; open-circuit chambers have been combined with direct calorimetry for studies on man (Dauncey, Murgatroyd & Cole, 1978) and primates (Dale *et al.*, 1967). Open-circuit chambers have also been widely used with poultry (Bønsdorff Petersen, 1969; Misson, 1974), pigs, sheep and cattle (Verstegen *et al.*, 1971; Vermorel *et al.*, 1973; Young, Kerrigan & Christopherson, 1975). A multichamber system for poultry has been built, and recently extended, at the AFRC's Poultry Research Centre in Scotland (Lundy, MacLeod & Jewitt, 1977; MacLeod, Lundy and Jewitt, 1985). At Leeds we have built two multichamber calorimeters for rats and these are discussed in detail below and in Chapter 8.

4.4.2*b* Ventilated hoods

The ventilated hood method has achieved widespread use for human and animal studies. It uses a plastic, perspex, or polythene canopy to cover the head, the upperpart of the body or the whole body; for small animal use the hood can cover a standard animal cage or even act as the cage itself (Armitage, Hervey & Tobin, 1979; Tobin & Hervey, 1984). Air enters the hood either through one or more specific inlets or through gaps between the hood and the subject; in most systems air is drawn through the hood by negative pressure created by a pump downstream so the danger of losing expired air is negligible (although common sense must prevail, on one of the earlier prototype cages at Leeds, the rats were found inserting their noses into one of the many inlets and expiring to atmosphere).

The ventilated hood can be traced back at least to Grafe (1909) and Benedict (1930). As with chambers there are far too many reports in the literature on ventilated hood systems to quote them all, but we suggest the reader might consult some of the following: Kinney *et al.*, 1964; Kappagoda, Stoker & Linden, 1974; Lister, Hoffman & Rudolph, 1974; Long *et al.*, 1979; Hampton, Stainthorpe & Tobin, 1982; Duggan, Tobin & Milner, 1984; Weissman *et al.*, 1985). Caldwell, Hammel & Dolan, 1966 have described a hood system for rats and placed inside a direct calorimeter chamber. We have found the ventilated hood to be highly successful for the measurement of energy expenditure of rats for periods of up to several months (Armitage *et al.*, 1984; Tobin & Hervey, 1984).

4.4.2c   *Masks and mouthpieces*

The third method for air collection uses a mask, mouthpiece or even an endotracheal tube attached to a line through which gas continuously flows; the subject inspires from and expires to the gas stream (Poole & Maskell, 1975; Wilmore, Davis & Norton, 1976; Schulze *et al.*, 1981). The use of a mouthpiece and nose clip increases minute volume by about 20–25 % compared to the ventilated hood (Weissman *et al.*, 1984) and is therefore best avoided, although the additional cost of breathing will not be great.

4.4.2d   *Ventilation and air-conditioning*

The air is drawn through the chamber, hood, or face assembly by a pump which is best placed at the end of the system. This creates a negative pressure throughout the calorimeter which eliminates leakage of expired air to atmosphere, although some care is necessary to ensure that loss does not occur by expiration directly out of the ventilated hood or the various face assemblies. The flow measurement is of the outlet airstream and any leak of fresh air into the system is unimportant, since it merely replaces part of the air entering at the inlet port (an error may be introduced if the composition of the inlet air and leaking air are different).

However ingress of air through a leak in the line between the sites of measurement of flowrate and gas concentration would have a serious effect on accuracy. Thus immediately after measurement of flowrate, the gas sample should be withdrawn through a side arm by means of an airtight pump, and supplied at positive pressure to the analysers; any leak after the pump is then to atmosphere and does not affect the gas concentration of the sample. Alternatively, air may be pushed through the system but to avoid errors due to leaks, this involves measuring inlet rather than outlet flowrate, and it also creates complications in the subsequent calculations (see Section 4.5).

The rate of air flow through the system should be sufficient to bring about a change of no more than 1 % in the concentrations of oxygen and carbon dioxide in the chamber or hood. Because such small changes are difficult to measure accurately, and require high-quality equipment and methods of calibration, it may be tempting to use a low flowrate (Garrow & Hawes, 1972) to produce greater changes in gas concentration. However, a rise in concentration of inspired carbon dioxide greater than about 1 % brings about an increase in respiratory effort and energy expenditure; above 4 % carbon dioxide the increase in respiratory effort becomes substantial and may be stressful to the subject (Mountcastle, 1974, p. 1450;

Weissman *et al.*, 1984). As a rule of thumb a satisfactory rate of ventilation is about five times the minute volume or alternatively at rest about 1 l/kg of bodyweight.

If animals are totally enclosed within the chamber or hood, this rate of air flow is insufficient to prevent the air becoming saturated, particularly if excreta are not removed from within the system, and some form of air-conditioning is required and this may involve partial recirculation of the airstream. For small animals passage of chamber air through a chemical drier is satisfactory. In the Leeds calorimeter (Tobin & Hervey, 1984), we use a simple air conditioner; the air in each cage is circulated by means of a rheostat-controlled fan through a tower holding a replaceable mesh container of silica gel or anhydrous calcium sulphate (Drierite). The rate of air flow is manually adjusted to maintain relative humidity in the range 50–60%. For large animals a commercial air conditioning unit might be necessary.

4.4.2e   *Gas mixing*

The expired air must be thoroughly mixed with the ventilating air prior to measurement of flow and concentration; this may be achieved by a circulating fan in the chamber or hood or by means of a mixing device in the outlet line. In the human calorimeter at Leeds, we (Hampton *et al.*, 1982) have used a mixing box designed at the Institute of Naval Medicine at Alverstoke, Hants and found it very satisfactory. At the Hannah we (Kendall, McLean & Smith, 1971) have passed the outlet air into an automatically switched reservoir which collects successive samples of gas over convenient periods before passing the mixed samples to gas analysers. The reservoir, which is made from two plastic basins, is divided into two sealed halves by a loosely folded thin-rubber diaphragm. As the sample fills one side of the reservoir it causes the diaphragm to move over until all the gas from the other side has been expelled downstream. The resulting pressure drop on the downstream side of the reservoir is used to trigger a simple electric circuit which switches the gas connections over so that the side previously emptied is filled, and the sample just accumulated is passed downstream. The variations in gas concentrations are thus integrated over each sampling period. If the volume of the reservoir is known the device may be used as an accurate flowmeter by noting the time taken between successive switching operations. The use of theatrical smoke is a good test of the mixing qualities of a system despite the difference in densities of air and the smoke.

4.4.2f   Measurement of flowrate

There are numerous methods by which flowrate has been measured in flow-through systems; most involve measurement of volume, in which case the temperature and pressure of the gas must be also determined. The main instruments used for measurement of volume have been: the wet gasmeter, the oil-filled wet gasmeter, the dry gasmeter, the rotameter, the orifice plate, the pneumotachograph, and the mass flow-meter. These are described in detail in Chapter 6. In our view the mass flowmeter is one of the best ways of measuring the flow of gas: as its name implies it measures the mass of gas and is therefore unaffected by temperature and pressure variations but it is affected by humidity. It gives an electrical output suitable for recorders or computers; we have found mass flowmeters very reliable, accurate, and precise.

The observed flowrate from the flowmeters will need to be adjusted to STPD (0 °C, 760 mm Hg, and dry). Except for the mass flowmeters this will involve measurement of temperature and pressure; and for all flowmeters humidity must be measured or the gas dried prior to measurement. Where the flowrate is less than 20 l/min drying the gas is probably the best solution. Drying may be achieved by chemical or physical means (for details see Chapter 7).

At rates of air flow greater than about 20 l/min the use of chemical drying agents becomes unsatisfactory particularly for measurements over several hours, because of the bulk and cost. Permapure driers may provide a solution. Drying the air by cooling is not a realistic alternative because of the complexity and cost. It is therefore necessary to measure the water vapour content of the gas in order to adjust the gas volume to STPD. Humidity may be measured by wet- and dry-bulb thermometers and other humidity sensors. Humidity sensors are available at modest cost with a digital read-out and often with an electrical output suitable to be fed to a recorder or computer. In those parts of the apparatus where gas is moist, great care must be taken to ensure that moisture does not condense in the line; heating tape may be wrapped around the tubing carrying the gas and the temperature increased slightly. There is one exception to the procedures described above. If the volume of gas is to be measured by a wet gasmeter, the gas must be first saturated, otherwise evaporation of water from the meter will lead to errors in its calibration and the measurement of volume (see Chapter 6).

4.4.2g   Methods of gas analysis

The methods of gas analysis used in flow-through systems are also diverse: at the simplest level, chemical analysis by the Haldane apparatus has been used though this is not practical for large numbers of measurements. Other methods include thermal conductivity i.e. the Noyens diaferometer, the oxygen electrode, the zirconium fuel cell and the quadrupole mass spectrometer. However, the majority of investigators have used the paramagnetic oxygen analyser and infra-red carbon dioxide and methane analysers. All of these are pressure sensitive (see Chapter 6) and so samples must enter at a constant water vapour content and with a constant pressure drop across the analyser. Many investigators have failed to describe how this has been achieved in their calorimeters and yet it is an important factor in accurate analysis.

The presence of undetected water vapour in the gas entering the analysers can introduce substantial errors, for example up to 25% (Beaver, 1973; see also Chapter 6). We believe that the calibration and sample gases should be thoroughly dried before entering the analysers, if necessary by adding a second drier to the line.

Absolute pressure regulators are available and these have been used in calorimeters, for example by Kinney *et al.* (1964). A combined electronic flow and absolute pressure regulator has been described by MacLeod *et al.* (1985). Whilst ideal, these are expensive to install and we have used a simple and cheap 'barostat' which has proved effective. Strictly speaking, our 'barostat' provides a constant pressure head, and the analyser is still affected by fluctuations in barometric pressure. However, it does eliminate pressure variations between sample and calibration gases. The apparatus is similar to that described by Janac, Catsky & Jarvis (1971). It comprises a perspex water-filled cylinder into which a side arm from the analyser sample line dips. At the end of the side arm there is an aquarium bubble disperser. The flow of sample or calibration gas is adjusted so that there is a constant stream of bubbles through the water; the pressure drop across the analysers is then equivalent to the height of water above the bubbler and is constant.

The problem of control of inlet pressure and moisture content is considered further in Chapter 7.

The means by which the measuring instruments in the flow-through calorimeter can be calibrated and the whole system checked are described in Chapter 7.

4.4.2h *A portable flow-through system – the metabolic rate monitor*

Nearly all ventilated flow-through systems have a fixed ventilation rate and consequently a varying gas concentration difference between outlet and inlet air. A notable exception is the Metabolic Rate Monitor (MRM) devised by Webb & Troutman (1970). The subject wears a mask covering the whole of the face; a flexible tube about 4 cm in diameter leads from the bottom of the mask to a 4 cm vane-axial type air blower powered by a 27 V d–c motor. The output of the blower is essentially linear with respect to input voltage; at 27 V the volume flow is 500 l/min. Air is drawn into the mask through two 15 cm long tubes mounted on top of the mask like a pair of horns. The blower housing incorporates an oxygen electrode whose output is compared with a fixed reference signal equivalent to some nominal 'preset' oxygen concentration. The differential output from the two voltages is amplified and this error signal is used to drive the blower. As the downstream oxygen concentration decreases, the error signal increases, the blower speeds up, and the volume flow increases, thereby returning the downstream oxygen concentration to its 'preset' value. The system thus produces an air flow volume which is directly proportional to oxygen consumption. Since the amplified error signal drives the blower, it too is directly proportional to oxygen consumption. The signal is displayed on the meter and is also fed to a recorder output. The instrument is set up so that the fixed oxygen concentration of the expired air gives a 1 % differential with respect to dry fresh air; thus oxygen consumption is $(0.01 \times V \, \text{l})$. The apparatus is calibrated by introducing a known quantity of nitrogen into the facemask, which is mounted on a dummy head, to replace a known quantity of oxygen (an allowance must also be made for water vapour which 'dilutes' oxygen concentration in the experimental situation).

The MRM is comfortable to wear and is fast responding. In over 100 measurements it achieved agreement with measurements made by a Douglas bag to within ± 0.1 l/min. Minor leaks in the facemask have little effect on the measurements because of the slight negative pressure within and this contrasts with need for air-tightness required with the K–M meter, Miser and Oxylog. Despite its simplicity the MRM has surprisingly not been widely used.

4.5   Calculation methods

In closed-circuit systems the primary measurements are oxygen consumption and carbon dioxide and methane production. These quantities may be substituted directly into an equation of the Brouwer type (see Table 3.8) as derived in Chapter 3. If the quantities of food intake and excreta are recorded and these materials sampled and analysed for carbon and nitrogen content, heat production may alternatively be calculated from the carbon–nitrogen balance (equation (3.13)). This alternative was particularly useful before the invention of paramagnetic oxygen analysers, as it eliminated the need for analysis of oxygen which was formerly a difficult measurement.

For open-circuit systems the primary quantities measured are ventilation rate and the composition of inlet and outlet air; the computation of oxygen consumption ($\dot{V}_{O_2}$), respiratory quotient (r), and metabolic rate ($\dot{M}$) appears more complex at first sight. The principle of the calculations is the same regardless of whether the ventilating airstream consists purely of respired air as in a mask, mouthpiece or tracheal cannula fitted with inspiratory and expiratory valves, or whether the subject breathes freely into a moving airstream as in a ventilated hood or respiration chamber. The schematic diagram of Fig. 4.15 illustrates the gas flows and concentrations of *any* open-circuit system.

Inlet air is often fresh air which may contain up to 5% by volume of water vapour. The remaining part, i.e. dried fresh air, has a constant composition and contains 20.95% oxygen, 0.03% carbon dioxide and 79.02% inert gases, which consist mainly of nitrogen and will hereafter be referred to as nitrogen.

Thus

$$FI_{O_2} = 0.2095; \quad FI_{CO_2} = 0.0003;$$
$$\text{and } FI_{N_2} = 1 - FI_{O_2} - FI_{CO_2} = 0.7902.$$

Fig. 4.15. Schematic diagram showing gas flows and concentrations in an open-circuit calorimeter.

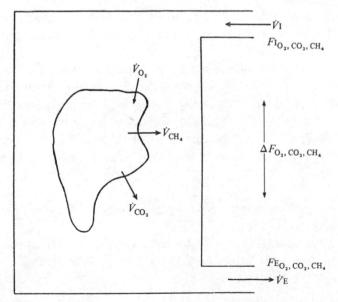

Because these values are constant for dry fresh air and because in most gas analysers the air is dried before analysis, gas concentrations are normally expressed as fractions of dry air. Hence for exhaust air or expired air the gas concentrations are $F_{E_{O_2}}$, $F_{E_{CO_2}}$ and for ruminant animals which also produce methane, $F_{E_{CH_4}}$.

Air flows by contrast are usually measured in moist air ($\dot{V}_m$) and are adjusted to STPD ($\dot{V}$) by the use of a correction factor ψ.

$$\dot{V} = \dot{V}_m \left(1 - \frac{P_w}{P}\right) \frac{P}{760} \cdot \frac{273}{(273 + T)} = \dot{V}_m \psi \qquad (4.10)$$

were P_w is the vapour pressure of moisture in the air.

4.5.1   Steady-state conditions

We will at first assume steady-state conditions, i.e. that sufficient time has elapsed for the outlet gas concentrations to equilibrate with the levels of gas exchange by the animal or subject.

The quantity of oxygen consumed is then

$$\dot{V}_{O_2} = \dot{V}_I F_{I_{O_2}} - \dot{V}_E F_{E_{O_2}} \qquad (4.11)$$

If the total volume of gases (CO_2 and CH_4) produced by the animal is equal to the volume of oxygen consumed (i.e. for non-ruminants when $r = 1$) then $\dot{V}_I = \dot{V}_E$ and equation (4.11) reduces to

$$\dot{V}_{O_2} \simeq \dot{V}_E (F_{I_{O_2}} - F_{E_{O_2}}) = - \dot{V}_E \Delta F_{O_2} \qquad (4.12)$$

This equation is sometimes used to estimate oxygen consumption, *but it is only correct when $r = 1$.* The magnitude of the error of using it when $r \neq 1$ is discussed later.

In general $\dot{V}_I$ and $\dot{V}_E$ are of similar magnitude but not equal; the same is true of $F_{I_{O_2}}$ and $F_{E_{O_2}}$. Consequently, extreme precision in estimates of these four variables would be necessary if $\dot{V}_{O_2}$ were to be calculated directly from equation (4.11). To avoid such difficulties it is preferable to devise methods whereby $\dot{V}_{O_2}$ can be calculated from the difference in concentration (ΔF_{O_2}) and only one flowrate. This may be achieved in several ways.

4.5.1a   Oxygen consumption from oxygen analysis and carbon dioxide absorption

Depocas & Hart (1957) derived formulae for various situations by making the substitution $\dot{V}_E = \dot{V}_I - \dot{V}_{O_2} + \dot{V}_{CO_2}$. Eliminating $\dot{V}_E$ from equation (4.11) then gives

$$\dot{V}_{O_2} = \dot{V}_I F_{I_{O_2}} - (\dot{V}_I - \dot{V}_{O_2} + \dot{V}_{CO_2}) F_{E_{O_2}} \qquad (4.13)$$

The dependence of this formula on knowing the value of $\dot{V}_{CO_2}$ may be eliminated if the sample of exhaust air to be analysed is first passed through a carbon dioxide absorber, so that the concentration of outlet air measured ($F'_{E_{O_2}}$) is that of CO_2-free gas. In that case $\dot{V}_{CO_2} = 0$ and equation (4.13) reduces to

$$\dot{V}_{O_2} = \dot{V}_I \frac{F_{I_{O_2}} - F'_{E_{O_2}}}{1 - F'_{E_{O_2}}} \tag{4.14}$$

The measurements involved are of $\dot{V}_I$, $F_{I_{O_2}}$, and $F'_{E_{O_2}}$ and no approximations have been made.

Alternatively Depocas & Hart (1957) suggested that the equation could be expressed as

$$\dot{V}_{O_2} = \dot{V}'_E \frac{F_{I_{O_2}} - F'_{E_{O_2}}}{1 - F_{I_{O_2}}} \tag{4.15}$$

This equation also involves no approximations, but practically it is less convenient because measurement of the CO_2-free outlet flowrate, $\dot{V}'_E$, involves absorbing carbon dioxide from the entire airstream instead of from only a sample. Except for small animals this may be impracticable.

Hill (1972) considered situations in which carbon dioxide is absorbed from both inlet and outlet airstreams before oxygen analysis. This leads without approximation to the equation

$$\dot{V}_{O_2} = \dot{V}'_I \frac{F'_{I_{O_2}} - F'_{E_{O_2}}}{1 - F'_{E_{O_2}}} \tag{4.16}$$

Since carbon dioxide concentration of inlet air can usually be assumed constant, $\dot{V}'_I$ in equation (4.16) can be replaced by $\dot{V}_I(1 - F_{I_{CO_2}})$.

In order to eliminate measurement of $\dot{V}'_E$ in equation (4.15), Wennberg (1975) made the substitution $\dot{V}'_E = \dot{V}_I(1 - F_{I_{CO_2}}) - \dot{V}_{O_2}$ which again without approximation leads to the solution

$$\dot{V}_{O_2} = \dot{V}_E \frac{F_{I_{O_2}} - F'_{E_{O_2}}(1 - F_{I_{CO_2}})}{1 - F'_{E_{O_2}} - (1 - r)[F_{I_{O_2}} - F'_{E_{O_2}}(1 - F_{I_{CO_2}})]} \tag{4.17}$$

This equation involves carbon dioxide concentration of inlet air and respiratory quotient but it is not greatly influenced by either value. Wenneberg showed that by inserting the values $r = 0.86$ and $F_{I_{CO_2}} = 0.0003$, the error of the resulting equation was always less than 1%.

$$\dot{V}_{O_2} = \dot{V}_E \frac{0.2093 - 0.9997 \, F'_{E_{O_2}}}{0.9707 - 0.86 \, F'_{E_{O_2}}} \tag{4.17a}$$

Equations (4.14)–(4.17) represent different methods for calculating oxygen consumption by absorbing carbon dioxide from one or both of the gas streams. Since carbon dioxide is not measured, neither respiratory quotient nor heat production can be calculated exactly from such procedures. Also the equations are not applicable to measurements on ruminant animals which produce appreciable quantities of methane. If heat production and respiratory quotient are to be estimated, then analysis of respired gases for both oxygen and carbon (and for ruminants, methane as well) is necessary.

4.5.1b   *Oxygen consumption when all respiratory gases are analysed*

The relationship between $\dot{V}_I$ and $\dot{V}_E$ is now obtained by equating the quantity of nitrogen in inlet and exhaust airstreams, i.e. the Haldane transformation

$$\dot{V}_I . F_{I_{N_2}} = \dot{V}_E F_{E_{N_2}}$$

or

$$\dot{V}_I(1 - F_{I_{O_2}} - F_{I_{CO_2}} - F_{I_{CH_4}}) = \dot{V}_E(1 - F_{E_{O_2}} - F_{E_{CO_2}} - F_{E_{CH_4}})$$

which may be rearranged in the form

$$\dot{V}_I = \dot{V}_E \left[1 - \frac{1}{F_{I_{N_2}}} (\Delta F_{O_2} + \Delta F_{CO_2} + \Delta F_{CH_4}) \right] \tag{4.18}$$

where

$$\Delta F = F_E - F_I$$

It should be noted that ΔF_{O_2} has a negative value because oxygen is consumed rather than produced. Substituting $\dot{V}_I$ from equation (4.18) into equation (4.11) then gives an accurate equation for oxygen consumption (McLean, 1972):

$$\dot{V}_{O_2} = - \dot{V}_E \left[\Delta F_{O_2} + \frac{F_{I_{O_2}}}{F_{I_{N_2}}} (\Delta F_{O_2} + \Delta F_{CO_2} + \Delta F_{CH_4}) \right] \tag{4.19}$$

The expression for oxygen consumption is seen here to consist of a main term (the product of outlet flowrate and oxygen concentration decrement between inlet and outlet air) and a correction term which involves a factor dependent only on the composition of inlet air $(F_{I_{O_2}}/F_{I_{N_2}})$ and $(\Delta F_{O_2} + \Delta F_{CO_2} + \Delta F_{CH_4})$.

When the volume of oxygen consumed is replaced by an equal volume of gas produced (i.e. for non-ruminants when $r = 1$) then $-\Delta F_{O_2} = \Delta F_{CO_2} + \Delta F_{CH_4}$, the correction term is zero and the equation reduces once more to equation (4.12).

When inlet air is fresh air $(F_{I_{O_2}}/F_{I_{N_2}} = 0.2095/0.7902 = 0.2651)$, then equation (4.19) reduces to

$$\dot{V}_{O_2} = - \dot{V}_E [\Delta F_{O_2} + 0.2651(\Delta F_{O_2} + \Delta F_{CO_2} + \Delta F_{CH_4})] \tag{4.19a}$$

The percentage error of using equation (4.12), i.e. of assuming $r = 1$, may be calculated by inserting a series of assumed values for ΔF_{O_2}, ΔF_{CO_2} and ΔF_{CH_4} into equation (4.19a) and plotting the resultant value of the error term, expressed as a percentage of $\dot{V}_{O_2}$, against respiratory quotient. Fig. 4.16 shows such a plot both for a non-ruminant ($\Delta F_{CH_4} = 0$) and a hypothetical ruminant animal ($\Delta F_{CH_4} = \Delta F_{CO_2}/10$). Over the practical range of respiratory quotient (see Chapter 3, Section 3.3.2) the error can be up to $\pm 6\%$. This potential error in the estimation of oxygen consumption cannot be avoided unless, in addition to measurement of oxygen, carbon dioxide is either absorbed or measured (Otis, 1964; Maxfield & Smith, 1967; McLean, 1972).

4.5.1c Carbon dioxide and methane production

Equations for carbon dioxide production $\dot{V}_{CO_2}$ and methane production $\dot{V}_{CH_4}$ may be derived in the same manner as for $\dot{V}_{O_2}$.

$$\dot{V}_{CO_2} = \dot{V}_E \left[\Delta F_{CO_2} + \frac{FI_{CO_2}}{FI_{N_2}} (\Delta F_{O_2} + \Delta F_{CO_2} + \Delta F_{CH_4}) \right] \tag{4.20}$$

$$\dot{V}_{CH_4} = \dot{V}_E \left[\Delta F_{CH_4} + \frac{FI_{CH_4}}{FI_{N_2}} (\Delta F_{O_2} + \Delta F_{CO_2} + \Delta F_{CH_4}) \right] \tag{4.21}$$

When inlet air is fresh air the correction term for carbon dioxide is negligible because $FI_{CO_2}/FI_{N_2} = 0.00038$ and for methane it is zero. $\dot{V}_{CO_2}$

Fig. 4.16. The percentage change with respiratory quotient in calorific value per litre of oxygen consumed ——, and the percentage error of assuming that oxygen consumption is equal to outlet volume multiplied by concentration difference between inlet and outlet air, for ruminant ...... and non-ruminant ———— animals.

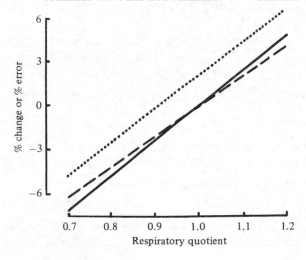

% change or % error

Respiratory quotient

and $\dot{V}_{CH_4}$ may therefore be calculated accurately from the simplified forms

$$\dot{V}_{CO_2} = \dot{V}_E \Delta F_{CO_2} \tag{4.20a}$$

$$\dot{V}_{CH_4} = \dot{V}_E \Delta F_{CH_4} \tag{4.21a}$$

4.5.1d *Respiratory quotient*

Respiratory quotient ($r = \dot{V}_{CO_2}/\dot{V}_{O_2}$) may be obtained by combining equations (4.19) and (4.20).

$$r = \frac{(1 - FI_{O_2})\Delta F_{CO_2} + FI_{CO_2}(\Delta F_{O_2} + \Delta F_{CH_4})}{-(1 - FI_{CO_2})\Delta F_{O_2} - FI_{O_2}(\Delta F_{CO_2} + \Delta F_{CH_4})} \tag{4.22}$$

When inlet air is fresh air this becomes

$$r = \frac{0.7905\,\Delta F_{CO_2} + 0.0003(\Delta F_{O_2} + \Delta F_{CH_4})}{-0.9997\,\Delta F_{O_2} - 0.2095(\Delta F_{CO_2} + \Delta F_{CH_4})} \tag{4.22a}$$

4.5.1e *Metabolic rate*

In Chapter 3 (equation (3.8)), it was shown that metabolic rate could be calculated using

$$\dot{M} = \alpha \dot{V}_{O_2} + \beta \dot{V}_{CO_2} + \delta \dot{V}_{CH_4} + \gamma \dot{N} \tag{4.23}$$

The values of the factors α, β, γ and δ are to some extent arguable. Using the Brouwer factors for farm animals (see Table 3.8) the equation is

$$\dot{M} = 16.18\,\dot{V}_{O_2} + 5.02\,\dot{V}_{CO_2} - 2.17\,\dot{V}_{CH_4} - 5.99\,\dot{N} \tag{4.23a}$$

where $\dot{M}$ is measured in kW, $\dot{V}$ in l/s, and $\dot{N}$ (urinary nitrogen) in g/s. Since in practice $\dot{V}_{O_2} \simeq \dot{V}_{CO_2} \simeq 10 \times \dot{V}_{CH_4}$, the oxygen term contributes about 75%, the carbon dioxide term about 25% and the methane term about -1% to the estimated value of $\dot{M}$. The urinary nitrogen term also contributes about -1%.

The calculation may be shortened in open-circuit calorimetry by substituting $\dot{V}_{O_2}$, $\dot{V}_{CO_2}$ and $\dot{V}_{CH_4}$ from equations (4.19), (4.20) and (4.21) into equation (4.23) (McLean, 1972), which gives

$$\dot{M} = \dot{V}_E[-(\alpha + \mu)\Delta F_{O_2} + (\beta - \mu)\Delta F_{CO_2} + (\delta - \mu)\Delta F_{CH_4}] + \gamma \dot{N} \tag{4.24}$$

where

$$\mu = (\alpha FI_{O_2} - \beta FI_{CO_2} - \delta FI_{CH_4})/FI_{N_2}$$

The value of μ depends only on the composition of inlet air and in practice always lies within the range 4.2–4.3. When inlet air is fresh air, $\mu = 4.29$ and, using the Brouwer factors, equation (4.24) becomes

$$\dot{M} = \dot{V}_E(-20.47\,\Delta F_{O_2} + 0.73\,\Delta F_{CO_2} - 6.46\,\Delta F_{CH_4}) - 5.99\,\dot{N} \tag{4.24a}$$

In this equation the relative values for the coefficients for oxygen, carbon dioxide and methane are quite different from those of equation

(4.23a). The oxygen term contributes nearly the entire value of $\dot{M}$. Fig. 4.17 which has been compiled from nearly 9000 individual determinations of heat production in the Hannah Institute calorimeter (McLean, 1986) shows that the percentage contribution to $\dot{M}$ of the terms involving ΔF_{CO_2} and ΔF_{CH_4} vary in relation to the time after feeding; also they are small and tend to cancel each other. If it is assumed that the urinary nitrogen term (not usually measured) contributes -1%, then the combined contribution of the terms in ΔF_{CO_2} and ΔF_{CH_4} and $\dot{N}$ has an average value of only $+0.21\%$. This small contribution may be accommodated by a corresponding adjustment to the coefficient of the oxygen term, resulting in the simplified equation applicable to ruminant animals

$$\dot{M} = -20.5 \, \dot{V}_E \Delta F_{O_2} \tag{4.25}$$

The error in using equation (4.25) instead of equation (4.24a) is at all times no greater than $\pm 1.2\%$. It should be noted that equation (4.25) differs from equation (4.12) only by the inclusion of the constant factor -20.5. However, whereas equation (4.12) provides only an approximate estimate of oxygen consumption ($\pm 6\%$), equation (4.25) gives a reasonably accurate estimate of heat production ($\pm 1.2\%$).

The small influence of the carbon dioxide term in the McLean equation (4.24a) in comparison with the Brouwer equation (4.23a) results from a fortunate coincidence. Fig. 4.16 shows plots against respiratory quotient of the percentage error which would occur if oxygen consumption were estimated from a simple flow times concentration relationship (equation (4.12)) instead of from the correct equation (4.19a). Also plotted in Fig.

Fig. 4.17. The daily variation in heat production ($\dot{M}$) of cattle and in the percentage contributions in the equation for $\dot{M}$ of terms involving carbon dioxide (CO_2), methane (CH_4) and their sum ($CO_2 + CH_4$). Vertical arrows indicate times of feeding.

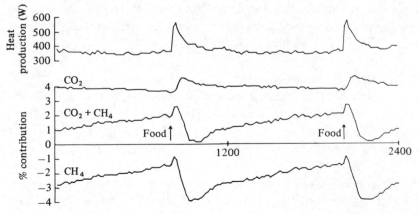

4.16 is the percentage change with respiratory quotient in the calorific value per litre of oxygen consumption, based on the values given in Table 3.2. The slopes of the two relationships are very nearly the same, and this has the fortunate consequence that the quantity $\dot{V}E\Delta F_{O_2}$ contains within itself very nearly the exact correction required for adjusting the calorific value per litre of oxygen consumption. Consequently the carbon dioxide term in equation (4.24a) is of only minor importance. This also shows that applying the calorific value given for the observed value of r in Table 3.2 to the quantity $\dot{V}E\Delta F_{O_2}$ (a practice not infrequently recorded in the literature) results in an error in the estimate of heat production of up to $\pm 6\%$.

Weir (1949) was the first to introduce a formula for calculating metabolic rate directly from flowrate and oxygen concentration decrement. His equation was based on slightly different calorific factors (see Table 3.8). He also proposed two alternative methods for introducing the correction for urinary nitrogen. Weir's equation (for man) is

$$\dot{M} = -21.09 \; \dot{V}E\Delta F_{O_2} - 9.08 \; \dot{N} \tag{4.26}$$

or

$$\dot{M} = -21.09 \; \dot{V}E\Delta F_{O_2}(1 - 0.082\,p) \tag{4.27}$$

where p is the proportion of the total heat production due to protein metabolism. Weir pointed out that the value of p is usually about 0.125 hence the protein correction amounts to about -1%. This partially accounts for the 2.9% difference between the factor 21.09 in Weir's equation and 20.5 in equation (4.25). The remaining 1.9% occurs partly because equation (4.25) allows for methane production, and partly because of different calorific factors (see Section 3.3.2) used in deriving the equations.

Equations (4.10)–(4.27) are all expressed in terms of gas concentrations and flowrates and apply to open-circuit systems. They are of course also applicable to total collection systems: flowrates ($\dot{V}$ l/s) must be replaced by volumes (V l) and the calculated results will also be in terms of volumes of gas in litres or quantities of heat in kilojoules.

4.5.2   *Dynamic conditions*

So far all formulae discussed have assumed static conditions. They are suitable for calculation of data derived from total collection systems or apparatus involving mouthpieces or facemasks with a small dead space in which the gas concentrations in exhaust air rapidly attain steady levels. However, in respiration chambers the fresh air ventilation rate per minute may be small compared with the volume of the chamber and steady-state conditions only attained after an hour or more. Fast response estimates of oxygen consumption and heat production can still be calculated by consideration of the rates of change of gas concentrations in the chamber (McLean & Watts, 1976). This type of calculation is also required for interpretation of results from confinement calorimeters.

Provided that the air in the chamber is stirred sufficiently to ensure rapid mixing of gases, then the mean concentration of oxygen inside the chamber is equal to that of the exhaust air ($F_{E_{O_2}}$). An additional term may then be included in equation (4.11) to allow for any accumulation of oxygen inside the chamber.

$$\dot{V}_{O_2} = \dot{V}_I F_{I_{O_2}} - \dot{V}_E F_{E_{O_2}} - V \frac{dF_{E_{O_2}}}{dt} \tag{4.28}$$

where V is the volume of the ventilated system. For confinement calorimeters $\dot{V}_I = \dot{V}_E = 0$, and the equation reduces to equation (4.1). The effect of the additional term is that in all the equations from (4.19) to (4.27) $\dot{V}_E$ is replaced by $(\dot{V}_E + V\frac{d}{dt})$. Thus the equation for oxygen consumption (4.19) becomes

$$\dot{V}_{O_2} = -\left(\dot{V}_E + V\frac{d}{dt}\right)\left[\Delta F_{O_2} + \frac{F_{I_{O_2}}}{F_{I_{N_2}}}(\Delta F_{O_2} + \Delta F_{CO_2} + \Delta F_{CH_4})\right] \tag{4.29}$$

and the equations for metabolic rate (equations (4.24a) and (4.25)) become

$$\dot{M} = \left(\dot{V}_E + V\frac{d}{dt}\right)$$
$$\times (-20.47\,\Delta F_{O_2} + 0.73\,\Delta F_{CO_2} - 6.46\,\Delta F_{CH_4}) - 5.99\,\dot{N} \tag{4.30}$$

and

$$\dot{M} \simeq -20.5\left(\dot{V}_E + V\frac{d}{dt}\right)\Delta F_{O_2} \tag{4.31}$$

In practice the gas concentrations are recorded at regular intervals τ seconds apart and the value of $(\dot{V}_E + V\frac{d}{dt})\Delta F$ is calculated from the nth and $(n+1)$th measurements as

$$\left(\dot{V}_E + V\frac{d}{dt}\right)\Delta F = \dot{V}_E \frac{[\Delta F_n + \Delta F_{(n+1)}]}{2} + V\frac{[\Delta F_{(n+1)} - \Delta F_n]}{\tau} \tag{4.32}$$

The differential terms may be included or excluded from all equations according to requirements. Over a prolonged experimental period they have no influence on the average values of oxygen consumption or metabolic rate obtained. Their only effect is to sharpen up the time response of the measurements. This was illustrated by McLean & Watts (1976) who calculated the results of a calibration trial in which nitrogen gas was injected into the Hannah Institute calorimeter to simulate oxygen consumption (see Chapter 7). Fig. 4.18 shows the rate of recovery of nitrogen calculated from ΔF_{O_2} both with and without inclusion of the differential term. It is evident that inclusion of the differential term greatly improves the speed of response, but introduces additional variability into the measurements. The variability arises because of the difficulty of accurately measuring the rate of change of oxygen concentration, and it is inevitably increased as the interval between measurements is shortened. This variability does not follow a normal Gaussian distribution because any inaccuracy in the determination of a single value of $F_{E_{O_2}}$ affects successive estimates of the differential term by equal and opposite amounts.

It is also possible to include differential terms in equations (4.14)–(4.17) resulting in the same improvement in response. However, a small approximation is involved here because the measured quantity $\mathrm{d}/\mathrm{d}t(\Delta F'_{O_2})$ differs slightly from $\mathrm{d}/\mathrm{d}t(\Delta F_{O_2})$. The error is very small.

Equations (4.19)–(4.32) assume that it is outlet flowrate $\dot{V}_E$ that is measured. Some observers prefer to measure inlet flowrate $\dot{V}_I$, a procedure which is particularly suited to static conditions of gas concentrations and for which equations (4.14)–(4.16) are applicable. If $\dot{V}_I$ is measured when gas concentrations are changing then a set of equations complementary to those given for $\dot{V}_E$ may be evaluated in exactly the same way by inter-

Fig. 4.18. Results of a recovery trial in which oxygen consumption was simulated by injection of nitrogen into the Hannah Institute calorimeter. (*a*) oxygen concentration difference, ΔF; (*b*) recovery of nitrogen calculated without inclusion of a differential term; (*c*) with a differential term.

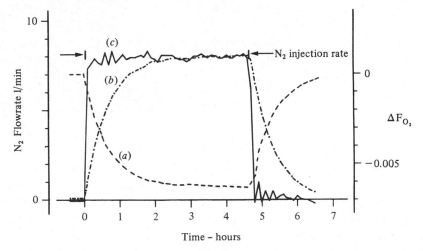

changing the letters ɪ and ᴇ. Equation (4.19) for oxygen consumption becomes

$$\dot{V}_{O_2} = -\dot{V}\text{ɪ}\left[\Delta F_{O_2} + \frac{F\text{ᴇ}_{O_2}}{F\text{ᴇ}_{N_2}}(\Delta F_{O_2} + \Delta F_{CO_2} + \Delta F_{CH_4})\right] \tag{4.33}$$

and equation (4.24) for metabolic rate becomes

$$\dot{M} = \dot{V}\text{ɪ}[-(\alpha+\mu')\Delta F_{O_2} + (\beta-\mu')\Delta F_{CO_2} + (\delta-\mu')\Delta F_{CH_4}] + \gamma\dot{N} \tag{4.34}$$

where

$$\mu' = \alpha F\text{ᴇ}_{O_2} - \beta F\text{ᴇ}_{CO_2} - \delta F\text{ᴇ}_{CH_4}/F\text{ᴇ}_{N_2}$$

In these equations the factors $F\text{ᴇ}_{O_2}/F\text{ᴇ}_{N_2}$ and μ' refer to the properties of exhaust air and must be measured rather than assumed. Since they are variable quantities the equations cannot be reduced to simplified forms complementary to equations (4.19a), (4.24a) and (4.25). To avoid such complications it is generally preferable to measure outlet rather than inlet airflow.

Brown *et al.* (1984) have proposed a generalised formula for the estimation of the consumption of any respiratory gas. They allow for the possibility that the STPD correction factor (ψ, see equation (4.10)) is variable, and that it is the measured flowrate $\dot{V}_m$ that is held constant rather than the STPD adjusted flow $\dot{V}$. This effectively introduces additional terms involving $d\psi/dt$ into equations (4.19)–(4.27). We have examined past records from our stored calorimetric data and found that the error of omitting these extra terms was less than 0.1 % under normal (steady temperature) conditions, and less than 1 % even when the calorimeter temperature was subjected to an imposed diurnal cycle (McLean, 1987).

4.5.3   *Practical aspects*

From the preceding sections it is clear that the crucial measurement in open-circuit indirect calorimetry is the product $\dot{V}\text{ᴇ} \times \Delta F_{O_2}$; terms involving carbon dioxide and methane are of minor importance. It is therefore desirable to employ the best possible flowmeter since any error in the measurement of $\dot{V}\text{ᴇ}$ will be reflected in a proportional error in the calculated value of oxygen consumption or of metabolic rate. Measurement of the oxygen concentration requires still more careful consideration. It is necessary not only to employ a very accurate oxygen analyser but also to ensure that it is utilised with maximum efficiency. The value of $F\text{ɪ}_{O_2}$ will normally be close to that of fresh air i.e. 0.2095. The value of $F\text{ᴇ}_{O_2}$ will depend on the type of calorimetry system. In those where ventilation relies on the respiratory action of the subject e.g. a mouthpiece or facemask with inspiratory and expiratory valves, the value of $F_{E_{O_2}}$ will usually be no greater than 0.1800 and the difference ($\Delta F_{O_2} = 0.2095 - 0.1800$) can be estimated fairly accurately. In pump-ventilated systems such as respiration chambers $F_{E_{O_2}}$ is often 0.2050 or higher, and the problem is one of estimating the small difference between two relatively large quantities. Obviously if $F_{I_{O_2}}$ and $F_{E_{O_2}}$ can each be measured with only $\pm 1\%$ accuracy (i.e. 0.2095 ± 0.0020 and 0.2050 ± 0.0020) then the difference ΔF_{O_2} will have a possible error approaching 100% (0.0045 ± 0.0040). It is clearly preferable to compute oxygen consumption from ΔF_{O_2} measured directly rather than from $F_{I_{O_2}}$ and $F_{E_{O_2}}$, or from $F_{E_{O_2}}$ and an assumed constant value of $F_{I_{O_2}}$.

Methods of arranging this are discussed under instrumentation in Chapter 6. The error in measuring the difference between two levels of oxygen consumption, say of a control and experimental treatment, can also be substantial.

4.6   Other indirect systems

All the methods of indirect calorimetry so far described in this chapter have involved some restriction on the freedom of the experimental subject. The use of facemasks or mouthpieces goes some way towards making measurements under normal living conditions, but their use requires some subject co-operation and they are thus mainly restricted to use with humans. The energy cost of many everyday activities has been successfully determined by such methods (Durnin & Passmore, 1967). Surgical preparations involving tracheal cannulation made measurements also possible on free-ranging and grazing animals. However, the surgical procedures are not without problems and, if only for ethical reasons, are to be avoided if alternative methods are available. This has prompted the search for alternative non-invasive methods of assessing gas exchange, or for correlations between heat production and some other more readily measurable physiological parameter. Non-invasive methods for gas exchange measurements include two based on estimation of carbon dioxide production by means of isotopic dilution.

4.6.1   Labelled carbon

In 1967 a group of Australian workers suggested that if $NaH^{14}CO_3$ was continuously infused into the body, the rate of elimination of labelled carbon in the form of carbon dioxide (they called it the CO_2-entry rate) could be calculated from the specific activity of carbon dioxide in one of the body fluids (Young *et al.*, 1967). They subsequently demonstrated satisfactory correlations between CO_2-entry rate and energy expenditure in cattle (Young, 1970) and in sheep (Corbett *et al.*, 1971; Engels, Inskip & Corbett, 1976). In these experiments CO_2-entry rate (l/h) was calculated as

$$\dot{V}_{CO_2} = \frac{\text{rate of infusion of activity } (\mu Ci/h)}{\text{specific activity of } CO_2 \ (\mu Ci/l)}$$

Energy expenditure was measured by a variety of different methods including tracheal cannulation and respiration chambers both with and without inclusion of an allowance for methane production. Carbon was infused as $NaH^{14}CO_3$ at rates of 1–10 $\mu Ci/h$ either subcutaneously, intraperitoneally or intravenously and the specific activity was measured in blood, urine, saliva or in expired carbon dioxide. After infusion for three hours in sheep and four to five hours in cattle, specific activity in the body fluids reached steady levels which were dependent on the level of metabolism.

It was found that the observed specific activity of carbon dioxide varies depending on the site of infusion and between different body fluids assayed. The calculated CO_2-entry rates were from 18 to 30 % higher than carbon dioxide production of grazing sheep measured by a tracheal cannulation technique. These differences were thought to be due to local differences in the production and elimination of carbon dioxide between the sites of infusion and sampling (Young, 1970; Corbett *et al.*, 1971). But Corbett *et al.* (1971) found that for any chosen pair of infusion and sampling sites good correlations could be obtained between energy expenditure and CO_2-entry rate. They regarded the CO_2-entry rate technique (CERT) as a practical index of energy expenditure (with an accuracy of about 10–15 %) provided that the same methods of infusion and sampling were used to obtain the empirical predictive equation and for the test application. The latest work indicates that saliva is the preferred fluid for sampling (Young – personal communication).

Whitelaw, Brockway & Reid (1972) did further detailed experiments, infusing sheep intravenously with $NaH^{14}CO_3$, and simultaneously estimating respiratory gas exchange in a respiration chamber. They found that CO_2-entry rate overestimated carbon dioxide production by only 2–4 % and they believed this small difference was due to retention of some radioactivity in bone. Whitelaw (1974) suggested that much of the variation found in the previous work might have been due to technical inaccuracies, and he stressed the importance of three basic requirements:

(1) Sufficient time must be allowed after onset of infusion before sampling begins, he recommended 12 h.

(2) The measurements must be made over a sufficient time for the effect of transient changes in body stores to be minimised, (this can be effected by sampling urine over a period by means of a bladder catheter, or blood over a period by a sampling device (Farrell, Corbett & Leng, 1970)).

(3) The equipment used for measurement of respiratory exchange must have an inherent accuracy better than that expected of CERT.

The method has been used to estimate the energy expenditure of grazing cattle by Havstad & Malechek (1982).

4.6.2   *Doubly labelled water*

This method was first suggested by Lifson, Gordon & McClintock (1955). It depends on their earlier finding that an isotopic equilibrium exists between the oxygen of respiratory carbon dioxide and the oxygen of body water (Lifson *et al.*, 1949). Lifson *et al.* (1955) measured the turnover in the body of water and carbon dioxide by monitoring the concentrations of isotopes in the urine after injection of doubly labelled water ($^2H_2{}^{18}O$) into mice. Since oxygen is eliminated as both carbon dioxide and water, its turnover rate is faster than that of hydrogen which is eliminated only as water. The difference in turnover rates provides a measure of the rate of production of carbon dioxide.

For simplicity we start by making four assumptions as follows:

(1) The body is in a steady state and acts as a single compartment with respect to labelled water.

(2) Labelled water is distributed only in the water compartment.

(3) The isotopes are lost only as water and carbon dioxide.

(4) Labelled water and carbon dioxide in urine and exhaled carbon dioxide have the same level of enrichment as in body water.

Given these assumptions it follows directly that

$$\dot{V}_{CO_2} = (W/2) \times (k_{18} - k_2) \times 22.41 \; 1/h \qquad (4.35)$$

where W is the size of the total pool of body water (mol) and k_{18} and k_2 are the fractional turnover rates of ^{18}O and ^2H respectively. They may be calculated from time-related measurements of the isotopic concentrations c_{18} and c_2 using $k = (1/c) \times (dc/dt)$. The factor 2 in the denominator of the equation represents the molar proportion of oxygen in carbon dioxide in relation to water.

In fact the four assumptions stated above are only approximately satisfied in practice. In particular there is a natural fractionation of the isotopes between different body fluids. Lifson *et al.* (1955) modified equation (4.35) by inclusion of the fractionation factors: f_1 (0.93) for ^2H_2O gas/^2H_2O

liquid; f_2 (0.99) for H_2 gas/$H_2{}^{18}O$ liquid; and f_3 (1.04) for $C^{18}O_2$ gas/ $H_2{}^{18}O$ liquid:

$$\dot{V}_{CO_2} = \left[\frac{W}{2f_3}(k_{18}-k_2) - \left(\frac{f_2-f_1}{2f_3}\right) \cdot \frac{\dot{m}}{18} \right] \times 22.41$$

or

$$\dot{V}_{CO_2} = \left[\frac{W}{2.08} \cdot (k_{18}-k_2) - \frac{\dot{m}}{624} \right] \times 22.41 \text{ l/h} \tag{4.36}$$

where $\dot{m}$ is the rate of water loss as vapour (g/h).

In addition allowance must be made for natural background levels of the isotopes, which are approximately 1 part in 150 for ^2H, and 1 part in 1000 for ^{18}O. The isotopic concentrations c_{18} and c_2 must be expressed in terms of excess concentration above these normal levels.

Lifson *et al.* (1955) measured c_{18} and c_2 in mice by means of mass spectrometry in two samples of blood and urine (for ^2H) and of expired carbon dioxide (for ^{18}O) separated by Δt ($= 1$–3 days), and calculated the turnover constants using $k = 2.3 \ (\log c)/\Delta t$. They measured total body water by carcass analysis at the end of the experiment. They found no significant difference between carbon dioxide production calculated by the doubly labelled water technique and total carbon dioxide collected from the 15 mice ($-3\% \pm 10\%$). They discussed the wider applications of the technique, but recognised that at that time it was not a practical alternative for measurement of heat production in large animals because of the cost of doubly labelled water.

Improvements in the precision of mass spectrometers have now made it possible to reduce greatly the required dosage of doubly labelled water, and the technique is emerging as a useful method of estimating metabolic rate in humans. Schoeller (1983) has calculated that with present levels of analytical accuracy for estimating isotopic enrichment (1.8×10^{-5} atom per cent excess [APE] for ^{18}O, and 3.3×10^{-5} APE for ^2H) the optimal doses for postneonatal humans are 0.3 g/kg total body water for $H_2{}^{18}O$ and 0.12 g/kg for ^2H_2O.

The doubly labelled water technique for adult humans has so far been validated against traditional methods in only four studies involving 14 subjects in total. The results are briefly summarised in Table 4.1.

Schoeller & van Santen (1982) estimated carbon dioxide turnover of four subjects from isotopic concentrations in urine samples taken on day 1 and day 14 after drinking labelled water. They calculated total body water (W) from isotopic dilution of urine samples collected 6 h after and saliva collected 3–4 h after drinking ^{18}O-labelled water. They also calculated energy expenditure as the difference between energy intake, estimated from dietary intake and food tables, and body energy storage, estimated from changes in total body water and total body mass during the period of the experiment. Finally, heat production, calculated from the carbon dioxide turnover, was compared to that estimated from intake and storage. Considering the complexity of these procedures, the agreement and precision obtained (Table 4.1) is quite remarkable. However, such estimations of energy intake and storage may be grossly inaccurate so that a further, more accurate, comparison was required to validate the method.

Schoeller & Webb (1984) compared the heat production of five subjects, estimated from the doubly labelled water technique with that obtained from continuous measurement using the Metabolic Rate Monitor (see Section 4.4.2h) of Webb & Troutman (1978). Total body water was calculated from ^{18}O enrichment of urine samples excreted after 12 h, and carbon dioxide turnover rate from isotopic analysis of urine samples taken on day 1 and day 5 after dosing. The agreement between the methods was again good enough for the doubly labelled water method to be sufficiently accurate for groups of subjects in the field.

Klein *et al.* (1984) measured the carbon dioxide production rates of a single subject living in a calorimeter. They calculated carbon dioxide turnover rates from differences in isotopic enrichment from successive pairs of urine samples collected throughout the 5 day experimental period. Although there were large differences between the heat production measured by the two methods on a day to day basis, the agreement over the whole period was exceptionally good.

Coward *et al.* (1984) adopted a different mathematical approach to the treatment of the measurements. They replaced W in equation (4.35) by D/I, where D is the total dose of labelled water given and I is the zero-time

Table 4.1. *The doubly labelled water technique*

Reference	Quantity compared	No. of subjects	Experimental period	%Difference (labelled H_2O – standard)	SD
Schoeller & van Santen (1982)	Energy intake – body storage	4	14	2.1	±5.6
Schoeller & Webb (1984)	Heat production	5	5	5.9	±7.6
Klein *et al.* (1984)	CO_2 production	1	5	1.8	±28
Coward *et al.* (1984)	CO_2 production	4	12	1.1	±1.6

intercept of the curve of excess isotope concentration against time. For a mono-exponentially decaying concentration, D/I is equivalent to the distribution space (the volume in which the isotope is distributed) of the isotope. Furthermore from a logarithmic plot of excess concentration against time, both the slope, which corresponds to k and the intercept, which corresponds to I, may be readily calculated. Values of D and I are of course different for each isotope, hence equation (4.36) becomes

$$\dot{V}_{CO_2} = [(1/2.08) \times (\{D_{18} \times k_{18}\}/I_{18} - \{D_2 \times k_2\}/I_2) - (\dot{m}/624)]$$
$$\times 22.41 \; l/h \quad (4.37)$$

Coward *et al.* (1984) collected frequent urine samples for isotopic analysis from each of four subjects during confinement in a whole-body calorimeter. They found the labelled-water method gave results (mean $\pm$ s.d.) 1.1 % ($\pm$1.6 %) and 4.1 % ($\pm$1.8 %) higher than those from the calorimeter for carbon dioxide and water respectively. Coward & Prentice (1985) used the same data to demonstrate that improved precision is obtained by (1) the slope/intercept method of calculation compared with methods based on two points and (2) by extending the period of measurement from 5 to 12 days.

The doubly labelled water method is still new and requires further validation. Its accuracy as a means of measuring carbon dioxide output appears to be within $\pm$2 %, and as pointed out by Coward & Prentice (1985) it is on this basis, and not as a means of measuring heat production, that it should be judged. This level of accuracy has so far been achieved only by the slope/intercept method of calculation for a 12 day trial. For estimation of heat production, the additional error due to the uncertainty of estimating respiratory quotient must be included and this will generally increase the overall uncertainty to at least 6 %. In a recent review, Coward *et al.* (1985) have considered possible applications and limitations of the method.

Schoeller & van Santen (1982) estimated the cost of the dose per subject in their study as $225. The method is totally non-invasive and it imposes no restriction on the subject. Despite its limited accuracy and its high cost it has considerable potential for applications such as estimating food requirements of populations, and large groups of free-ranging animals. It should also be stressed that the method employs stable isotopes only and involves no radioactive substances.

4.6.3 Measurements that correlate with heat production
Many attempts have been made to estimate heat production from measurement of some other correlated physiological parameter. These

parameters have included the rate of insensible perspiration (Benedict & Wardlaw, 1932), pulmonary ventilation (Durnin & Edwards, 1955) and heart rate.

Heart rate in particular has been extensively investigated as a possible predictor of heat production in both man (e.g. Booyens & Hervey, 1960; Malhotra, Sen Gupta & Rai, 1963; Andrews, 1971; Christensen *et al.*, 1983) and animals (Webster, 1967; Yamamoto & Ogura, 1985). Correlations which are good enough for predictive purposes within specified limits or conditions have been reported for individual subjects or animals: but there appears to be no way of avoiding the need for some form of prior calibration of each subject against a method involving measurement of oxygen consumption.

It is also possible to estimate heat production using a diary technique. The subject or an observer must compile a detailed record of either the activities or of the types and quantities of food eaten over an extended period. It is then possible to compute the total energy expenditure from the activity record using previously determined values for the energy cost of the activities and the time spent on each activity (Durnin & Passmore, 1967). Alternatively, the energy intake may be calculated from the quantities of different foods consumed, and their energy content which may be obtained from food tables such as those of Paul & Southgate (1978); it is then assumed that a state of energy balance exists in the subjects and that energy intake equals energy expenditure.

We regard all of these methods, whether involving correlates or diary techniques, as being secondary derivatives of indirect calorimetry. They can be useful in providing approximate estimations of heat production under circumstances where true indirect calorimetric methods cannot be put into practice; but they must by definition be less accurate, and are usually in practice very much less accurate, than the calorimetric methods used in their validation.

4.7   Summary

Confinement respiration chambers of the 'open and shut' type are especially suitable for measurement of gaseous exchanges of large animals. They depend only on accurate measurement of gas concentrations. The rate of gas consumption is calculated from the rate-of-change of gas concentration multiplied by the volume of the system.

Closed-circuit respirometers are best suited to small animal applications, but may also be used in conjunction with masks or mouthpieces for a cooperative subject such as man. Also some elaborate though very successful closed-circuit chambers have been built for man and large animals. In all these systems, oxygen is measured as the volume required to replenish that consumed in the chamber, and carbon dioxide production is determined from the weight gain of absorbers.

In total collection and open-circuit systems, heat production is calculated from the products of the rate of air flow (or volume) and the difference in gas concentrations between air leaving and entering. They are suitable for all types of subject, and various portable systems of these types have been devised. Using open-circuit systems it is possible to calculate heat production with $\pm 1.2\%$ accuracy and fast response from measurements of flow and oxygen concentration difference only; but high quality instrumentation and particularly careful attention to calibration is needed.

The main advantage of methods involving isotopic tracer techniques is the freedom conferred on the subject, but they are expensive and provide only approximate estimates of heat production, because they are based only on the measurement of carbon dioxide.

Note added in proof

Since writing our original manuscript a new instrument, the Deltatrac Metabolic Monitor (manufactured by Datex Instrumentarium Corp., PO Box 357, SF-00101, Helsinki, Finland) has become commercially available.

The Delatrac Metabolic Monitor, an open-circuit flow-through system, is designed specifically for clinical measurements on hospital patients, either breathing into a ventilated canopy which encloses the head, or mechanically ventilated by a respirator. Its design includes:

(1) A new type of differential oxygen cell.

(2) Reliance on a constant airflow generated by an inbuilt fan and a restrictor, rather than on a flow measuring device.

(3) Sample gas conditioning to ambient air conditions by means of Permapure tubing.

(4) New calculation methods made possible as a result of the differential measurement of oxygen concentration.

The limited tests which we have so far been able to perform on the Deltatrac Metabolic Monitor do appear to confirm the manufacturer's claimed accuracy of approximately $\pm 5\%$ for oxygen consumption, carbon dioxide production and respiratory quotient, provided that the instrument is re-calibrated daily using standard gas mixtures.

5

Direct calorimeters

Direct animal calorimeters measure the total heat generated inside them and partition the heat into its two components, evaporative and non-evaporative. Non-evaporative or sensible heat is heat given off from an animal by radiation to surrounding surfaces, by convection to the surrounding air and by conduction to any objects with which the animal is in contact. Evaporative heat loss occurs because the conversion of liquid water into vapour requires heat energy. The latent heat of vaporisation is the heat required to vaporise unit mass of water. When water is vaporised in the respiratory passages during normal respiration and panting, or at the skin surface during perspiration, the latent heat of vaporisation is derived mainly from the animal. This heat loss by the animal is transferred to the air in the form of increased humidity; the enthalpy of the air (which is a measure of its energy content and depends on temperature, humidity and pressure) is increased.

5.1.1 The psychrometric chart

The relationships between temperature, humidity and enthalpy of air are often illustrated in the form of a psychrometric chart. Fig. 5.1 is a scaled down version of a psychrometric chart for standard barometric pressure. The heavy line is a plot against temperature of the mixing ratio (i.e. humidity expressed as weight of water per weight of dry air) for air saturated with water vapour; it represents the maximum amount of vaporised moisture that air may contain.

Suppose that air passing along a duct initially has temperature T_1 and humidity r_1 represented by point A. Saturated air at the same temperature would have humidity r_s. Air at point A is therefore $(r_1/r_s) \times 100\%$ saturated, and this quantity is known as its percentage relative humidity.

Air humidity may be expressed in various other ways including specific humidity (mass of vapour per unit mass of the air-vapour mixture), absolute humidity (mass of vapour per unit volume of the mixture) and water vapour pressure. The properties of the psychrometric chart described here apply approximately but not exactly to all of these.

If the air is cooled from point A its temperature falls without change in humidity until it reaches point D which corresponds to saturation at some temperature T_3. Any further cooling beyond this point results in the formation of dew, as liquid water condenses out of the air. T_3 is known as the dewpoint temperature of air at any condition represented by a point along the straight line passing through A and D. Dewpoint temperature is yet another term used to express air humidity. If the air is now cooled further to point B it attains a dewpoint temperature of T_2 and a humidity r_2. When it is reheated to its original temperature it reaches point C, still with humidity r_2 and the relative humidity is now $(r_2/r_s) \times 100 \%$. Air that has undergone the cycle of temperature changes from T_1 through T_2 and back to T_1 again has thus given up moisture, and its heat content or enthalpy has been reduced by the latent heat required to vaporise the condensed water.

Enthalpy of air increases with increasing temperature and with in-

Fig. 5.1. Scaled-down version of a psychrometric chart.

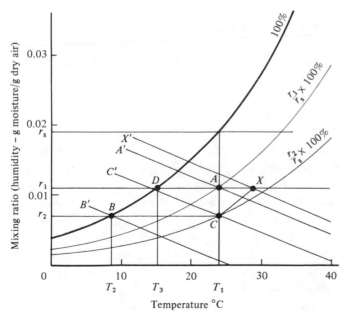

creasing humidity, and it is possible to draw lines of constant enthalpy on the chart such as a *AA'*, *BB'* and *CC'*. When a small object such as a thermometer bulb is covered with a wetted wick and suspended in a moving airstream, moisture evaporates from the wick cooling it, causing the temperature of the thermometer bulb to fall. Eventually, the fall in temperature is arrested when the rate of heat gain from the surrounding air equilibrates with the rate of heat loss by evaporation. The temperature depression of the wet bulb is greater for dry air than for wet air and is zero for saturated air. It is found that lines of constant wet-bulb temperature on the psychrometric chart coincide almost exactly with lines of constant enthalpy.

Psychrometric charts are usually drawn with lines of constant enthalpy and/or wet-bulb temperature and also with curved lines of constant relative humidity spaced at regular intervals. If any two measures of the condition of the air are known (dry-bulb temperature, wet-bulb temperature or enthalpy, dewpoint temperature or mixing ratio, relative humidity) the others may be read from the chart. Psychrometric charts are useful in illustrating the enthalpy changes that take place within direct calorimeters. Because enthalpy also increases with increasing pressure the values on a psychrometric chart refer to some nominated pressure, however in most applications concerning animal calorimetry pressure changes are negligible.

5.2   Types of direct calorimeters

Nearly all direct calorimeters may be classified under one or other of four headings: isothermal, heat-sink, convection or differential. These forms are illustrated schematically in Fig. 5.2.

5.2.1   *Isothermal calorimeters*

Isothermal calorimeters (Fig. 5.2(*a*)) are chambers lined on all surfaces with a barrier layer of insulating material and surrounded by a constant-temperature water jacket. As sensible (i.e. non-evaporative) heat produced inside the chamber passes through the barrier layer into the water jacket it causes a rise in the temperature on the inner surface of the barrier layer, the outer surface remaining at the temperature of the water jacket. Measurement of either the mean temperature rise of the barrier layer or of the mean temperature gradient across it provides an indication of the overall heat flow through the surfaces. In the design of isothermal calorimeters the objective is to ensure so far as possible that all the sensible heat output from the animal passes through the surfaces and is not convected out of the chamber in the form increased temperature of the ventilating air. On the other hand, evaporative heat output from the animal is carried out of the chamber by the ventilating air and may be estimated by measuring the increase in air humidity. Evaporative and non-evaporative heat losses from the animal are thus distinguished.

In the first isothermal calorimeters (d'Arsonval, 1886; Rubner, 1894) the insulating barrier was a closed air gap and changes in its mean temperature were estimated from changes either in the volume or pressure of the air confined in the gap. Murlin & Burton (1935) measured the temperature gradient across a 2″ thick 'kapok-like' lining around their calorimeter and Prouty, Barrett & Hardy (1949) built a calorimeter for small animals in which the heat flow across an air gap was measured by thermocouples attached to the walls, floor and ceiling. But the major breakthrough leading to accurate and fast responding isothermal calorimetry was the introduction by Benzinger & Kitzinger (1949) of the gradient-layer calorimeter with a network of thermocouples interwoven through a plastic sheet. Fig. 5.3 taken from Benzinger *et al.* (1958) illustrates the gradient-layer principle. Heat flow across any small area of the insulating layer lining the surface of the chamber is proportional to the temperature gradient across the layer at that point. It follows that if the layer is

Fig. 5.2. Types of direct calorimeter. (*a*) isothermal; (*b*) heat-sink; (*c*) convection; (*d*) differential.

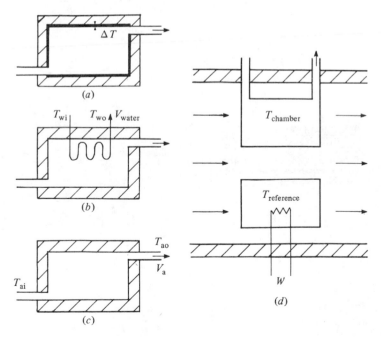

of uniform thickness and thermal conductivity, the mean temperature gradient across the entire layer is proportional to the integrated heat flow through all parts of the surface, regardless of where the heat is generated within the chamber. The mean temperature gradient may be conveniently measured by an array of series-connected thermocouples spaced at regular intervals or by resistance thermometers. Evaporative heat loss may also be measured by the gradient layer principle as shown by Fig. 5.4, also taken from Benzinger (1958). In addition to the calorimeter chamber two plate-meters P_1 and P_2 are included in the ventilation circuit. These each consist of two water jacketed metal plates lined with heat sensitive layers identical to that lining the chamber, and mounted face to face so that the air, passing up and down through channels formed between the two plates, attains the temperature at which they are operated. P_1 is operated at chamber temperature T_1 and P_2 at the desired dewpoint temperature T_2. The enthalpy changes undergone by the ventilating air are best understood by reference

Fig. 5.3. The Gradient Principle of Calorimetry. A 'gradient layer', shown below, lining completely and uniformly the entire inner surface of a cavity (above) records correctly the total heat from a source within, regardless of its location and the size or shape of the cavity. (Reproduced from Benzinger *et al.*, 1958.)

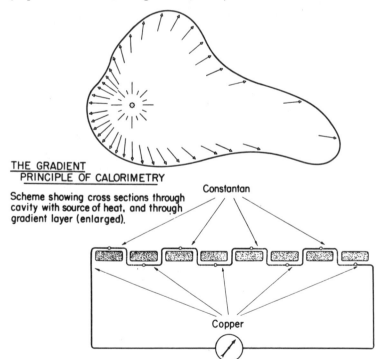

THE GRADIENT
PRINCIPLE OF CALORIMETRY

Scheme showing cross sections through cavity with source of heat, and through gradient layer (enlarged).

Constantan

Copper

to Fig. 5.1. Air is first saturated with water vapour in a saturator (S) operated at temperature T_2 (point B), it then gains heat through the sensitive layer of P_1 to enter the calorimeter chamber at temperature T_1 and with dewpoint temperature T_2 (point C). The latent and sensible heat produced by the animal are capable of raising the enthalpy of the air to point X, but the sensible fraction passes through the sensitive layer of the chamber so that air leaves the chamber still at temperature T_1 but with enhanced humidity (point A). Finally, during its passage through P_2 the air is again cooled, and the moisture which was added from the animal is condensed out so that the air leaves the heat-sensitive system at its initial condition, i.e. saturated at T_2 (point B). The combined heat output through all three sensitive layers is the total heat output of the animal. The non-evaporative fraction (the enthalpy difference $XX'-AA'$) is measured by the layer of the chamber, and the evaporative fraction (the enthalpy difference $AA'-CC'$) by the combined layers of the platemeters. What in fact occurs is that, in addition to evaporative heat output from the animal, P_2 measures

Fig. 5.4. The Gradient Calorimetry of Circulating Air. Ventilatory air, forced through the calorimeter in an open circuit, exchanges heat at equal rates of opposite sign in two plate-shaped gradient calorimeters, P_1 and P_2, if no heat is added in the calorimeter. Any added heat from the subject induces a net excess potential in the 'platemeter', P_2, which is recorded with P_1 and P_2 in series electrically. Saturation at temperature, T_2, of the air entering and leaving, is ascertained by a saturator (S) and by condensation of excess vapour in P_2. This circuit serves to remove vapour and to exchange the respiratory gases. (Reproduced from Benzinger *et al.*, 1958.)

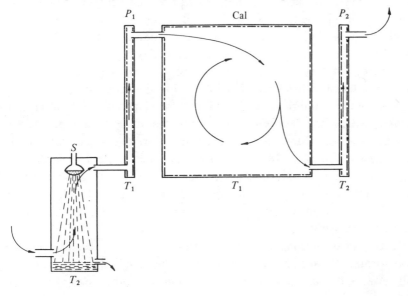

the heat evolved as the ventilating air is cooled from T_1 to T_2 (the enthalpy difference $CC'-BB'$), but the latter is exactly cancelled by the (negative) output from P_1 which measures the heat taken up as the air in it is heated from T_2 to T_1. The response of the entire system to changes in heat load can be very rapid, it is not affected by fluctuations in the ventilation rate and the only measurements to be recorded are the voltage outputs from the gradient layers.

5.2.2   Heat-sink calorimeters

Heat-sink calorimeters (Fig. 5.2(*b*)) are ones in which heat loss through the surfaces is prevented. Heat generated by the subject is removed from the chamber by a liquid-cooled heat exchanger whose cooling power is regulated to ensure equality between the temperatures of air leaving and air entering the chamber. The rate of heat removal by the heat exchanger is estimated from the flowrate and temperature rise of the coolant. In the design of heat-sink calorimeters an important objective is to avoid condensation of water vapour on the surfaces of the heat exchangers; this ensures correct partitioning between non-evaporative heat which is removed by the heat exchanger and evaporative heat which is convected out of the chamber by the ventilating air. The earliest heat-sink calorimeters (Atwater & Rosa, 1899; Armsby, 1904; Atwater & Benedict, 1905) were made of two concentric metal shells with a thin air gap in between. Heat flow through these surfaces was prevented by manipulating the outer shell temperature (by means of electric heaters or cooled pipes) to be the same as that of the inner shell; also the efficiency of the heat exchangers inside the chamber could be altered by moving the position of metal screens which partially obscured their radiating surfaces. These were probably as accurate and fast responding as any heat-sink calorimeters built since; but they were extremely complex in construction and operation. The technique was greatly simplified in the series of animal calorimeters built at Cambridge (Hill & Hill, 1914; Capstick, 1921; Deighton, 1926) in which the heat exchanger consisted of cooling pipes at the chamber surfaces. More recent heat-sink calorimeters (Garrow *et al.*, 1977; Dauncey *et al.*, 1978) have relied on cellular materials of very low thermal conductivity and capacity to prevent heat flow through and heat storage in the walls; they have also used heat exchangers of high efficiency operating only slightly below the air temperature and so avoiding condensation problems.

Heat-sink calorimeters are often described as adiabatic because no heat flows through the surfaces of the chamber. This terminology is acceptable only because it is unlikely to cause confusion with other forms of animal calorimeter. However, a heat-sink calorimeter is not truly adiabatic since heat is removed from the chamber by the heat exchanger. The name heat-sink calorimeter is preferred here as being a more precise description.

5.2.3 Convection calorimeters

Convection or air calorimeters (Fig. 5.2(*c*)) rely on measurement of the mean temperature rise of air ventilating a chamber containing the animal. Heat is prevented from escaping through the walls either by insulation or by external circulation of air at the same temperature or by a radiation shield. The method depends on very accurate comparison of the mean temperatures of the ingoing and outgoing airstreams. Early calorimeters of this type proved difficult to operate and were not very accurate. But the method has been revived and improved with the successive development of three human calorimeters in recent years (Visser & Hodgson, 1960; Tschegg *et al.*, 1979; Snellen, Chang & Smith, 1983).

5.2.4 Differential calorimeters

In differential calorimeters (Fig. 5.2(*d*)) there are two identical chambers, one containing the animal and the other an electrical heater whose (measured) heat output is adjusted to produce identical temperature increases in both chambers. It is particularly suited for measurements on small animals.

It will be apparent from the following sections that the descriptions isothermal, heat-sink and convection are by no means distinct or even accurate. A gradient-layer calorimeter is not strictly isothermal because a small variation occurs both across the thickness of the layer and over the surface; also a correction has to be made because some heat is lost by convection. Similarly, in convection and heat-sink calorimeters corrections have to be made for heat loss through the walls. However, most direct calorimeters are usually designed with the intention that the bulk of the heat is measured either as it flows through a heat-sensitive surface or via a heat-sink or as convection, and may be classified accordingly.

In some calorimeters a 'compensating heater' is included inside the chamber and its heat output is controlled so that the combined non-evaporative heat output from the animal and the compensating heater is held constant. This avoids alterations in the amount of heat stored in the structure and furnishings and can therefore lead to improved time response of both gradient-layer and convection calorimeters. The same also occurs in heat-sink calorimeters where variations in the rate of removal by the heat exchanger automatically compensate for alterations in the heat output of the animal. The heater inside the dummy chamber of a differential calorimeter may also be said to compensate for changes in the animal's heat output, and the dummy chamber is sometimes referred to as a compensating chamber. The term compensation calorimeter can thus be used to mean almost any type and indeed has been applied to all of them; it is a confusing description and should be avoided.

The majority of direct animal calorimeters built in recent years operate on either the gradient-layer or on heat-sink principles. These are each considered in the following sections first by a description of a particular calorimeter, followed by a more detailed treatment of design aspects for calorimeters of that type.

5.3   Gradient-layer calorimeters

5.3.1   *The Hannah Research Institute large animal calorimeter*

The gradient-layer calorimeter at the Hannah Research Institute was built for studies on the physiology of body-temperature regulation in large farm animals. Many of the construction methods followed those used in the slightly smaller calorimeter (Pullar, 1969), built at the Rowett Research Institute, Aberdeen, to accommodate sheep. The description that follows is an updated version of the original report on the Hannah calorimeter (McLean, 1971). Fig. 5.5 is a schematic diagram of the calorimeter. The ventilation system comprising the humidifier, dehumidifier, heater, fan and flowmeters provides the chamber with a continuous supply

Fig. 5.5. Schematic diagram of gradient-layer calorimeter. *V*, air flow-control valves or dampers; *R*, rotameter flowmeters; *T*, thermal flowmeters.

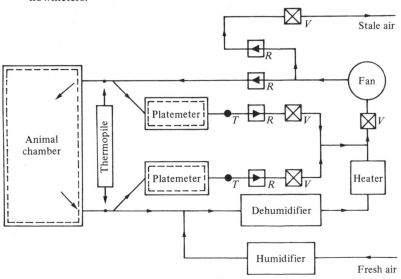

of conditioned air; it has no heat-measuring function. Most of the sensible heat output from the animal passes through and is measured by the gradient layer surrounding the animal chamber; evaporative heat is measured by the platemeters. Only a small proportion (approximately $\frac{1}{16}$) of the air ventilating the chamber is passed through the platemeters, whose heat sensitive layers are therefore designed to be 16 times as sensitive as that of the layer surrounding the chamber. The main purpose of the reduced flow through the platemeters is to limit their physical size. The platemeters are operated at a temperature T_2 just below that of the dewpoint of air entering the chamber. Air leaves each platemeter saturated at T_2. The difference between the electrical outputs of the two platemeters represents the difference in enthalpy of the air leaving as compared with the air entering the chamber; this is mainly the evaporative fraction of the animal's heat output. If the air leaving the chamber differs in temperature from the air entering then this represents sensible heat from the animal which has not been detected by the gradient layer surrounding the chamber, but has been detected instead by the platemeters. The thermopile which measures the temperature difference between the two airstreams is included to record this error in partition, and to correct for it. The thermopile sensitivity is adjusted to balance that of the chamber and platemeters; its output is added to the former and subtracted from the latter.

Four essential conditions must be satisfied for accurate calorimetric measurements:

(1) The two platemeters must be matched in sensitivity, temperature of operation and rate of flow of air samples, and complete cooling and dehumidification of the air samples to saturation at the operating temperature must be ensured.

(2) The samples drawn through the two platemeters must be truly representative aliquots, so far as temperature and humidity are concerned, of the air entering and leaving the chamber.

(3) The operating temperature of the platemeters must not exceed the dewpoint temperature of the air entering the chamber.

(4) The ratio of sample flowrate to total chamber ventilation rate must be fixed in relation to the sensitivity ratio of the platemeter and chamber gradient layers.

5.3.1a *The animal chamber*

The animal chamber (Fig. 5.6), measuring 2.13 m high $\times$ 2.6 m $\times$ 1.52 m internally, is built upon a framework welded from 76×5 mm mild steel angle and tee sections. The rectangular cooled panels that form

Fig. 5.6. The inside of the calorimeter chamber showing the tubular stall for restraining the animal. The original faeces-well which was later removed, can be seen in this picture.

the heat-sensitive floor, walls and ceiling are bolted to the inside of the framework. Each panel (Fig. 5.7) consists of two 6.3 mm thick aluminium plates which are bolted to face each other but held 23 mm apart by aluminium spacer-bars. Cooling water circulates through a series of channels formed between the two plates by the spacer-bars. The whole assembly is held together by screws passing through the plates and spacer-bars, and is sealed with neoprene-cork jointing. The screw-heads are countersunk on the inside surface of the chamber. The heat-sensitive surface is preformed on each panel in successive layers embedded in epoxy

Fig. 5.7. Cut away drawing of a heat sensitive panel. A, aluminium back plate; B, spacer-bars; C, neoprene–cork joint; D, aluminium face plate; E, resin-bonded-paper sheet or cotton cloth; F, Tufnol strips, every second strip wound spirally with constantan and half of each turn short-circuited by copper (shown in black); G, cotton cloth; H, aluminium foil; J, bolt holes for fixing to framework; K, bolt holes for assembly of the panel; L, hole for entry of circulating water.

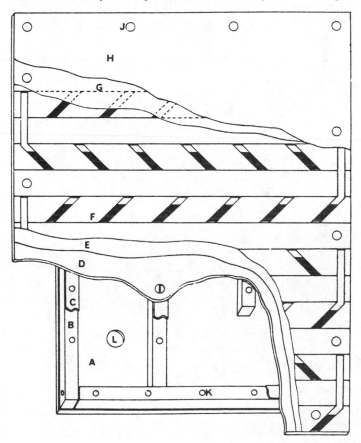

resin (Araldite). The first layer is either a 0.15 mm thick sheet of resin-bonded paper or a layer of cotton cloth for electrical insulation. The second layer, which is the heat-sensitive layer, consists of resin-bonded paperboard (Tufnol) strips 38 mm wide and 0.8 mm thick. These are laid side by side so as to form a continuous sheet covering the whole surface. Every second strip is wound spirally with constantan tape 1 cm wide by 0.08 mm thick at a pitch of 76 mm. Every half-turn of each constantan spiral is short-circuited by a length of copper tape 13 mm wide by 0.025 mm thick, the joints between the copper and constantan being hand-soldered, one on each surface of the Tufnol strip. The wound strips thus have series-connected copper–constantan thermocouples, formed every 76 mm along their length, which compare surface temperatures on either side of the strip. The construction of strips of this kind was originally described by Cairnie & Pullar (1959). Each wound strip is series-connected to the adjacent wound strips, so that a network of thermocouples is

Fig. 5.8. The calorimeter chamber showing the animal entrance door suspended from its travelling carriage, and the door into the airlock for use by personnel.

formed, with one couple for every 58 cm^2 of the whole surface, making approximately 4500 thermocouples in all. The heat-sensitive layer is covered by a layer of fine cotton cloth impregnated with epoxy resin to form electrical insulation. The final layer is a protective covering and consists of 0.1 mm thick aluminium foil, except over the floor and the lower parts of the walls, which have a hard-wearing but thermally conducting surface made of 0.7 mm thick perforated aluminium sheet embedded in epoxy resin with a sprinkling of sand added to the floor panels before the resin has completely set; this provides a non-skid surface.

Formation of the heat-sensitive layer on each panel had to be done with the panel lying horizontal, i.e. before it was mounted on the steel framework. In the design of the heat-sensitive panels it was therefore necessary to ensure that all the bolt holes for fixing the panels to the framework were drilled at points where there was to be no thermocouple winding. After the panels had been secured in place, the gaps in the Tufnol layer over the countersunk screw holes were filled with an insulating compound.

The entire rear-wall panel of the chamber forms the main door (Fig. 5.8); it is suspended from an overhead travelling-carriage and can be sealed to the rest of the chamber by spring-loaded catches compressing an expanded rubber joint. A small hinged door for entry of personnel via an airlock is situated near the front of the chamber. Sixteen heat-sensitive panels are used in all to form the chamber and doors. They are arranged in such a way that the inside of the chamber is completely lined with heat-sensitive surfaces, the small gaps between adjacent panels being filled with thermal insulation. There are a number of other small openings in the walls and ceiling of the chamber for observation, service entries, ventilation and illumination. These openings are mostly made 58 cm^2 in area or twice that size, and one or two thermocouples are left out of the heat-sensitive array at these points. The observation ports are double-glazed with perspex. The entire structure is insulated on the outside with 15 cm expanded polystyrene and finally covered in plywood. When originally built the chamber was fitted with a gradient-layer lined well, which could be sealed off from the rest of the chamber, for collection of animal excreta. This proved unsuccessful and was later removed.

5.3.1*b* *Ventilation*

The chamber is ventilated by a centrifugal fan which maintains a total flow of 2500 l/min against 15 cm H$_2$O. Approximately 90 % of the total flow is recirculated through the chamber, dehumidifier and heater (Fig. 5.5). The humidity of air entering the chamber is fixed by the operating temperature of the dehumidifier, which is a water-cooled condenser. The air is then passed through the 2 kW heater-bank, fan and variable-area flowmeters (rotameters) before re-entering the chamber. The heater-bank is controlled so that air entering the chamber is within ± 0.2 °C of chamber temperature. The heater-bank is placed close to the dehumidifier so as to minimise the possibility of condensation in the connecting ducting at high operating humidities. The variable temperature profile of the air emerging from the heater-bank and mixing with the sample-return flow is eliminated by the stirring action of the fan. The thermopile consists of 28 pairs of copper–constantan junctions placed in the ducts so as to compare inlet and outlet air temperatures.

Air is allowed to escape from the system through a rotameter and control valve at a rate of approximately 250 l/min and a similar amount of fresh air enters via the humidifier. The purpose of the humidifier, which is a small tank of heated water open to the fresh air duct, is to ensure that fresh air mixing with the main airstream does not reduce its dewpoint temperature below the temperature of the dehumidifier. The chamber is effectively at atmospheric pressure due to the negligible pressure drops in the ducts connecting the chamber, dehumidifier and fresh air inlet. The stale-air exhaust duct is led through the same outside wall of the building as that through which fresh air enters. Each duct opening is enclosed in a perforated metal cage approximately 1 m square and 0.5 m deep mounted on the outside wall. These cages are needed to reduce the effects of wind gusts on the air circulation of the calorimeter to an acceptable level.

The main air-circulation system is insulated throughout with 10 cm expanded polystyrene and covered with plywood.

5.3.1c   *Air sampling*

Samples of the circulating air are drawn at 150 l/min from points just before entry and just after exit from the chamber. The samples are passed through identical circuits each consisting of a platemeter, thermal flowmeter, rotameter and flow-control valve before rejoining the main airstream at the suction side of the fan.

Each platemeter consists of two water-cooled panels 2.44 m high by 0.45 m wide. The panels are mounted vertically facing one another but separated by a 22 mm gap, through which the air flows. The air passes up and down through a series of channels cut out from a sheet of expanded rubber which is sandwiched between the heat-sensitive surfaces of the two panels (Fig. 5.9). Drainage tubes with water-traps are fitted at the bottom of the airways to allow escape of condensed water. The panels are of the same construction as those forming the walls and ceiling of the animal chamber except that the gradient layers have a greater density of thermocouples, one to every 3.6 cm² of surface, giving approximately 5000 thermocouples in all for each platemeter. The basic sensitivity of the gradient-layers of the platemeters is therefore approximately 16 times greater than that of the chamber panels.

The rates of airflow through the main air-circulation and sampling systems are adjustable by means of valves and dampers at points V (Fig. 5.5). Since all air flows throughout the entire system are generated by the one prime air mover, once set they remain in fixed ratios to one another for any given temperature and humidity condition. Indication of all air flows is provided by the rotameters, but since these are affected by changes in air density it is necessary to reset all flows at levels appropriate to each temperature and humidity condition. No further control of air flow is required.

The thermal flowmeters consist of heaters of the same electrical resistance mounted in each air-sample line and both carrying the same heating

Fig. 5.9. Cross-section through the airways of a platemeter.

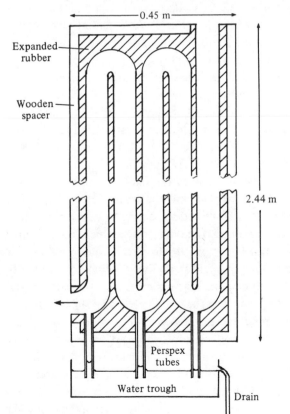

current. Comparison of the temperature increase in each airstream caused by these heaters provides a delayed but more precise measurement of the difference between the two sample flowrates than is possible with the rotameters.

5.3.1d   *Water cooling system*

Cooling water is circulated through the panels of the calorimeter chamber by all-plastic pumps from a 1.8 m³ tank. The panels are arranged in series in four circuits and the total flow is 200 l/min. The tank is fitted with a refrigeration coil, immersion heaters, a temperature-sensing element and various baffles. An additional stirring-pump is used to circulate water over the immersion heaters. In order to avoid corrosion and galvanic action the tank and the refrigeration coil are plastic coated, and PVC piping is used throughout. The only metals exposed to the circulating water are aluminium and stainless steel. The gradient layer of the calorimeter chamber responds immediately to any change in water temperature; an increase of 0.001 °C/min gives an output signal equivalent to a heat load inside the chamber of -2.8 W. Because of this, maintenance of minimal rate of change of water temperature is the essential requirement of the controller rather than maintenance of temperature within a fixed differential band. The controller used employs the same type of sensing element as that of Cairnie & Pullar (1959). This consists of a block of aluminium surrounded by a gradient layer, the whole being immersed in the water tank. The output from the gradient-layer surrounding the block provides a sensitive indication of the instantaneous rate of change of water temperature. This output is fed to a solid-state proportional controller that regulates the power supplied to a 1 kW immersion heater so that the rate of change of water temperature does not exceed ± 0.0005 °C/min. For efficient operation the controller must always detect a positive demand for heating power, free from sudden changes. These conditions are ensured by continuous running of the refrigerator which has a cooling capacity of up to 2 kW. In order to avoid waste, a manually adjustable expansion-valve is fitted to the refrigerator, so that cooling power may be set at a level appropriate for the operating temperature of the calorimeter.

The system for circulating water to the platemeters and dehumidifier is similar to that for the chamber except that control of water temperature is less precise. This is acceptable because most causes of error signal in the electrical output, such as fluctuations in water temperature, are eliminated due to the gradient layers of the two platemeters being connected in opposition. Temperature is controlled to ± 0.1 °C by means of a proportional controller operated from a thermistor immersed in the tank. The tank capacity is 1.1 m³ and a total of 100 l/min of water is circulated first through the platemeters and then through the dehumidifier. This arrangement ensures that the condensing temperature of the platemeters cannot exceed that of the dehumidifier.

5.3.1e *Instrumentation*

The outputs of the chamber, platemeters and thermopile are connected together as shown schematically in Fig. 5.10. Voltage analogues of sensible, evaporative and total heat losses appear across terminals 1 and 2, 2 and 3 and 1 and 3 respectively. These were originally monitored by a 2.5 mV multi-point potentiometric recorder. An additional similar circuit (omitted from the simplified diagram of Fig. 5.10) produces a voltage representing heat loss through the floor which is also available for recording. Alteration of the resistances connected across each unit, allows adjustment of the output sensitivity of each unit to 46.5 W/mV, which gives a full-scale range of 0–116 W (100 kcal/h) on the recorder. Additional switched resistances allow a choice of other less sensitive output ranges.

Although still retained to provide a visual record, the multipoint recording system has since been superseded by a computer-controlled data logger. In addition to monitoring the direct heat outputs of the gradient layers, the logger also records various temperatures of the calorimeter system and of the animal as well as ventilation rates and gas concentrations for indirect estimation of heat production. The data logger is based on a DEC PDP11/05 computer and a precision digital voltmeter and is programmed to scan up to 100 input channels and to process, display and

Fig. 5.10. Schematic diagram of calorimeter output circuit. The shunt resistances R, R_{pi} and R_{po} have values chosen to balance the output sensitivities of the two platemeters, the thermopile and the chamber.

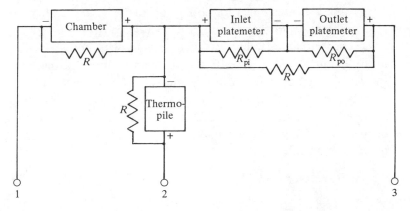

store all information at 10 min intervals. It is also programmed to control illumination in the chamber, availability of food and water to the animal and to operate alarms in the event of malfunctions. Apart from the visual and audible alarms operated by the computer, there are a number of automatic safety features built into the air- and water-temperature control systems to protect against overheating or failure of water or air circulation. Particularly important is a battery-operated reserve fan which is brought into operation in the event of failure of the public electricity supply.

5.3.1f   *Initial calibration*

For initial calibration the voltage outputs of the chamber (*Ch*), thermopile (*Th*) and platemeters (*Pm*) were each measured individually. The outputs from these units are:

$$Ch = (\dot{Q}_{\mathrm{N}} - \dot{Q}_{\mathrm{C}} - \dot{Q}_{\mathrm{L}})/k_{\mathrm{ch}} \tag{5.1}$$

$$Th = \dot{Q}_{\mathrm{C}}/\dot{V}k_{\mathrm{th}} \tag{5.2}$$

$$Pm = (\dot{Q}_{\mathrm{E}}\dot{v}/\dot{V} + \dot{Q}_{\mathrm{C}}\dot{v}/\dot{V} + \Delta\mathrm{p})/k_{\mathrm{pm}} \tag{5.3}$$

where $\dot{Q}_{\mathrm{N}}$ and $\dot{Q}_{\mathrm{E}}$ are the rates of non-evaporative and evaporative heat production in the chamber; $\dot{Q}_{\mathrm{C}}$ is the rate of non-evaporative heat convection out of the chamber; $\dot{Q}_{\mathrm{L}}$ is the rate of heat leakage through any gaps in the heat-sensitive layer; $\dot{V}$ and $\dot{v}$ are chamber and platemeter flowrates; k_{ch}, k_{th} and k_{pm} are constants (sensitivities); $\Delta\mathrm{p}$ is a term included to represent the effect of any mismatch between the two platemeters either due to unequal sensitivities or unequal sampling flowrates.

Initial calibration, as performed with the chamber controlled at 25 °C and the platemeters at 10 °C, proceeded in logical steps as follows:

(1) The chamber was not ventilated, the air inlet and air outlet were sealed off ($\dot{Q}_{\mathrm{C}} = 0$) and the output (*Ch*) was measured for various known levels of electrical energy ($\dot{Q}_{\mathrm{N}}$) dissipated inside the chamber. Plotting the results gave a straight line with slope equal to the basic chamber sensitivity (k_{ch}) and intercept equal to rate of heat leakage ($\dot{Q}_{\mathrm{L}}$).

(2) The ventilation system was then put into operation and with no load in the chamber ($\dot{Q}_{\mathrm{N}} = \dot{Q}_{\mathrm{E}} = 0$), the air-temperature controller was offset by various amounts to create a series of different levels of $\dot{Q}_{\mathrm{C}}$. Since chamber, thermopile and platemeters all respond to variations in $\dot{Q}_{\mathrm{C}}$ their sensitivities were thus compared. The output circuit (Fig. 5.10) could then be made up with values of R chosen so as to reduce all three sensitivities to the desired value k:

$$k = k_{\mathrm{ch}} = k_{\mathrm{th}}\dot{V} = k_{\mathrm{pm}}\dot{V}/\dot{v} = 46.5 \text{ W/mV (40 kcal.h}^{-1}.\text{mV}^{-1}) \tag{5.4}$$

Hence
$$\dot{Q}_{\mathrm{N}} = (Ch + Th)k + \dot{Q}_{\mathrm{L}} \tag{5.5}$$
$$\dot{Q}_{\mathrm{E}} = (Pm - Th)k - \Delta\mathrm{p}\,.\,\dot{V}/\dot{v} \tag{5.6}$$

(3) Still with no load in the chamber ($\dot{Q}_{\mathrm{N}} = \dot{Q}_{\mathrm{E}} = 0$) the value of $(Pm - Th)$ and therefore of $\Delta\mathrm{p}$ was made zero by adjusting R_{pi}. Only a minor adjustment was necessary, which was unlikely to have affected overall platemeter sensitivity, but stage 2 was repeated as an additional precaution.

(4) As a final check stage 1 was repeated but this time with the chamber ventilated. This completed the initial calibration at 25 °C.

The calibration of the platemeters and thermopile is slightly affected by the operating temperature and humidity because of alterations in the physical properties of air. To avoid the need for compensating adjustments to the resistances R and R_{pi}, it was more convenient to adjust flowrates $\dot{v}$ and $\dot{V}$ during stage 2 of initial calibrations at other temperatures. After a number of initial calibrations had been done in this way, graphs were plotted showing the required flowmeter settings at different operating conditions.

The leakage $\dot{Q}_{\mathrm{L}}$ was found to be approximately 0.17 W for each 1 °C difference between chamber and ambient temperatures.

5.3.1g *Routine calibration and operation*

For routine calibration and operation of the calorimeter some refinements are built into the computer programme to improve accuracy. Once the calorimeter has been set running at any required temperature and humidity condition and with the predetermined flowrates, a series of routine calibration checks are made as follows:

(1) With no load in the chamber the zero levels of sensible and evaporative heat output are checked. If necessary the zero for evaporative heat is corrected by means of a small adjustment to inlet platemeter flowrate. Any difference between platemeter flowrates ($\Delta\dot{v}_x$) is noted.

(2) The air temperature controller is offset by a few degrees. This has a transient effect on the outputs of the chamber platemeters and thermopile at the end of which their outputs should have all altered by the same amount. Any differences in their responses is used to determine the ratios of thermopile to chamber sensitivity (θ) and platemeter to chamber sensitivity (φ) at the pre-set flowrates $\dot{V}_x$ through the chamber and $\dot{v}_x$ through the platemeters.

(3) Air temperature control is now restored and the chamber sensitivity is checked against a known quantity of heat generated electrically within the chamber. Any error factor (ψ) in the chamber sensitivity is noted.

These checks provide all the information required for adjustment of future measurements to accord with the latest calibration figures, and to allow for slight changes in flowrates from the original settings which may subsequently occur due to changes in barometric pressure, electricity supply voltage, etc. Changes from the preset flowrates are important because the sensitivity of the thermopile and platemeters are dependent on flowrates as shown by equation (5.4). The sensitivity correction factors to be applied to the chamber, thermopile and platemeter outputs are ψ, $\theta \dot{V}/ \psi \dot{V}_x$ and $\varphi \dot{V} \dot{v}_x / \psi \dot{V}_x \dot{v}$, respectively, where $\dot{V}$ and $\dot{v}$ are current levels of chamber and platemeter flowrates recorded by the data logger. There is also a small zero correction to be made to the platemeter output if the difference in flows between the two platemeters has drifted from the calibration value of $\Delta \dot{v}_x$.

Finally, a further correction to the platemeter output is necessary to account for the fact that platemeter sensitivity to evaporative heat from an animal is slightly different from the sensitivity observed during calibration with a non-evaporative heat load. The error which arises because the water is recondensed to the platemeter temperature (T_2) rather than to the air temperature (T_1) is discussed in Section 5.3.4b.

All of these small adjustments to the calorimeter outputs are built into the computer software and are applied automatically.

5.3.1h   Performance

The calorimeter can operate at any temperature in the range 10 to 40 °C, with platemeter temperatures, i.e. dewpoint temperature of air entering the chamber, from 8 to 30 °C. Prolonged operation at air temperatures below 10 °C causes the output to become unstable. The instability is associated with a gradual breakdown of the electrical insulation of the gradient layers and is probably due to moisture trapped in the epoxy resin at the time of construction. The hardening compound of epoxy resins is hygroscopic and gradient layers should be constructed in a dry atmosphere.

Over the operating range of temperature and humidity the accuracy of total heat loss measurement is approximately $\pm 1\%$ or ± 3 W, whichever is the greater.

Three tracings of output records from the calorimeter are reproduced in Fig. 5.11. They represent heat emission from (*a*) a 250 kg steer standing and lying down at 15 °C, (*b*) a 90 kg man lying, standing and doing moderate exercise at 27 °C and (*c*) a calibrating heater and moisture source at 20 °C.

Speed of response to a step change in heat output is seen from Fig.

5.11(*c*). 80% of the change is recorded within 3 min and 90% within 10 min but the full response takes nearly an hour. The delay is largely attributable to the heat capacity of the tubular metal stall inside the chamber (Fig. 5.6).

5.3.2 Gradient layers

The design of gradient layers involves compromise. The requirements are as follows:

(1) Sensitivity must be adequate to provide output signals suitable for the measuring equipment.

Fig. 5.11. Tracings of output records from the Hannah calorimeter. (*a*) Rates of total ($\dot{Q}$), evaporative ($\dot{Q}_E$) and non-evaporative heat emission ($\dot{Q}_N$) and heat passing through the floor ($\dot{Q}_F$) from a 250 kg steer whilst standing and lying on the floor at 15 °C. The bars denote standing periods. (*b*) Rates of $\dot{Q}$, $\dot{Q}_E$ and $\dot{Q}_N$ from a 90 kg man lying down, standing and doing moderate exercise. (*c*) Rates of $\dot{Q}$, $\dot{Q}_E$ and $\dot{Q}_N$ from a 0.29 kW electric heater and in addition with a wet cloth hung up inside the chamber. An operator entered the chamber for a few seconds at points marked X for insertion and removal of the cloth.

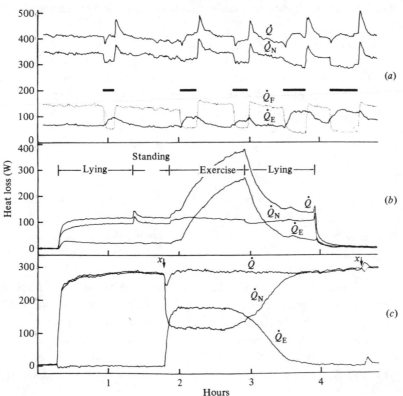

(2) Spacing between the temperature sensing points must be close enough to ensure accurate estimation of the mean temperature gradient.

(3) Speed of response to changes in heat load must be reasonably rapid.

(4) Internal surfaces must be washable, resistant to abrasion and corrosion and must absorb thermal radiation.

Mathematical analysis of the first three of these is possible and necessary at the design stage in order to decide on constructional details. But many approximations are needed and it is impossible to compute the eventual performance with precision. This must be found later by calibration of the finished instrument.

5.3.2a   *Basic sensitivity*

Fourier's Law states that once a steady state has been attained, the rate of heat flow ($\dot{Q}$) through a uniform sheet of area A, thickness D and thermal conductivity K is given by

$$\dot{Q} = KAT_D/D \qquad\qquad (5.7)$$

where T_D is the temperature difference between the two surfaces of the sheet.

The quantity $D/K = T_D/(\dot{Q}/A)$ represents the basic sensitivity of an insulating layer in terms of the temperature gradient generated per unit of heat flux through it. It is found in practice that the magnitude of the heat flux per unit area of chamber surface is similar regardless of the size of the calorimeter. Thus, for example, the Hannah calorimeter designed for cattle which emit heat at a rate of 0.5–1 kW has an internal surface area of 26 m², giving an average flux of 20–40 W/m²; whereas the area of the calorimeter described by Carter *et al.* (1975) for rats which emit 2–4 W is 0.132 m², giving 15–30 W/m².

If we assume $\dot{Q}/A = 25$ W/m² then, for a typical gradient-layer material (say thickness $D = 1$ mm, thermal conductivity $K = 0.25$ W. m⁻¹.°C⁻¹) the temperature gradient is of the order of $T_D = 0.1$ °C.

5.3.2b   *Spacing of measuring points*

The heat flow through different parts of the chamber surface, and therefore the magnitude of the temperature gradient, varies considerably over the surface of the chamber. It is necessary to ensure that the number of points at which the temperature gradient is measured is adequate for accurate estimation of the mean. We will consider two extreme cases.

Firstly, suppose that all the heat generated in the chamber radiates from a point source a distance x from one of the surfaces and that the distance between measuring points is λ. It may readily be calculated that the maximum fraction of the total radiation impinging on an area of the surface which does not include a measuring point is less than 15% for $x = \lambda$ and 8% for $x = 2\lambda$. Since in practice the heat source is very much more widespread than a point source it is evident that, provided the animal is kept at least a distance λ away from the surface of the chamber, there should be adequate averaging of heat exchanges by radiation.

Secondly, suppose that the animal is lying on the chamber floor so that, say, 50% of its heat loss is by conduction to the floor; we may assume that approximately one quarter of the animal's surface will be in contact with the floor and that the surface area of the chamber will be about five times that of the animal, i.e. 50% of the total heat loss will pass through about $\frac{1}{20}$ of the surface of the chamber. It should be possible to define this heat flow with an accuracy of 1% of the total by using 50 measuring points. Hence $20 \times 50 = 1000$ measuring points for the whole surface should be sufficient. If the calorimeter is roughly cubical there will be not less than $\frac{1000}{6} \simeq 11 \times 11$ measuring points per side, and hence the linear distance between measuring points (λ) needs to be not more than $\frac{1}{11}$ of the linear dimension of the chamber. This condition is easily compatible with keeping the animal a distance λ from any of the walls.

5.3.2c Speed of response

When a heat load is suddenly applied a delay occurs before the temperature gradient is built up because some heat is required to warm the gradient layer itself. Since the average temperature of the layer must increase by $T_D/2$ the total heat per unit area required to warm it up is

$$Q = AD\rho cT_D/2 \tag{5.8}$$

where ρ is density and c specific heat of the layer material. If we were to assume that all the heat impinging on the layer is initially retained in the layer to warm it up and none flows out from the other side, then the time taken to accumulate this heat would be found by dividing equation (5.8) by equation (5.7):

$$\text{Time} = Q/\dot{Q} = D^2\rho c/2K \tag{5.9}$$

In reality heat does flow out from the other side of the layer and the time for equilibration is somewhat greater than indicated by equation (5.9). A more rigorous mathematical analysis of this problem is given in Appendix IV (part 1) and its result is summarised in Fig. 5.12.

Fig 5.12 shows that the surface temperature of the layer builds up according to an exponential-like curve with a time constant (τ) given by

$$\tau = 4D^2 \rho c / K \pi^2 \qquad (5.10)$$

In every period of duration τ, a true exponential (indicated by the lighter curve of Fig. 5.12) rises by 63 % towards the final level. The temperature gradient across an insulating layer rises rather more quickly reaching 80 % of its final level in the first period τ, but thereafter its equilibration approximates very closely to a true exponential rise. Appendix IV (part 1) also confirms that the eventual temperature gradient is $T_D = \dot{Q}D/KA$ (cf. equation (5.7)).

The analysis shows that high thermal sensitivity is obtained from a thick layer (D large) of low thermal conductivity (K small), but these conditions also lead to a slow response. Since τ is proportional to D^2 it is particularly important to keep D as small as possible; speed of response is also improved by using layer material of low density (ρ) and specific heat (c). Also, the material should be incompressible to prevent D from altering. A list of relevant physical properties of materials frequently used in the construction of calorimeters is given in Appendix V. For the Hannah calorimeter the calculated time constant of the basic insulating layer (0.8 mm Tufnol) is $\tau = 2.04$ s.

The above analysis refers only to an unprotected insulating layer. In practice the insulating layer is usually faced with a protective metal foil. The temperature build-up of a thin foil is analysed in Appendix IV (part 2) where it is shown (equation (A4.6)) that it follows a simple exponential

Fig. 5.12. Time response of a gradient layer to a change of heat flow through it. The lighter line shows the time response of a simple exponential function.

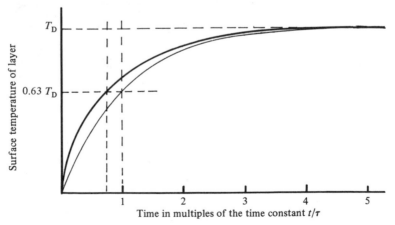

Time in multiples of the time constant t/τ

curve. For the Hannah calorimeter there are two time constants; 1.43 s for the 0.1 mm thick aluminium foil covering the walls and ceiling and 5.3 s for the 0.7 mm thick aluminium covering of the floor. The combined delay of both an insulating layer and its metal covering will tend to be near to the greater of the two.

A delay of considerably greater magnitude and practical importance is usually caused by the presence of objects of high heat capacity inside the chamber. Feed bins, drinking bowls and cages or other restraining struc-tures required in animal chambers usually have a larger heat capacity than the layer itself and can introduce delays measurable in minutes or even hours rather than seconds. The limiting factor here is not so much the slow equilibration of heat within the object as the slow transfer of heat to it from the heat source in the chamber. All items contained in the chamber should be kept as light as possible and have a high value of thermal diffusivity ($\kappa = K/\rho c$, see Appendix V). The animal stall in the Hannah calorimeter made from aluminium piping accounts for most of the delay in thermal response (Fig. 5.11(c)).

Delay in the response of calorimeters due to the time taken for the gradient layer, the chamber surfaces or the chamber contents to equilibrate in temperature may be reduced if the temperature changes are minimised. This can be achieved in a gradient-layer calorimeter if a heating element inside the chamber is so controlled that the voltage output of the sensitive layer remains constant. The power fed to the heater then represents a correction to be subtracted from the calorimeter output.

Sherebrin & Burton (1964) wound a heating wire uniformly over the inner surface of their gradient-layer calorimeter and controlled the power supply to the heater by a servo-mechanism which maintained a constant heat flow through the gradient layer. This reduced the time for 98 % response of their chamber (60 cm long × 25 cm diam.) from 12 min to 2.5 min. A similar system used by Bradham & Carter (1972) for their dog calorimeter (92 × 61 × 61 cm) reduced the response time from 7 h to 6 min; although it is not clear why the response time of this gradient-layer instrument was so poor in the first place.

5.3.2d   *Resistance-thermometer layers*

Resistance thermometers or thermocouples may be used to meas-ure the mean temperature difference T_D between the two surfaces of the gradient layer. The former is made by running a fine wire to and fro in equally spaced parallel lines all over the surface of the chamber. Two such wire networks of identical resistance are fixed one on either side of the insulating layer. The difference in electrical resistance (ΔR) between the two wires is

$$\Delta R = AR_0 aT_D \tag{5.11}$$

where R_0 is the resistance (at 0 °C) per unit area of the surface for each wire and a is the temperature coefficient of resistivity. Spinnler *et al.* (1973) used resistances in the form of printed copper circuits on either side of a 2.4 mm thick layer of epoxy resin. They divided the resistance of each side into two halves and connected them in the form of a Wheatstone bridge, as shown in Fig. 5.13. For this arrangement the voltage output of the bridge (V) is given by

$$V = i\Delta R/4$$

and combining with equations (5.7) and (5.11) gives

$$V = iR_0 aD\dot{Q}/4K \tag{5.12}$$

Provided R_0 is constant, i.e. that the resistance of the circuit is spread uniformly over the surface, the voltage output is proportional to the total heat flow $\dot{Q}$. It does not depend on the surface area of the chamber. The sensitivity of the layer may be varied by altering the current (i) fed to the Wheatstone bridge, but a limit is set by the amount of heat ($i^2R_0/4$) which is generated per unit area on each side of the insulating layer itself. The heat generated on the inside surface has to pass through the layer and is therefore measured by it. To keep this heat to a minimum the current must be small, but in order to avoid loss in sensitivity this means that the resistance per unit area must be kept high.

Spinnler *et al.* (1973) used a relatively thick layer material ($D = 2.4$ mm,

Fig. 5.13. Connections of resistances of gradient layer to form Wheatstone bridge. (Modified from Spinnler *et al.*, 1973.)

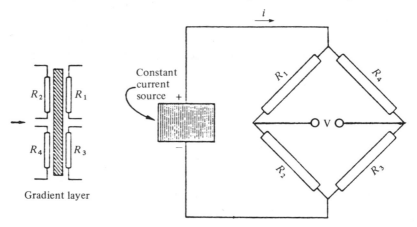

$K = 0.4$ W m^{-1} °C^{-1}) with the printed circuits made of copper ($R_0 = 334\Omega$, $a = 0.0041$). They then adjusted the current so as to obtain an output of $V = 1$ mV from a heat load of $\dot{Q} = 25$ W. Equation (5.12) shows the current required as 19.4 mA, which would produce $i^2 R_0/4 = 0.31$ W/m^2 and $iR_0/2 = 3.25$ V/m^2 in the gradient-layer winding. In fact the current, voltage and power dissipation in Spinnler *et al*.'s apparatus are slightly greater than those figures because of the additional power load of the recording system.

Fine nickel wires were used as resistance thermometers in their gradient-layer calorimeters for small animals by Young & Carlson (1952) and Sherebrin & Burton (1964). Nickel provides a higher resistance network over a small area than a similar copper network. Spinnler (private communication) has also constructed a calorimeter for babies in which printed circuits of nickel are used.

5.3.2e Thermocouple layers

Copper-constantan is the obvious choice of material for thermocouple layers. Its thermoelectric power (ε) is nominally 40 μV/°C, but it varies depending on the composition of the constantan alloy. The voltage output (V) from a heat sensitive layer of a chamber with n thermocouples per unit area is therefore given by

$$V = nA\varepsilon T_D$$

combining this with equation (4.7) gives

$$V = \frac{n\varepsilon D\dot{Q}}{K} \tag{5.13}$$

Equation (5.13) is the same as equation (5.12), but with $iR_0 a/4$ replaced by $n\varepsilon$. To obtain the same output sensitivity (1 mV for 25 W) from the same layer material (2.4 mm epoxy resin), as used by Spinnler *et al*. (1973), would require approximately $n = 100$ thermocouples per square meter of surface. The voltage output is again independent of the area of the chamber and the only requirement is that the thermocouples are uniformly distributed (i.e. n is constant).

Because the heat flow per unit surface ($\dot{Q}/A$) is similar regardless of the size of the chamber, the total number of thermocouples for any required voltage output is also similar for all sizes of chamber. Many of the thermocouple gradient-layer calorimeters that have been described have a total of 4000–5000 thermocouples and a typical output of about 10 mV. This was necessary in the period between 1950 and 1970 for measurement with the potentiometric strip-chart recorders then available. With modern computer-based data-logging equipment the sensitivity of the recording device is no longer a critical factor in deciding on the number of thermocouples required.

Some loss of sensitivity may occur in thermocouple gradient-layers because the metal tape has to bridge across the insulating layer twice for each thermocouple (see Fig. 5.14). As well as direct heat leakage through the metal bridges, there is a possibility that lateral heat conduction along the metal towards the bridges may result in a reduced temperature gradient at the measuring points. Mathematical analysis (see Appendix IV, part 3) shows that the fractional reduction in sensitivity due to lateral conduction from a measuring point situated at a distance L from a bridge is

$$\theta_1 = \frac{2}{e^{L/l} - e^{-L/l}} \tag{5.14}$$

where

$$l = \sqrt{(DD'K'/K)} \tag{5.15}$$

is a characteristic length dependent on the thickness D' and D and thermal conductivities K' and K of the metal and insulating layer materials respectively. As an example, for a 0.025 mm thick copper tape, bridging a 1 mm thick Tufnol sheet, $l = 6.3$ mm. The relationship between θ_1 and L/l is summarised in Table 5.1; it shows that the distance L requires to be about five times the value of l to keep the loss of sensitivity due to lateral

Table 5.1. *The factor* (θ_1) *by which sensitivity is reduced due to lateral heat conduction towards metal bridges through a gradient layer for different values of the ratio* L/l *(see text)*

L/l	θ_1
1	0.65
2	0.27
3	0.10
4	0.01

Fig. 5.14. Section through an idealised gradient layer with all-copper metal bridges passing through the insulation.

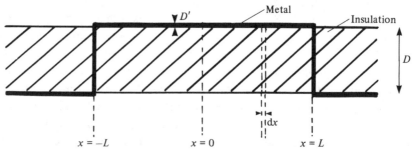

heat conduction along the metal tape down to 1 %, if $L = 3 \times l$ sensitivity is reduced by 10% and if $L = 2 \times l$ by 27%.

For constantan, which has a thermal conductivity K' 16 times less than copper, the value of l is only $\frac{1}{4}$ that for a copper tape of the same thickness. The layer windings should therefore be arranged so that the ratio L/l and therefore θ_1, is the same for both the copper and constantan arms of the thermocouple. This usually means that the junction between the two metals is formed about four times further from the copper bridge than from the constantan one. The optimum position will depend on the thicknesses of both tapes.

The fraction of the total heat flux which passes directly through the copper bridges (θ_2) rather than through the insulating layer is calculated in Appendix IV (part 4):

$$\theta_2 = nlD''(l+D')/(l+D) \qquad (5.16)$$

where D'' is the width of the tape and n is the number of thermocouples per unit area of surface. Since l is normally large compared with both D and D', equation (5.16) may usually be approximated to $\theta_2 = nlD''$. The corresponding value for constantan will again be about $\frac{1}{4}$ that for copper. The fractional reduction in sensitivity due to direct conduction of heat through copper–constantan bridges is therefore $1.25\theta_2$.

The predicted sensitivity of the completed layer may be estimated by applying the correction factors θ_1 and θ_2 to equation (5.10).

$$\text{Sensitivity} = (V/\dot{Q}) = n\varepsilon D(1-\theta_1).(1-1.25\,\theta_2)/K \qquad (5.17)$$

The length of copper tape per thermocouple is $2L$ and its electrical resistance is $2L\rho'/D'D''$, where ρ' is resistivity of copper. For constantan with resistivity 25 times as high the corresponding values are length $L/2$ and resistance $12.5L\rho'/D'D''$. The combined length $2.5L$ is obviously related to the distance between measuring points and therefore to $1/\sqrt{n}$. Hence

$$L = m/2.5\sqrt{n} \qquad (5.18)$$

where m is a constant factor, usually in the range 1–4, which depends on the geometry of the thermocouple array. The overall resistance (R_c) of the entire thermocouple network for the chamber is therefore

$$R_c = \frac{5.8mA\rho'\sqrt{n}}{D'D''} \qquad (5.19)$$

The properties of a heat-sensitive layer based on thermocouples are thus determined from equations (5.13)–(5.19).

As an illustration of how the equations are used for design purposes, Table 5.2 lists the values for three gradient layers: the chamber-surface layer of the Hannah calorimeter, its platemeters and a small flask calorimeter designed to measure heat

Table 5.2. *Design summary for gradient layers of the Hannah calorimeter chamber and platemeters and for a small calorimetric flask. All three are based on $D = 0.8$ mm thick Tufnol strips wound with $D' = 0.025$ mm copper tape and constantan, for which the characteristic dimension $l = 5.5$ mm*

	Purpose of layer	Animal chamber	Platemeters	Flask calorimeter
A	Surface area (m²)	26.2	1.78	0.037
n	Thermocouples per area (m⁻²)	172	2760	10330
nA	Total number of thermocouples	4506	4912	382
V	Uncorrected voltage output (mV for 20 W/m²)	11.5	12.6	0.97
m	Array factor $= 2.5L\sqrt{n}$ (see Fig. 5.13)	1.41	2.24	3.91
L	Length of copper strips (mm)	43.0	17.1	15.1
D''	Width of copper strips (mm)	13.0	5.0	2.5
L/l	(see equation (5.14))	7.82	3.11	2.75
θ_1	Lateral flow loss factor	0.001	0.089	0.128
θ_2	Direct flow loss factor	0.010	0.080	0.150
$\|$	Corrected voltage output (mV for 20 W/m₂)	11.4	10.3	0.69
R_c	Total electrical resistance (Ω)	153	173	24
τ	Basic time constant (s)	2.04	2.04	2.04
τ_F	Time constant of aluminium (s)	1.43 (wall) 5.3 (floor)	1.43	5.3

(For physical properties see Appendix V.)

of fermentation in a sample of cow's rumen fluid. All three are designed for a typical load of $\dot{Q}/A = 20$ W/m² and are based on an insulating layer of $D = 0.8$ mm thick Tufnol and copper tape $D' = 0.025$ mm thick, together with constantan (for physical properties see Appendix V). This choice of materials determines the values of the characteristic dimension l (5.5 mm) and the basic layer time constant (2.04 s) which are common to all three layers. The form of the thermocouple array is a crucial design factor. For the animal chamber the main consideration was to limit the overall resistance (R_c) of the network covering a large area. A wide copper tape was used and it was wound around Tufnol strips of width $\lambda/2$, with a blank strip between each wound one (Fig. 5.15(a)). This arrangement kept the factor m and hence L as small as possible, but the ratio L/l was still large enough to prevent any heat loss by lateral conduction (θ_1). For the platemeters a compromise was necessary. Both the width of the Tufnol strips and the pitch of the windings were set at λ (Fig. 5.15(b)), giving an intermediate value for m. The ratio L/l was still large enough to limit the sensitivity loss factor θ_1 to only 0.09 without the overall resistance becoming excessive. For the flask calorimeter, although a more sensitive measuring instrument was available, the prime consideration was to boost output voltage by crowding as many thermocouples as possible into the small surface area; but the value of L still had to be kept large enough to avoid increasing θ_1. The Tufnol strips were made as wide as possible within the limits of keeping the

Fig. 5.15. Geometrical form of a single thermocouple winding for each of three gradient layers. (a) animal chamber; (b) platemeters; (c) small flask calorimeter. Full lines indicate copper (shaded) and constantan tape on top of Tufnol; broken lines indicate tape below. λ, λ' and λ'' are dimensions of the thermocouple grid; D'' is width of the copper tape.

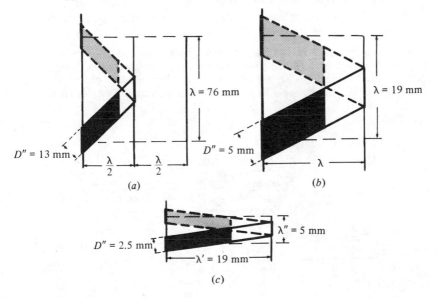

separation between strips λ' small compared with the linear dimension of the flask, and the copper was wound in as tight a spiral as possible. This resulted in a rectangular grid of thermocouples with dimensions λ' and λ'' (Fig. 5.15(c)) and $\lambda'\lambda'' = 1/\sqrt{n}$. The resultant value of m was large.

5.3.2f Construction

Following Benzinger's (1949) introduction of the gradient-layer concept, it was first developed at the research laboratories of the American Society of Heating and Ventilating Engineers by Huebscher, Schutrum & Parmelee (1952) who employed thermocouple chains made from copper–constantan ribbon. They were made by welding a bar of copper to a bar of constantan and rolling it into a thin ribbon of which half the width was copper, and half constantan. The ribbon was woven in basketweave fashion over and under strips of insulating material and at every cross-over point a deep notch was cut out of the ribbon from one side or the other so that only one of the two metals formed the bridge passing through the insulating layer. The method which is illustrated in Fig. 5.16, taken from Hammel & Hardy (1963), was employed in a number of calorimeters built in USA. Huebscher *et al.* (1952) used a vacuum press to bond the woven layer between sheets of cellulose acetate which were in turn covered with aluminium foil. The alternative process of winding constantan tape spirally around strips of insulation and soldering copper shorting strips to every half-turn is less costly in materials but very laborious. Care has to be taken to ensure sound soldered joints and to avoid spreading the solder along the unshorted constantan tape.

Epoxy resin is an excellent material for encapsulating the various layers,

Fig. 5.16. Formation of a gradient layer from copper–constantan centre-welded ribbon interwoven with woven glass tape. The thermojunctions are formed by cutting deep notches out of alternate sides of the ribbon. (Reproduced from Lawton, Prouty & Hardy, 1954.)

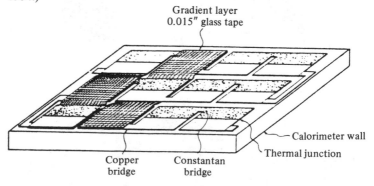

and the mixture five parts of Araldite MY753 to one part Hardener HY956† has been found ideal. After mixing, the resin must be kept stirred to avoid very rapid setting which can occur almost explosively due to the exothermic nature of the reaction. The panel to be layered should be mounted horizontally and water at 50 °C passed through its jacket. At this temperature the resin at first becomes very viscous but sets hard within 20 min. The various strata (insulating cloth, thermocouple layer, insulating cloth, metal foil) are formed one at a time with at least 4 h in between for curing the resin whilst the temperature is maintained. The process should take place in a dry, well-ventilated atmosphere and care must be taken to avoid excessive breathing on the unset resin which at this stage is slightly hygroscopic. All materials used must be clean and completely free from grease. The resin does not adhere to polythene which may therefore be used for covering any boards or weights employed. Once cured the resin has a high-impact strength but any surplus is readily filed, scraped or drilled.

Spinnler (personal communication) for his calorimeter at Zurich fixed the gradient layers in position under vacuum. In order to do this a sheet of polyethylene was laid over the gradient layers and taped to the edges of the aluminium plate. This created a bag over the gradient layer from which the air could be pumped out; the outgassing was facilitated by 'massaging' the gradient layers by hand through the polyethylene. The procedure was effective in eliminating air pockets under the gradient layer. The gradient layers of the platemeters were also covered with PVC foil and no difficulty was found in utilising the chamber at low dewpoints (down to 2 °C). These construction methods used at Zurich appear to offer a considerable improvement over those which we employed at the Hannah Institute.

5.3.3 *Animal chambers*

Most large gradient-layer calorimeters that have been described are of rectangular shape, probably because it is easier to construct the heat sensitive layers on flat surfaces. Tiemann's (1969) calorimeter for humans is cylindrical, which has advantages in ensuring a uniform distribution of air movement in the chamber.

5.3.3a *Air leaks*

Chambers must be airtight and the only problem areas in this respect are the seals around doors and hatches. There does not appear to be any satisfactory alternative to some form of compressible jointing

---

† Ciba Geigy Plastics and Additives Co., Duxford, Cambridge CB2 4QA.

material such as rubber, cork or plastic. Most of these under near-permanent compression tend to lose elasticity, and adjustment of securing catches is usually necessary from time to time. An inflatable tube can also be used for the seal between closed surfaces. Because of the difficulty of sealing large openings it is desirable to design the ventilation system so that the pressure inside the chamber is very close to barometric pressure. A simple test to measure leaks around doors or, indeed, in any part of the system, is to seal off air inlet and outlet ports so as to isolate the chamber or section to be tested, and to pump air in or out through a small rotameter whilst at the same time measuring the pressure change using a water manometer. The pumping speed, starting from a low level, is slowly increased until a detectable steady pressure is attained and the leak rate at this pressure may then be read from the rotameter.

5.3.3b   Heat leaks

Windows, service entry points, door seals and joints between heat-sensitive panels are all potential areas where heat may leak in or out of the chamber undetected by the heat-sensing layer. Heat leakage must be avoided as far as possible by the use of double- or triple-glazing with clear plastic, cork or rubber bungs, thermally insulating filler compounds, etc. It is virtually impossible to exclude heat leakage completely from an operational animal chamber, and its extent must be estimated during calibration by a method such as that described in Section 5.3.1f. It should be noted that any gap in the gradient layer such as a window, through which heat exchange is completely prevented, does not adversely affect the measurement of heat loss. In theory even a whole wall may be left uncovered by the gradient layer and, provided that it is perfectly insulated, all of the heat loss will be diverted through the remaining (heat-sensitive) surfaces of the chamber. A corollary to this is that it is not necessary for the internal finish or furnishings of the chamber surfaces to be uniform. For example, a mat placed on the floor, though it may reduce heat flow through the floor by increasing the thermal insulation, only diverts that part of the heat flow to other surfaces; it in no way alters either the total heat flow or the sensitivity of the floor layer itself.

5.3.3c   Illumination

An electric light bulb or other electrical device operating inside the chamber produces heat which is automatically measured as part of the heat load. In the Hannah calorimeter the electricity supply to all such devices is measured by a digital wattmeter which is monitored by the data logger. It is, however, better to avoid intermittent operation of electrical equipment in the chamber because many devices have considerable heat capacity and slow warm-up time. A useful method of providing a limited amount of illumination is to duct light into the chamber from outside through a solid perspex tube which passes through a wall or ceiling panel.

5.3.3d Animal-restraining structures

Animal chambers have to include provision for restricting the movement of the subjects and for collection of excreta. Pullar (1969) placed his animals in a mobile cage which could be rolled in and out of the chamber. The thermal inertia due to the high heat capacity of the cage involved considerable loss in speed of response, but excreta was easily collected in receptacles placed below the cage. At the Hannah Institute we chose to have our animals standing on the chamber floor and we built in fixing points for a lightweight stall formed from tubular aluminium (Fig. 5.6). The bolts securing the fixing points had to pass through the floor and ceiling panels of the chamber and were a potential source of heat leakage; this was reduced by sleeving the bolts in plastic and covering their heads inside the chamber with plastic domes. We also found that the perforated aluminium sheet embedded in epoxy resin provided a floor surface that was hardwearing, thermally conducting and resistant to the corrosive effects of spilled excreta. The addition of a sprinkling of sand before the resin had completely set was essential as without it the final surface was extremely slippery, especially when wet.

In chambers built for human subjects (Benzinger *et al.* 1958; Tiemann, 1969; Spinnler *et al.*, 1973) provision has been made for suspending a nylon net hammock or chair for the subject to rest on.

5.3.4 Ventilation
5.3.4a Ventilation rate

Humidity control is the most important factor in determining the level of ventilation in an animal chamber. The greatest problem is encountered with a physically active subject in a warm environment; here the heat loss is high and it is all evaporative. For every watt of evaporative heat loss the vapour produced is 25 mg/min. If the ventilation rate is V l/min the air passing through the chamber will increase in humidity by $25/V$ mg/l per Watt of evaporated heat. At 40 °C saturated air contains 51 mg/l of moisture, hence the increase in relative humidity is approximately $50/V$ %/W. To limit the relative humidity increase to 10% for an exercising man producing 250 W thus requires a ventilation rate of 1250 l/min. At lower temperatures the proportion of heat loss by evaporation is reduced, but so also is the capacity of air for moisture (see Fig. 5.1); the increase in relative humidity of air passing through the chamber is likely to be only slightly less than at 40 °C.

The ventilation rate in relation to the volume of the chamber also determines the speed of response to changes in evaporative heat output. The time taken for 95% response in the humidity of outlet air is approximately equal to the time for three complete changes of the chamber air (e.g. 6 min for a 2.5 m³ chamber ventilated at 1250 l/min). Although a high level of ventilation is desirable for improving both speed of response and humidity control it also involves increasing the size of the platemeters.

5.3.4b *Platemeters*

In the straight-through ventilation system used by Benzinger and illustrated in Fig. 5.4, the measurement of heat losses is independent of ventilation rate. Provided that the air passing through each unit (P_1, P_2 and the chamber) leaves that unit at its operating temperature (i.e. that the units are 100% efficient as heat exchangers) then the measurement and partition of heat losses is accurate. Many gradient-layer calorimeters employ this system (e.g. Benzinger *et al.*, 1958; Spinnler *et al.*, 1973), often an additional platemeter is included between the chamber and P_2 (Benzinger & Kitzinger, 1949; Pullar, 1969). The additional platemeter operating at chamber temperature serves as an extension of the chamber and ensures that the air leaves it at the correct temperature. Alternatively, internal ducting can be designed so that the exit air is equilibrated with the chamber temperature before being exhausted (Hammel & Hardy, 1963).

There is little to be gained from a detailed mathematical analysis of heat exchange along the air pathways inside platemeters. The complex geometry and the large number of variables to be considered rule out this approach. At the Hannah Institute we have adopted an empirical approach based on experimental trials with a crude prototype.

Two water-cooled plates were mounted face to face with air channels formed in between by strips of expanded rubber sandwiched between the plates, in the manner of Fig. 5.9. Warmed air at temperature T_0 was passed into the air channel and air temperature was recorded by thermocouples placed at intervals along its length. The air pressure drop was also noted. The temperature along the length (x) of the channel decreased towards the water temperature (taken as zero) according to an exponential relationship

$$T = T_0 \, e^{-x/L}$$

The distance L represents the path length for the temperature to alter by 63 % towards its final value; it was found to depend on the velocity of air flow (v) and the separation between the plates (b) according to the relationship

$$L = 144vb/(1+v)$$

Also the pressure drop per meter length of air path (P) was given by

$$P = 0.67v(1+1.29v)$$

These values were used to calculate the path length and the total area of platemeter surface required as well as the total pressure drop for a platemeter of 99.9 % efficiency (i.e. $T/T_0 = 0.001$) which corresponds to $x = 7L$. Some examples of these calculations are given in Table 5.3. It must be emphasized that the empirical equations and the values given in Table 5.3 apply only to the conditions of the tests on which they are based ($v = 1.5$ to 5 m/s and $b = 2$ to 5 cm) and extrapolation to other conditions may be very unreliable.

Generally it is preferable to keep the gap between the plates small, but at very high ventilation rates this becomes impracticable as both the air velocity in the channel and the pressure drop become too great; these conditions also generate excessive noise. Platemeters needed for ventilation rates as high as 2500 l/min (the ventilation rate of the Hannah cow calorimeter) would have had to be unacceptably large in order to keep the pressure drop down to 50 mm H_2O which was considered the acceptable limit. These considerations led to the decision to use an air-sampling system (Fig. 5.5) for that calorimeter. The platemeters conditioning the samples were ventilated at only 150 l/min resulting in a great reduction in size.

Another point to be borne in mind during platemeter design is the possibility of errors arising from heat loss at the edges of platemeters. At the boundary of an air pathway defined by a strip of expanded rubber there is an abrupt transition between no heat transfer and heat transfer from the air to the aluminium-foil lining of the layer. Some heat is conducted along the foil under the rubber and it is essential that the heat-sensitive layer extends sufficiently far under the rubber that this heat is included in the measurement. The problem is considered in Appendix IV (part 5) where it is shown that for a pathway width w at the edge of the platemeter the proportion of the heat exchange escaping past the layer is $1/2we^{-L/l}$, where $l = \sqrt{(DD'K'/K)}$ is a characteristic length analogous to that described in Section 5.3.2e but with D' and K' now referring to the thickness and thermal conductivity of the aluminium foil. L is the width of the section of heat-sensitive layer extending under the rubber. Take, for example, a layer of 0.8 mm Tufnol, covered by 0.1 mm aluminium foil and divided into pathways $w = 5$ cm wide. The value of l for this layer material

Table 5.3. Some examples of calculated platemeter dimensions required to condition different levels of ventilation with 99.9% efficiency (for 99.0% efficiency the values in the last three columns may all be reduced by one-third)

Total ventilation (l/min)	Separation of plates (cm)	Width of channel (cm)	Air velocity (m/s)	Path length (m)	Total area (m²)	Pressure drop (mm H$_2$O)
150	2	5	2.5	14.3	0.86	10
250	2	5	4.2	16.1	0.97	29
500	2	10	4.2	16.1	1.77	29
1000	2	20	4.2	16.1	3.4	29
1000	5	20	1.7	31.3	6.5	11
2500	2	37	5.5	16.9	6.2	50
2500	5	23	3.5	38.9	9.3	50

is 0.82 cm. If the layer extends only 1 cm under the rubber then the proportion of the total heat escaping past the layer is 2.4 %; for $L = 2$ it is only 0.7 %. Most of the heat exchange in a platemeter takes place in the first meter or less of the pathway; it is only in this region that the rubber must be extended well over the sensitive layer to ensure against heat escaping it.

In addition to reducing the size of the platemeters a further advantage of the sampling system is that heat flow through both platemeters is in the same direction (outward towards the water jacket) and the difference signal is obtained by connecting the electrical circuits in opposition. Whereas in the straight-through flow system the platemeter heat flows are in opposite directions and the electrical circuits are not reversed. This means that most forms of output error signal arising in the platemeters (e.g. due to fluctuations in coolant temperature or inefficient cooling) are self-cancelling. The requirement for accurate temperature control as well as refrigerant and fan capacity are all reduced in comparison with the straight-through flow system.

There are, however, considerable disadvantages, the most important being that the measurements become dependent on flowrates. The important quantities are not the absolute levels of flow but the ratios of platemeter ventilation to chamber ventilation and the equality between the flowrates in the two platemeters. Fortunately, these ratios once set are unlikely to alter in the short term provided all air movement in the system is generated from a single fan. But the ratios are slightly affected by changes in the operating conditions of temperature and humidity or by accumulation of dust or debris in the ducting.

The heat-sensitive layers of platemeters in a sampling system have to be of greater sensitivity than those in a straight-through flow system and a dehumidifier and heater have to be included. The dehumidifier in the Hannah calorimeter was in fact first built as a trial platemeter and was used without a gradient layer for the tests described earlier in this section. After it had been established that this unit, measuring 2.4×1.2 m, could at most be only 98 % efficient as a heat exchanger, it was used instead as the dehumidifier; thus satisfying the condition that in a calorimeter employing a sampling system the dehumidifier must not be more efficient than the platemeters (see condition 3, Section 5.3.1).

Whether sampling or straight-through ventilation is employed there is an inherent error in the platemeter method of measuring evaporative heat loss because water is not recondensed to air temperature.

Consider the evaporation of m g of water injected into the chamber at T_1. $Q_E = mL_{T1}$ Joules of heat are absorbed from the chamber and m g of vapour are added to the circulating air (where L_{T1} is the latent heat of vaporisation at T_1).

The quantity of heat detected by the outlet platemeter but not detected by the inlet platemeter is the heat of cooling mg of vapour from T_1 to T_2 plus the heat of condensing it at T_2. i.e.

$$[(T_1 - T_2)c_p + L_{T2}]m$$

which by the law of Hess is the same as

$$[(T_1 - T_2)c_w + L_{T1}]m$$

where c_p and c_w are specific heats of water in vapour and liquid form respectively. But the heat detected by the chamber is $-mL_{T1}$. Therefore the total heat detected is

$$m(T_1 - T_2)c_w = e$$

This is the permanent error of the system. The overestimate of evaporative and therefore of total heat loss may be expressed as

$$e = \frac{(T_1 - T_2)c_w}{L_{T1}} \times Q_E \simeq \frac{(T_1 - T_2)\,Q_E}{580} \tag{5.20}$$

The error is the same regardless of the conditions of evaporation. If the water first cools or warms before evaporating at a lower (e.g. wet cloth) temperature or higher (e.g. skin) temperature than chamber air, then the platemeters still record the amount of heat $(mL_{T1} + e)$. The heats of changing the temperature of the water in the first place and then changing it back again to T_1 are equal and opposite and cancel out in the output of the chamber; however they may not be detected at the same time. If the calibration of the calorimeter for evaporative heat loss is done using a dry heat source as the standard (as described for the Hannah calorimeter) then the inherent evaporative error is not detected during calibration and must be allowed for during data processing.

In some gradient-layer calorimeters, sensing elements comparing humidity between outlet and inlet airstreams are used instead of platemeters for measuring evaporative heat loss (Tiemann, 1949; Hammel & Hardy, 1963; Caldwell, Hammel & Dolan, 1966). This saves considerable effort and expense but leaves the evaporative heat measurement dependent on flow and humidity, both of which are difficult to measure accurately. It also involves converting the observed humidity increase to heat loss by applying a multiplying factor. This procedure, which is also necessary with heat-sink and convection calorimeters, can lead to more serious errors which are discussed in Section 5.7. Quattrone (1965) used a straight-through flow system but omitted the saturator and first platemeter shown by Fig. 5.4. They were replaced by a simple unit that conditioned the temperature but not the humidity of inlet air to chamber temperature. Evaporative heat loss was estimated from the increase in output of the remaining (outlet) platemeter compared with that observed when there was no heat load in the chamber; the use of a multiplying factor was thus avoided but the measurement was still flow-dependent. Quattrone stated that it was essential to eliminate the saturator from the flow system, otherwise condensation would have occurred in the calorimeter at the low ventilation rate necessitated by the oxygen- and carbon dioxide-detecting equipment. These problems might, however, have been overcome by partial recirculation of a larger ventilatory flow.

5.3.4c *Thermopiles*

A thermopile is often included to correct for any sensible heat ($\dot{Q}_c$) convected out of the chamber

$$\dot{Q}_c = V\rho c \, \Delta T \tag{5.21}$$

If the thermopile is made from copper–constantan (thermoelectric power ε) with n pairs of junctions then its output (Th) is $n\varepsilon\Delta T$. Hence

$$Th = n\varepsilon\dot{Q}_c / V\rho c \tag{5.22}$$

Alternatively the sensible-heat convection may be measured with resistance thermometers. By analogy with equations (5.12) and (5.13) the resistance thermometer output (Tr) is

$$Tr = iR_0 \, a\dot{Q}_c / 4V\rho c \tag{5.23}$$

The estimate of sensible-heat convection from the chamber, whether by thermocouple or resistance thermometer, unlike the chamber output, is dependent on the chamber ventilation rate. It is a correction voltage to be added to the chamber output and subtracted from the platemeter output to improve accuracy of partition between evaporative and non-evaporative heat loss; it has no influence on the total heat output.

5.3.4d *Fans, heaters, ducting, etc.*

Nearly all direct calorimeters are required to act also as indirect calorimeters. As a result of the high ventilation rate required for humidity control the concentration differences of oxygen and carbon dioxide between chamber inlet and outlet are normally too small for accurate measurement. To overcome this the bulk of the main air flow is usually recirculated round a closed loop with only a small percentage of it being exhausted to waste and replaced with fresh air. Indirect calorimetric measurements may then be made by monitoring the difference between oxygen and carbon dioxide concentrations in stale air and fresh air (see

Chapter 4). For the direct measurement, prevention of leaks is only important in the chamber and platemeters; but for indirect calorimetry, leaks at any point in the system can cause errors. High pressure differentials in the ducting system are therefore to be avoided. The various items comprising the ventilation system must be so arranged as to minimise pressure drops, especially to ensure that the chamber itself operates at close to barometric pressure. In flow-through systems this can be accomplished by using two fans, one pushing air into the chamber and one drawing it out (Spinnler *et al.* 1973). For sampling systems the chamber pressure can be kept very close to barometric pressure by ensuring that the fresh-air inlet duct is of wide bore and joins into the recirculation ducting close to the outlet from the chamber.

To operate against the pressure drop across platemeters and flowmeters, which may total as much as 150 mm water, a centrifugal fan or rotary compressor is required; the former are prone to leaks at the shaft, the latter are usually leak-free but tend to be noisy.

The heater bank should be placed immediately after the dehumidifier and before the fan; this arrangement prevents condensation of moisture from the saturated air emerging from the dehumidifier; it also ensures good thermal mixing of the air leaving the heater by the fan. In the absence of thermal mixing variable temperature profiles may be set up across the width of the duct and they may persist along long lengths of ducting. Elimination of these temperature profiles by thermal mixing is the key to ensuring that air samples drawn into platemeters are true aliquots of the main airstream. Small heaters warming the air emerging from the sampling platemeters are also necessary to prevent condensation.

It is a wise precaution to fit the heater bank with a drain incorporating a U-trap. Normally water does not accumulate in the heater bank, but in the event of a power supply failure, after which ventilation is resumed but the heaters remain tripped in the off state, condensation can occur in the heater bank. This can have disastrous effects when the power supply is restored to the heaters.

Flowmeters incorporated in the circuit may take various forms. These are discussed in Chapter 6.

Heat exchanges between the air in the recirculation ducting, other than that forming interconnections between the chamber and platemeters, are unlikely to have any adverse effect on calorimetric measurements. However, it is best to insulate the ducting if only to prevent condensation of atmospheric moisture on the outside.

5.3.5   *Temperature control*

The output of a gradient-layer calorimeter is proportional to the temperature gradient, not to the absolute level of temperature. In fact there is a slight dependence on the operating temperature because many of the physical properties appearing in the sensitivity equations (5.12) and (5.13), including D, a, ε and K are slightly influenced by temperature. The combined effect is that the sensitivity of the chamber gradient-layer increases by about 1 % for every 10 °C increase in operating temperature, for both thermocouple and resistance thermometer types of layer. Other than this the operating temperature has no influence on performance. Indeed a chamber could be successfully operated with opposing walls at entirely different temperatures. There would be a flow of heat in at one wall, but this would be exactly balanced by heat flowing out at another with no net heat exchange. In practice this means that there is no special requirement for the cooling system to create a uniform temperature distribution over the whole surface; but the temperature must not vary with time. When the water temperature changes it induces a flow of heat through the chamber walls. The amount of heat flow, and therefore the magnitude of the error signal, depends on the rate of change of water temperature and the heat capacity of the calorimeter contents including the gradient layer itself; it is best determined by practical experiment. In the Hannah calorimeter with the metal stall, feed bin and drinking bowl in place, an output equivalent to 100 W is observed if the temperature rises by 1 °C per hour. Hence to keep the error signal to below 1 W requires a temperature stability of 2.7×10^{-6} °C/s.

A temperature control system designed to maintain a very low rate of change of water temperature rather than minimal deviation from a set point was described by Cairnie & Pullar (1959) and is illustrated schematically by Fig. 5.17. The water supply tank contains a refrigerator coil and three heating elements A, B and C. Heater A is fed with current from a controller which can operate in two alternative modes; mode 1 under the control of a temperature sensing device (e.g. thermistor bridge) and mode 2 under the control of a gradient layer surrounding a large block of aluminium immersed in the tank. The temperature is first brought to near the desired operating level using manual control of either the refrigerator or boost heater C. The refrigerator is then left running at a fixed level to ensure that there is a demand for reheating and the system is left to settle under the influence of the controller in mode 1 (thermistor control). The current indicated by meter M now fluctuates as control action takes place about a mean level $(\bar{\imath})$. $\bar{\imath}$ may drift for some hours after a change in operating temperature as the calorimeter system equilibrates to the new thermal demand. After this the level of $\bar{\imath}$ may be altered to mid-scale on meter M by a single adjustment of the variable transformer supplying power to heater B; any added power to B being quickly reflected by an equivalent reduction in the power supplied by the controller to A. Finally the controller is switched to mode 2 (block control) where alterations in current output to A are proportional to the output signal from the block. A system of this type can provide the very stable temperature control that is necessary for a gradient layer.

It may be advantageous to introduce baffles to direct the water movement in the tank, and good stirring is necessary especially over the heating elements. For optimum control the block should be placed close to the outlet that feeds water to the calorimeter. Precautions must be taken to avoid galvanic action between the aluminium of the calorimeter and other metals. The refrigeration coil, pumps and piping should be of aluminium, plastic or stainless steel; copper and brass should *not* be used.

In calorimeters with straight-through ventilation systems temperature control of the platemeters must conform to the same standards as that of the chamber. If the air-sampling system is employed accuracy of temperature control of the platemeters is less stringent. It is, however, possible to operate some gradient-layer calorimeters with changing temperatures. Control experiments must first be done with no heat load in the chamber to establish a time-series relationship between chamber output and tem-

Fig. 5.17. Schematic diagram of a controller designed to minimize the rate of change of water temperature. A, controlled heater; B, variable heater; C, boost heater; D, thermistor.

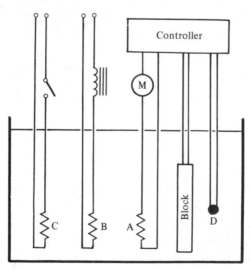

perature. The relationship may then subsequently be used to correct the calorimeter output for the effects of changing temperature. This method has been successfully used with the Hannah calorimeter to investigate thermal responses of animals to step changes (McLean, Downie & Jones *et al.*, 1983*a*) and cyclical changes (McLean & Stombaugh *et al.*, 1983*b*) in the temperature of their environment.

5.4   Heat-sink calorimeters
5.4.1   *The Dunn calorimeter*
The heat-sink calorimeter at the Dunn Clinical Nutrition Centre of the Medical Research Council at Cambridge was built to study energy regulation and obesity problems in humans. A brief description of it written by Murgatroyd (1984) forms the basis of the description given here.

5.4.1*a*   *The structure and ventilation*
The internal floor of the calorimeter (Fig. 5.18) measures 4.29 × 2.93 m and the internal volume is 29.5 m³. It is divided into a furnished living room and a wash room. The calorimeter stands on joists 0.5 m above the floor of an outer shell which is insulated and air conditioned. The temperature of the shell space may be controlled to within 0.1 °C over the range 10 to 35 °C.

Fig. 5.18. Schematic diagram of the Dunn heat-sink calorimeter. Arrows indicate direction of air-flow.

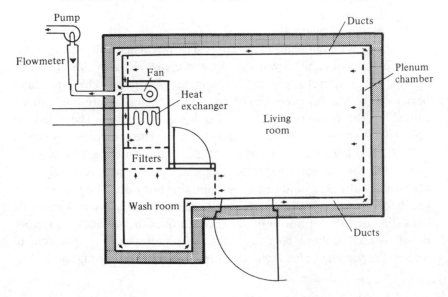

The calorimeter itself is constructed of expanded polystyrene 150 mm thick. This is self-supporting; the only additional structural members being aluminium 'T' section bearers within the ceiling and the wooden door and window frames. The whole of the outside and the end walls of the inside are clad with 3 mm thick expanded PVC sheet. The internal surfaces of the floor, ceiling and side walls are completely covered with rigid PVC rectangular 120 mm × 25 mm tube sections to provide a durable inner-wall surface with a 25 mm duct between it and the polystyrene. At the ends of the calorimeter, perforated screens form the inner walls of plenum chambers. Air is collected from behind one screen, filtered and cleaned by an activated carbon filter, pressurised by the recirculation fan, passed over the heat exchanger and put back into the calorimeter living space through the second perforated screen which it reaches after passing through the duct formed behind the inner wall surfaces. The inner-wall surface is thus decoupled from the wall structure and any radiant heat reaching it is taken up without delay by fast-moving air behind it.

All non-evaporative heat losses from the subject inside the calorimeter are thus absorbed by the recirculating air and then given up to the heat exchanger. Heat loss through the polystyrene walls is prevented by controlling the temperature of the water delivered to the heat exchanger so that the mean temperature gradient across the walls is zero.

Some of the circulating air is drawn from a point near the discharge of the fan and sucked through a rotameter flowmeter before being exhausted to waste. This air is replaced by fresh air at the same temperature which leaks into the calorimeter from the shell space.

5.4.1b   *Water circulation*

The water-circulation system is shown schematically in Fig. 5.19. A pump circulates chilled water at approximately 1 l/min through three heaters, the heat exchanger and a rotameter flowmeter. The smoothing heater (up to 925 W) is controlled to even out temperature fluctuations caused by the refrigerator and sensed by a thermistor. The reference heater adds a constant 100 W to the water flow. The control heater (up to 1450 W) regulates the temperature of the water entering the heat exchanger inside the calorimeter so that the mean temperature gradient across the walls, floor and ceiling is maintained at zero. The temperature gradient across the walls of the chamber is monitored by a thermopile consisting of 135 thermocouples distributed over the surfaces; its output is fed to a controller which regulates the power supply to the control heater. Proportional plus integral controllers are used throughout.

5.4.1c *Estimation of heat losses*

Non-evaporative heat loss ($\dot{Q}_{\mathrm{N}}$) from the subject is given by

$$\dot{Q}_{\mathrm{N}} = \dot{Q}_{\mathrm{EX}} + \dot{Q}_{\mathrm{W}} + \dot{Q}_{\mathrm{C}} - \dot{Q}_{\mathrm{F}} \tag{5.24}$$

where $\dot{Q}_{\mathrm{EX}}$ = heat removed by the heat exchanger; $\dot{Q}_{\mathrm{W}}$ = heat flowing through the chamber surfaces; $\dot{Q}_{\mathrm{C}}$ = heat convected out of the chamber by the ventilation; $\dot{Q}_{\mathrm{F}}$ = heat produced by the circulating fan.

$\dot{Q}_{\mathrm{EX}}$ is the product of water temperature increase across the heat exchanger (ΔT_{EX}) and the water flowrate. But rather than measuring the flowrate directly, $\dot{Q}_{\mathrm{EX}}$ is estimated by comparing ΔT_{EX} with the temperature increase (ΔT_{REF}) across the 100 W reference heater:

$$\dot{Q}_{\mathrm{EX}} = \frac{\Delta T_{\mathrm{EX}}}{\Delta T_{\mathrm{REF}}} \times 100 \tag{5.25}$$

ΔT_{EX} and ΔT_{REF} are measured by four-thermocouple thermopiles situated before and after each heater unit. Because of the efficiency of the temperature control systems $\dot{Q}_{\mathrm{W}}$ and $\dot{Q}_{\mathrm{C}}$ are both very small; however they are measured and applied as corrections. $\dot{Q}_{\mathrm{W}}$ is obtained from the mean temperature gradient across the insulation measured by the thermopile, and the thermal conductance of the walls. $\dot{Q}_{\mathrm{C}}$ is obtained from the temperature difference between inlet and outlet ventilating air, the air mass-flowrate and specific heat. $\dot{Q}_{\mathrm{F}}$ is continuously monitored by a wattmeter.

Evaporative heat loss from the subject is estimated (see Section 5.7) from the flowrate (approximately 200 l/min) and humidity increase of the exhaust air; air humidity being measured by a dewpoint hygrometer.

Fig. 5.19. Schematic diagram of the water circulation to the heat exchanger. T_{EX}, T_{REF} and T_{W} are temperature gradients across the heat exchanger, reference heater and walls respectively.

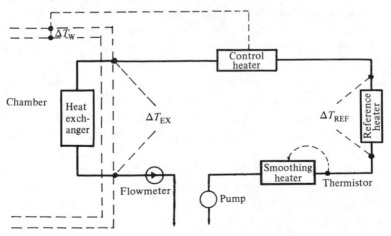

Data handling is carried out using a Solartron Integrated Measurement System interfaced to a Hewlett Packard 1000 system computer. This is programmed to make the measurements necessary for calculating heat loss at one minute intervals, and it outputs the mean heat loss results at preselected intervals – usually 10 min.

5.4.1d   Calibration

Calibration of sensible heat loss is made by using electrical heat sources in 60 W increments up to 480 W. The total power input at each calibration point is measured by a wattmeter of 0.5% accuracy. A linear response is obtained, repeatable to within ± 1%, but the measured heat is 3.5% less than the heat input. The reasons for this have so far proved elusive and so at present results are corrected with a scaling factor.

The combined heat loss calibration is carried out by infusing water onto an electrical heater at a constant rate. The response time of evaporative heat measurement is longer than expected, probably due to the adsorption of water onto the plastic materials of the wall structure. This does not affect either the accuracy or response time of the total heat measurement as calibrations have shown that the 'latent heat' of adsorption is interchangeable with the sensible heat loss.

In use, corrections are made to the heat loss measurements to take account of heat gains and losses arising from food and drink, washing water, etc. Correction is also made for heat radiated into the calorimeter by the external light sources. The total correction from these sources is in the region of 7% of a typical 24 h heat loss.

5.4.2   Other heat-sink calorimeters

The Dunn calorimeter was not designed primarily for high speed of response. Emphasis was placed on providing ample space and comfort for the volunteer subjects who spend periods of up to 13 days inside it. Nevertheless the response time of the heat measurement system, with the calorimeter fully furnished, is only 25 min for 98% response. The good performance of this calorimeter is largely due to the design of the individual air and water temperature control systems, where emphasis was placed on low heat capacities, and optimum response from proportional-with integral controllers. In this respect it represents a great advance on an earlier prototype built by the Dunn group and fully described by Dauncey, Murgatroyd & Cole (1978). In that calorimeter the water temperature delivered to the heat exchanger was controlled to minimise convective heat loss via the ventilating air stream ($\dot{Q}_c$) rather than conductive heat loss through the walls ($\dot{Q}_w$). No provision was made to circulate the incoming air around a wall cavity. $\dot{Q}_c$ and $\dot{Q}_w$ were both subject to considerable variation and introduced delays into the measurement.

Jacobsen, Johansen & Garby (1982) defined an ideal heat-sink calorimeter as a chamber with zero heat capacity and perfect mixing of air in the chamber and its surroundings, both being maintained at a common reference temperature. In such a calorimeter sensible heat loss from the subject at any instant would be equal to the heat removed in the sink at that time. They described the human heat-sink calorimeter at Odensee, Denmark and analysed its departure from the ideal in terms of a series of time constants describing delays and heat-storage capacities of various parts of the system. These factors were then incorporated in correction terms and programmed into the software of the computerised measuring system of the calorimeter. The resulting system had a precision for sensible and evaporative heat loss of 1.5 and 2 % and time constants for 95 % response of 14 and 7.5 min respectively. This time response is, however, only marginally better than that of directly measured output from the Dunn calorimeter.

Pickwell (1968) included a compensating heater in his heat-sink calorimeter, built to accommodate ducks, in which the temperature of the water supply to the heat exchanger was held constant. He suspended an electrical heating coil to hang in front of the heat exchanger and controlled its power supply so as to keep the temperature of the calorimeter box (monitored by a single thermocouple) constant. By setting the coolant flow at a suitable rate the calorimeter temperature could be regulated at any desired level. Pickwell described this calorimeter as a 'compensating heat-exchange calorimeter'; it is, however, no more compensating than an ordinary heat-sink calorimeter, the compensating action being merely transferred from the heat exchanger to the heating coil.

5.4.3 Suit calorimeters

A form of heat-sink calorimeter which imposes little restriction on the subject is the suit calorimeter described by Webb, Annis & Troutman (1972). The heat exchanger takes the form of a water-cooled garment made from a network of fine (1.6 mm ID) vinyl tubing which fits snugly over the body (Fig. 5.20). Water is circulated at 1.5 l/min, via a common manifold located at waist level, through the tubing which is divided into six individual circuits for the four limbs, torso and head. The spacing of the tubing is closest over skin areas normally associated with high levels of heat loss. Sensible heat exchange from either the skin or from the tubing network to the environment is minimised by thermally insulating outer garments which are permeable to water vapour, and by

Fig. 5.20. Photograph of a subject wearing diamond-pattern water-cooling garment. (Reproduced from Webb *et al.*, 1972.)

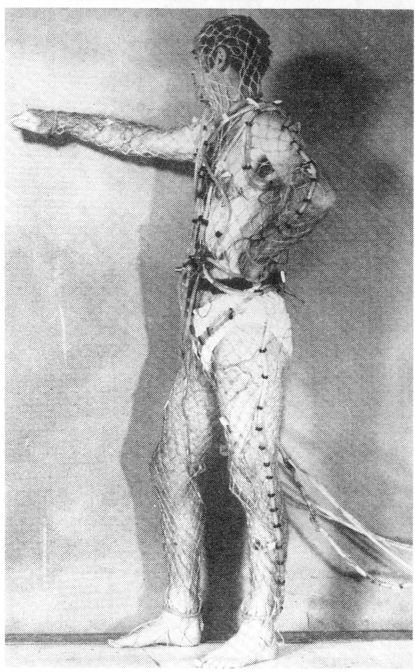

maintaining the environmental temperature at 30 °C. The complete assembly weighs less than 10 kg.

Heat given up to the water ($\dot{Q}_w$) is estimated from the mass-flowrate ($\dot{m}$), specific heat (c_w) and temperature difference between water outlet (T_0) and inlet (T_i):

$$\dot{Q}_w = \dot{m}c_w(T_o - T_i) \tag{5.26}$$

Evaporative heat loss is estimated from hourly measurements of the subject's weight, recorded on a balance accurate to 5 g. Normally the suit calorimeter is worn in conjunction with the metabolic rate monitor (Webb & Troutman, 1970) described in Chapter 4. This allows estimation of (respiratory) convective heat loss and also correction of the weight loss measurements to allow for the difference in weight between carbon dioxide produced and oxygen consumed.

A unique feature of this calorimeter is the control of the rate of heat removal by the water circulation so that the subject is always maintained in thermal comfort. This is achieved by controlling the temperature of the water inlet to the suit (T_{wi}) so as to satisfy the condition

$$T_{wi} = 35.0 - 6(T_o - T_i) + 1.1 \, (33.8 - T_s) \tag{5.27}$$

The cooling power is thus controlled so that it matches the flow of heat from the body to the skin surface, but with a modifying influence from a feedback signal which tends to limit the fall in water inlet temperature as the mean skin temperature (T_s) falls. The constants contained in equation (5.27) were selected by the designers on the basis of their earlier experiments with water-cooled suits to determine the amount of cooling required to suppress sweating (Webb & Annis, 1968).

According to the authors (Webb *et al.*, 1972) the suit maintains the subject in a state of biothermal neutrality. He is never chilled, never sweats significantly and is in a state of continuous thermal comfort, even when exercising vigorously. The subjects gladly tolerate automatic control while asleep, at rest or working for experimental periods of over 24 h. A detailed description of this calorimeter and subsequent refinements has been published recently (Webb, 1985).

The suit calorimeter produces an abnormal type of environment for the subject. Sensible heat loss from the surface is by conduction whereas in normal environments it is by radiation and convection. During exercise the sensible heat loss is greatly increased by keeping the skin cool, whereas a normal environment sets a limit to the sensible heat loss, and sweating is induced. In the suit the proportion of the total heat lost by evaporation was less than 38 % even for an active subject (Webb *et al.*, 1972). The suit calorimeter is a useful tool for studying physiological aspects of exercise and of subjects performing tasks in adverse environments from which they must be artificially protected; but it cannot, of course, be used to study physiological or nutritional effects due to changes in the normal environment. It is also a portable calorimeter in two senses: it allows the subject to perform different tasks while he carries the calorimeter around with him; it also may be packed in a suitcase and transported from place to place.

An earlier suit calorimeter was described by Young, Carlson & Burns (1955). It employed the gradient-layer principle with resistance thermometers in the form of wire grids located on either side of an insulated layer of underclothing. The heat-sensitive layer was divided into sections allowing partitional calorimetry of different body regions. This suit was heavily insulated and suitable for studies on subjects in Arctic regions, for which it was designed. It could not be used in warm environments and even in cold conditions became very uncomfortable due to the accumulation of sweat when the subject was active.

5.5   Convection calorimeters

Convection calorimeters are designed with the objective of ensuring that all sensible heat output from the subject is transferred to the ventilating air and measured as convective heat loss, i.e. as the product of air mass-flowrate, specific heat and temperature increase of the ventilating air.

Detailed plans for a proposed convection calorimeter for humans were described by Visser & Hodgson (1960). The subject was confined in a horizontal cylindrical chamber having double aluminium walls with an air gap between. Ventilating air was blown in and circulated tangentially round the air gap before entering the chamber proper through slots in the inner aluminium shell. It was claimed that this arrangement would ensure that any heat radiated by the subject to the inner shell would be reabsorbed by the circulating air, and heat loss through the outer shell would be virtually eliminated. This calorimeter was used experimentally by Snellen & Baskind (1969) who confirmed its accuracy but only after it had been enclosed in an outer wooden box ventilated with a fraction of the conditioned air.

The technique was developed further in a calorimeter intended to accommodate a seated human subject which was built at the Health Sciences Centre, St John's, Newfoundland, and which has been fully described by Snellen, Chang & Smith (1983). The calorimeter vessel, shown in cross section in Fig. 5.21, consists of three concentric upright cylinders. The two outer cylinders are formed from 1.2 and 1.7 mm thick aluminium plates with 5 cm polyurethane insulation applied to the outsides. The inner cylinder consists of six 0.2 mm thick curved sheet metal panels supported inside the middle cylinder on teflon spacers, and arranged so that they overlap, leaving six vertical slots along the length of the cylinder. Conditioned air from a common inlet plenum is diverted by an adjustable baffle to pass in roughly equal quantities around each of the two cavities (approximately 8 cm wide) formed between the cylinders. The air in the outer cavity passes all the way round into an exhaust plenum from where it is returned to the air-conditioning unit. The air in the inner cavity passes tangentially through the slots formed between the overlapping plates into the central chamber of the calorimeter where the subject is confined. All three cylinders are sealed at the ends with circular sections of similar construction (except for extra strength in the floor sections) so that the cavities and air circulation are continuous over all surfaces. Air from the inner chamber is exhausted through a pipe in the roof covered at both ends with caps mounted on spacers. The inner cylinder acts as a radiation shield and gives up its absorbed heat to the air in the inner cavity. Since air is circulating at the same temperature through both cavities no heat should escape from the central chamber except by convection via the ventilating air.

The entire vessel splits open along a rib parallel to the long axis, one half

Fig. 5.21. Cross-section of a convection calorimeter vessel. (Reproduced from Snellen *et al.*, 1983.)

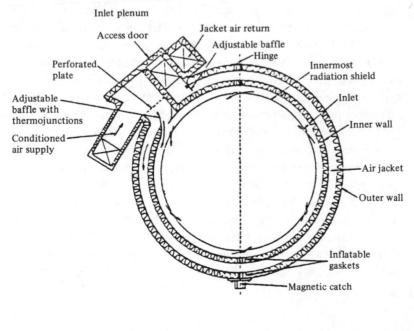

being fixed and the other hinged to allow entry of the subject. When shut the two halves are held together by electromagnetic catches and sealed together by inflatable gaskets at the striking faces.

The quantities of evaporative and non-evaporative heat lost by the subject are estimated from the differences between outlet and inlet air humidities and temperatures. The air temperature difference is measured by a thermopile with 192 pairs of copper–constantan junctions. Humidity difference is measured by variable-resistance hygroscopic film sensors. Air flowrate is kept constant but not continuously monitored. Instead the computer-outputs derived from the thermopile and humidity sensors are related to heat and water vapour outputs in the calorimeter vessel by direct calibration. For non-evaporative calibrations known rates of electrical heat produced in the chamber are measured by a wattmeter. For evaporative calibration water from a continuous infusion pump is vaporised and introduced into the chamber at known rates.

The high ventilation rate through the calorimeter proper (12.5 m³/min) gives a time constant for 63 % response to changes in evaporative heat loss of 1.5 min. For non-evaporative heat loss the response is not a simple exponential and involves two time constants of 0.86 and 7.4 min. Accuracy of the measurements is ± 1 W for non-evaporative and ± 4 W for evaporative heat loss.

The convection principle has also been used in a somewhat larger calorimeter, also intended for human subjects whilst lying, standing or exercising, and built at Vienna (Tschegg *et al.*, 1979). This calorimeter includes a compensating heater inside the chamber whose heat output is controlled so that the temperatures of outlet and inlet ventilating air are equal. The air flow is in the opposite direction to that of Snellen *et al.*'s (1983) calorimeter in that it is the exhaust air rather than the inlet air which is circulated round an inner cavity surrounding the chamber. This cavity is in turn surrounded by an outer cavity or shell space in which the air temperature is independently controlled to be equal to that of the inner cavity. Response time is of the order of 15 min and accuracy ± 2 W for both evaporative and non-evaporative heat losses.

Other convection calorimeters include that of Carlson *et al.* (1964) who relied on cellular insulation to minimise heat flow through the surfaces of their rectangular chamber which was also designed for humans. An interesting feature of this calorimeter is that the insulation was porous so that conditioned air could be passed through it to speed up thermal equilibration when the temperature of the chamber was altered. Kelly, Bond & Heitman (1963) converted an existing psychrometric room into a convection calorimeter for pigs by building a ventilated chamber of smaller dimensions inside it. Heat flow through the 'high heat capacity' walls of the inner chamber was minimised by controlling the air temperature on the outside to approximately chamber temperature. In these calorimeters the object of preventing heat flow through the walls was only partially achieved and heat leakage through the surfaces was monitored with heat flow discs by Carlson *et al.* (1964) and by Kelly *et al.* (1963) and applied as a correction to the convection measurement. That major problems were encountered can be judged from the statements of the authors. Carlson *et al.* (1964) stated that 'the dependence on air temperature, velocity and air humidity measurements make operational accuracy a constant problem'. Kelly *et al.* (1963) also list a number of disadvantages.

The advantages claimed for the convection method are simplicity of construction and speed of response to change in heat output of the subject. However, it depends on extremely accurate measurement of both air flow and temperature and humidity differences. The calorimeters of Visser & Hodgson (1960) after modification, Tschegg *et al.* (1979) and Snellen *et al.* (1983), which include ventilation of wall cavities, appear to be the only ones of this type which have operated wholly successfully.

5.6   Differential calorimeters

The differential principle, in which the temperature rise of the chamber caused by the subject is matched with that of a duplicate chamber containing an artificial heat source, was first used by Noyons (1927) in a calorimeter for humans. It was developed by Benedict & Lee (1937) whose description of their instrument is very detailed. Their main concern was to achieve rapid response by building twin chambers of very low heat capacity. The chambers consisted of brass cylinders of 130 l volume which, including the built-in resistance thermometers, weighed only 3 kg each. The animal, a goose, was placed in one chamber and a heating element in the other, the compensating chamber. The chambers were mounted side by side in a conditioned airstream and elaborate precautions were taken to ensure that the external conditions of air movement and temperature were the same for each. Only the animal chamber was ventilated internally, but convection of heat from the chamber was avoided by controlling the temperature of inlet air to be the same as that of outlet air. Evaporative heat loss was estimated from the humidity increase of the ventilating air. Control of the electrical heat delivered to the compensating chamber so that the temperature of the two chambers remained equal was effected manually.

Deighton (1939) built a similar system for poultry. He included a wetted muslin wick in his compensating chamber which could be variably exposed to match the evaporative heat output in the animal chamber.

This type of calorimeter because of its lightweight construction is only suitable for relatively small caged animals, but its advantage lies in its speed of response (of the order of 1 min). Hawkins (1958) described a device for automatic control of the heat supply to the compensating chamber of a differential calorimeter for mice.

5.7   Evaporative heat loss

Calculation of the rate of evaporative heat dissipation from measurements of the sweat rate has been a subject of considerable controversy.

The latent heat of vaporisation of pure water (L_T) varies with temperature as shown in Table 5.4. But when sweat or saliva is vaporised heat is required not only for the vaporisation which occurs near to the temperature of the skin or respiratory passages but also to change the condition of the vapour to that of the surrounding air. Hardy (1949) proposed a formula for the 'heat of vaporisation of sweat' which included not only the latent heat of vaporisation but also the heats of both warming and expanding the vapour to ambient temperature and vapour pressure. The heat of expanding the vapour according to Hardy could be quite large and typically could add a further 10 % onto the heat of vaporisation. It was not until some years later that Monteith (1972) and Wenger (1972) independently pointed out that Hardy's formula, which in the meantime had become widely accepted and quoted in textbooks (e.g. Blaxter, 1962; Mount, 1968; Kerslake, 1972) was thermodynamically incorrect. Wenger (1972) confirmed by thermodynamic reasoning that the heat (i.e. the

Table 5.4. *The latent heat of vaporisation of water*

Temperature (°C)	Latent heat of vaporisation (J/g)	(cal/g)
0	2494	595.9
10	2471	590.4
20	2448	584.9
30	2426	579.5
40	2403	574.0

increase in enthalpy, h_e) of converting unit mass of liquid water at temperature T_s to vapour at T_a is

$$h_e = L_T + c_p(T_a - T_s) \qquad (5.28)$$

where $c_p (= 1.91 \text{ J/g }°C^{-1})$ is specific heat of water vapour at constant pressure. In practice the specific heat term has very little influence on h_e because for sweating skin $(T_a - T_s)$ is usually small and may be either positive or negative. Wenger also showed that the effect of dissolved salts present in sweat exerts negligible influence on the heat required for its vaporisation. There remained, however, experimental evidence from various workers including Snellen (1966), Mitchell *et al.* (1968) and Snellen, Mitchell & Wyndham (1970) which appeared to show that the observed heat of evaporating sweat was considerably greater than that given by equation (5.28). Wenger (1972) discounted these observations on statistical grounds, but Monteith (1972) believed that some undetected systematic calorimetric error must be involved.

In fact the error probably lies in the implicit assumption made by nearly all workers and stated unequivocably by Kerslake (1972) that the latent heat of vaporisation is all derived from the subject and none from the surrounding air. For example, Snellen *et al.* (1970) measured the amount of water ($\dot{m}$) given up by their subject; they also measured heat losses using Visser & Hodgson's (1960) convection calorimeter, in which all radiative and conductive heat losses from the subject are converted into convective heat gain by the circulating air. However, if some of the heat of vaporising sweat is drawn from the air inside the calorimeter rather than from the subject then this would be observed as an apparent decrease in the non-evaporative heat loss ($\dot{Q}_N$) from the subject. Snellen *et al.* (1972) calculated evaporative heat loss ($\dot{Q}_E$) from the difference between heat production ($\dot{M}$, estimated by indirect calorimetry) and $\dot{Q}_N$, i.e. $\dot{M}-\dot{Q}_N$. Any underestimation of $\dot{Q}_N$ would be reflected as an overestimation of $\dot{Q}_E$ and also of the 'heat of evaporation of sweat' which they calculated as $\dot{Q}_E/\dot{m}$. Other high values for the heat of evaporation of sweat derived by Snellen (1966), Mitchell *et al.* (1968) and inferred from the measurements of Nielsen (1966) may be accounted for by similar reasoning.

In passing it is worth noting that underestimation of non-evaporative heat loss could be a contributory factor to the finding by Snellen (1966) and Snellen *et al.* (1970) that the straight line relating non-evaporative heat loss to the skin-to-air temperature gradient failed to pass through the origin.

At the time when these measurements were made they appeared to lend support for Hardy's (1949) equation which also predicted a high value for the heat of evaporation of sweat. However, the measurements of all these observers are also consistent with the heat of vaporisation given by equation (5.28), provided that it is assumed that a proportion (about 7%) of the heat is derived from the air rather than the subject. This implies that part of the latent heat of vaporisation itself is drawn from the air and not just the heat of the 'secondary processes' (Kerslake, 1972). There is no reason why this should not be so. The fact that the vaporisation takes place near the skin surface in no way precludes part of this heat being drawn from the air; indeed it would be surprising if none were.

Snellen *et al.* (1970) did not observe a similarly high value when they measured the heat required to vaporise water from open trays placed in their calorimeter; and the expected latent heat of vaporisation was also found in similar calibration experiments using the St John's calorimeter (Snellen *et al.*, 1983). In these instances, however, there is no alternative heat source, and the heat must ultimately be all derived from the air – either at the site of evaporation, or in subsequent warming of the vapour or in preventing a fall in temperature of the water tray. In this special case the evaporative heat must be equal and opposite to the non-evaporative heat. The wet-bulb thermometer is another example of the same special case. From its linear response Kerslake (1972) concluded that the heat taken up per unit mass of water evaporated must be nearly constant and equal to the latent heat of vaporisation. The contrast between these two physical examples, where the expected latent heat of vaporisation is observed, and the biological measurements giving apparently high values, may have served to perpetuate the belief that the disparity has some biological explanation. However, a statement by Hardy (1949), that he had obtained a high value in his (unpublished) observations on an electrically heated cylinder partly covered with an evaporating surface, is also revealing. In this physical example there is again an alternative heat source and it is no longer possible to equate heat of vaporisation with non-evaporative heat gain.

Uncertainty as to the source from which the heat of vaporisation is derived does not in any way affect its total quantity (h_e); but it does throw serious doubt on the accuracy of estimating overall heat balance whenever evaporative heat loss is calculated from the quantity of water vaporised. It is not the evaporative heat loss itself which is in error, but the non-evaporative loss which will appear to be reduced if some of the heat of vaporisation is drawn from the air. The problem is by-passed in the gradient-layer-type of calorimeter complete with platemeters, in which the heat losses are measured as heat fluxes and no assumptions are made regarding the source of the heat of vaporisation. The gradient-layer calorimeter thus avoids the issue of the source of the heat of vaporisation of sweat. It estimates the total heat loss by the animal correctly as the heat flow through the chamber surfaces plus the enthalpy increase of the circulating air; and it partitions the total loss between the heat required to increase the humidity (evaporative) and the remainder (non-evaporative). But this may not be the same partition as occurs at the skin surface.

Unless it can be assumed that all of the heat of vaporising sweat is derived from the animal, it is a mistake to combine an evaporative heat loss factor whose value depends on skin temperature with a fixed non-evaporative factor. But this is the usual practice in most calorimeters which operate by measuring the humidity increase of the circulating air. It would be better, in these calorimeters, to employ the latent heat of vaporisation of water at calorimeter temperature as the evaporative factor. This would avoid error in the estimate of total heat loss although leaving the partition between evaporative and non-evaporative open to criticism.

There is clearly a need for further experimentation: (1) to confirm that the calculation procedures using factors based on surface temperature can lead to apparently high heats of vaporisation, even in purely physical systems; and (2) to determine what proportion of the true heat of vaporisation of sweat is derived from the air under different circumstances (e.g. nude, hirsute or clothed skin) and with different levels of air movement.

The term 'heat of evaporation of sweat' in fact requires to be more precisely defined than heretofore. Should it be confined only to that portion of the latent heat of vaporisation which is actually derived from the subject? This would be a true description but, if used to estimate evaporative heat loss, and combined with the observed non-evaporative heat loss which may also be an underestimate, it would compound the error in total heat loss. Also, it surely should not be defined as the value necessary to provide overall heat balance in calorimetric experiments as derived by Snellen *et al.* (1970). This would merely overestimate evaporative heat loss to compensate for the underestimation of non-evaporative heat loss. We favour the definition given by equation (5.28) because this is the true quantity of heat required to convert each gram of liquid sweat or saliva into vapour at ambient conditions. But it must be emphasised that this definition does *not imply* that all the heat originates from the subject.

We recognise that the views expressed in this section may arouse further controversy, but we believe that this just emphasises the need for further work to be undertaken to determine the source from which the heat of vaporisation is drawn under different circumstances. To what extent is it affected by clothing or fur or by the air movement over the skin?

5.8   Other energy exchanges

In most instances the evaporative and non-evaporative heat losses detected by a calorimeter represent all the heat loss from the subject, and the two together should add up to the total rate of heat production. But there are exceptions. Heat from electric light bulbs, motors and instruments in the chamber are other obvious heat sources, already mentioned, which have to be taken into account. If the subject exercises inside the chamber the additional work is usually dissipated immediately as heat and is measured by the calorimeter. This applies, for example, to exercise on a bicycle ergometer, where the extra heat appears at the braking device; but if the shaft of the ergometer passes through the chamber wall and the brake is situated outside, the heat is of course not detected by the calorimeter. When retrievable work is done in the chamber, such as lifting weights onto a shelf, the energy imparted to the weights is stored in mechanical form so long as they remain on the shelf, and the heat detected by the calorimeter is correspondingly less.

Energy may also be stored as heat inside the chamber, particularly in the form of increased body temperature of the subject (see next section); such heat is not detected by the calorimeter. An example of heat storage which often has to be taken into account in animal calorimeters is the heat exchanges of food, drinking water and excreta. When food or water is consumed by an animal it gains heat as it is warmed from calorimeter temperature to the temperature of the animal; this represents heat loss by the animal which is not detected by the calorimeter. It is by no means negligible and it often stimulates a corresponding and readily observable increase in heat production by the animal to maintain thermal balance in the body. The reverse effect occurs when the animal defaecates or urinates. The excreta, assuming they remain inside the chamber, cool down giving up heat which is detected by the calorimeter. Over a prolonged period the heat of warming food and water and the heat of cooling excreta cancel each other out and there is no net error in the heat output as measured by the calorimeter. But in the short term imbalances do occur. The heat is lost from the tissues of the animal shortly after food or water is consumed, but it does not appear on the calorimeter output record until after excretion of the waste products. Corrections based on the weights and specific heats of the food and waste materials may be made if the feeding and watering habits of the animals are monitored (McLean *et al.*, 1982).

5.9 Body heat storage

The amount of heat stored in the body tissues may in theory be estimated from measurement of change in mean body temperature multiplied by the body mass and mean specific heat. The figure 3.47 J/g is widely accepted as the mean specific heat of mammalian body tissues; but estimation of mean body temperature poses considerable problems.

The major part of the body remains close to the temperature of arterial blood, and the so-called deep-body temperature can be relatively easily measured by a single thermometer placed in or near the mouth, axilla, rectum or eardrum. In most homeothermic species the deep-body temperature varies by no more than ± 1 °C under conditions of normal health. However, the peripheral regions of the body, though representing a smaller proportion of the total body mass, undergo much greater changes in temperature, especially in relation to changes in environmental warmth. They can thus make a large contribution to changes in mean body temperature. Mean body temperature is therefore estimated from a weighted mean of deep-body and mean skin temperatures.

5.9.1 Mean skin temperature

Mean skin temperature may be found by measuring the temperatures of many different skin regions and weighting them according to their relative proportions of the total surface area. Clearly, accuracy is improved by employing a large number of regions; but a limit has to be set for convenience and practicality. For mean skin temperature of humans a formula proposed by Hardy & Du Bois (1938*a*) has gained wide acceptance as being accurate if not convenient. The body surface is divided into seven regions and the mean surface temperature of each is estimated as the arithmetic mean of the temperatures at one to four individual measuring sites, there being 15 measuring sites in all. The weighting accorded to each region is shown in Table 5.5. A number of workers have proposed formulae using smaller numbers of measuring sites and these have been reviewed and tested against the full Hardy & Du Bois formula by Mitchell & Wyndham (1969) and by Lund & Gisolfi (1974). Two formulae which appear to give reasonably good consistency from relatively few measuring sites (50–70% frequency of agreement to within ± 0.2 °C with the full equation) are the six-point formula of Palmes & Park (1947) and the four-point formula of Ramanathan (1964); the weightings used in these formulae are shown in Table 5.5.

From a study of the surface temperature distribution of cattle exposed to fluctuating thermal environments, McLean *et al.* (1983*a*) devised a weighting formula for that species; this is also shown in Table 5.5.

Table 5.5. *The skin regions employed and the weighting ratios accorded to them in formulae for estimating mean skin temperature. In the formulae of Hardy & Du Bois and McLean mean temperature of some of the individual regions is estimated from measurements at more than one site*

Species	Humans			Cattle
References	Hardy & Du Bois (1938a)	Palmes & Park (1947)	Ramanathan (1964)	McLean et al. (1983a)
Skin regions and weighting ratios	Head 0.07	Forehead 0.14		Ear 0.04
	Arms 0.14	Forearm 0.11	Arm 0.3	Dewlap 0.02
	Hands 0.05	Palm 0.05		
	Foot 0.07			Lower limbs 0.12
	Leg 0.13		Leg 0.2	Upper limbs 0.32
	Thighs 0.19	Lateral thigh 0.32	Thigh 0.2	
		Back 0.19		Upper trunk 0.25
	Trunk 0.35	Chest 0.19	Chest 0.3	Lower trunk 0.25

5.9.2   Mean body temperature

Burton (1935) correlated changes in body-heat content of human subjects in basal and postabsorptive states with rectal and mean skin temperatures and derived a formula for changes in mean body temperature (ΔT_b):

$$\Delta T_b = 0.65 \times \Delta T_r + 0.35 \times \Delta T_s$$

Hardy & Du Bois (1938*b*) made a similar study on subjects exposed to environmental temperatures in the range 22–35 °C and found the optimum weighting ratio to be 0.80:0.20. In later experiments (Hardy, Milhorat & Du Bois, 1941) they concluded that the weighting factors were temperature dependent, the ratio altering from 0.5:0.5 to 0.8:0.2 over the environmental temperature range 22–36 °C. Livingstone (1968) showed that, in addition to depending on environmental temperature, the ratio should be adjusted according to the length of exposure to an altered environment. A rapid decrease in the mean skin-temperature coefficient was required to accommodate the initial rapid cooling of the skin (during the first five minutes), thereafter the mean skin-temperature coefficient gradually rose again during the following 60 min.

For cattle, McLean *et al.* (1983*b*) found 0.86:0.14 to be the deep-body to mean skin-temperature weighting ratio which correlated best with calorimetric determinations of body-heat storage.

For a human subject or large animal a 1 °C change in mean body temperature is equivalent to the total amount of heat produced in a period of the order of one hour. A 1 % imbalance between heat production and heat loss would thus be expected to manifest itself as a change in mean body temperature of only 0.01 °C/h; such a small rate of change is only realistically measurable over a period of several hours. On the other hand during a condition likely to create a large imbalance between heat production and heat loss (e.g. vigorous exercise), readjustments to the thermal distribution in the body must be expected, and the weighting factors for estimation of mean skin and mean body temperatures are more open to question. Other factors such as temporary heat storage in the muscle mass, changes in specific heat with differing levels of hydration and changes in body weight add further uncertainty to the measurements. It is therefore possible to make only very crude estimates of body-heat storage from measurements of changes in body temperatures, especially in the short term.

6

Instrumentation

The accuracy of a calorimetry system can be no better than the accuracy of the measuring instruments it employs. But it is equally true that accurate measuring instruments cannot improve upon imprecise basic data: that is to say there is no advantage in employing a digital voltmeter that can resolve to four digits to measure the voltage output of a transducer that is accurate to only three. The two must be matched. Instruments for a wide variety of physical measurements are commercially available. However, where budgetary considerations preclude the use of sophisticated equipment it is often possible to achieve high levels of precision with relatively inexpensive devices. In this chapter we will concentrate on both extremes of sophistication and simplicity.

Table 6.1. *Measurements required in different forms of calorimetry*
*, fundamental or primary measurements; +, subsidiary measurements; ×, intermediate.

		Temperature	Humidity	Pressure	Gas concentration	Gas flowrate	Gas volume	Weight	Voltage
Indirect	Confinement	+	+	+	*		*		
	Closed-circuit	+	+	+	×		*	*	
	Open-circuit	+	+	+	*	*			
Direct	Gradient layer	+				+			*
	Heat sink	*	*	+		×			
	Convection	*	*	+		*			
	Differential	*	*	+		×			

The basic measurements required depend on the type of calorimeter and they are summarised in Table 6.1. The asterisks indicate the fundamental measurements for each type of calorimeter whereas the plus signs refer to secondary measurements used in correction terms for making adjustments to the primary data. Multiplication signs refer to measurements that fall somewhere in between the other two categories.

Most of the measurements must be recorded repeatedly at preset time intervals; malfunctions must be readily recognisable and the recorded data subjected to immediate or subsequent computation. This effectively means that all instruments have to produce an electrical output in analogue form which is applied to a chart recorder providing a visual record, or in analogue or digital form to a computerised logger. Instrumentation may also be required for environmental control, for timing of experimental procedures and for operation of fail-safe features.

6.1   Instruments for basic measurements
6.1.1   Voltage
In the gradient-layer calorimeter all heat exchanges are transformed in the heat-sensitive panels into voltages, which are thus the primary measurements as shown in Table 6.1. However, as just described, nearly all the instruments for measuring the other parameters listed in Table 6.1 also produce voltage analogue outputs, so that voltage measurements are necessary in practically every form of calorimetry.

A high quality digital voltmeter (DVM) with ± 1 μV resolution and automatic ranging is virtually an essential piece of laboratory equipment.

6.1.2   Temperature
Temperature measurement is required in all forms of calorimetry either in combination with humidity and pressure for STPD corrections, or in combination with humidity and flowrates as a fundamental measurement in most forms of direct calorimetry. Even measurements by gradient-layer calorimetry are slightly dependent on temperature. Also humidity is usually derived from a basic measurement of temperature (e.g. wet-bulb or dewpoint).

Temperature may be measured with thermocouples, thermistors or resistance thermometers. Thermocouples are circuits made up from two dissimilar metals: when the two junctions between the metals are held at different temperatures a voltage is generated which is approximately proportional to the difference in temperature. In thermocouple thermometers, one junction (the reference junction) is held at a known fixed temperature and the temperature of the other may then be estimated from the voltage generated in the circuit. The resistance thermometer relies on the fact that the electrical resistivity of some metals increases with increasing temperature (usually by a few parts per thousand per degree C). A Wheatstone-bridge circuit, made up with one arm containing a resistance probe made from suitable metal, is thus sensitive to the temperature of the probe. A thermistor thermometer is a form of resistance thermometer employing a probe made from a semi-conductor material with a high negative temperature coefficient of resistance (usually about -4% °C). All of these are available in a wide variety of commercial instruments, with either digital or analogue display and output. Absolute accuracy of ±0.1 °C is adequate for most calorimetric applications, although ±0.025 °C is possible, and ±0.01 °C for differential measurements which are fundamental to heat-sink, convection and differential calorimetry. Interchangeability and consistency between sensor probes in multiprobe instruments is an important feature. Where a large number of sensors are required, or where frequent replacement is necessary, copper–constantan thermocouples are particularly suitable since they may be made up as required. It is necessary to ensure uniformity of thermoelectric characteristics between different samples of constantan wire, particularly between wires of different gauge. This test may be carried out by making up a thermocouple circuit with three junctions from the two constantan wires to be compared and a copper wire (Fig. 6.1). If junctions X and Z are held at the same temperature, then provided the two constantans (A and B) have the same thermoelectric properties there will be no change in voltage output when junction Y is subjected to large changes in temperature. Once this has been established a multi-thermocouple system may be made up with heavy gauge wires of copper and constantan in parallel terminating in fine-gauge wires which

Fig. 6.1. Circuit for comparing the thermoelectric properties of two samples of constantan.

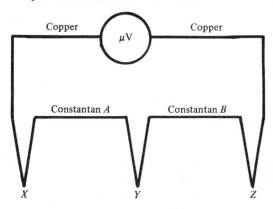

form the sensitive measuring junction at the tip. With this arrangement the fine-gauge wires do not cause excessive heat loss by conduction from the measuring point, and high electrical resistance of the circuit is also avoided. If the lightweight tip wire should break it can be replaced easily. All junctions return to a common constant-temperature bath or thermostated block with accommodates the reference junctions. Each thermocouple circuit so formed thus terminates in a pair of copper wires. These are switched in turn to the measuring instrument (voltmeter). Errors can occur if the electrical connections of the circuit are not carefully formed. However, ± 0.1 °C overall accuracy is easily achieved with silver-soldered joints provided all wires to be joined are first cleaned and 'tinned', a good mechanical contact is made and the quantity of solder used is limited to the minimum necessary to secure the joint. It is still desirable to insulate the switching unit thermally and protect it from draughts.

6.1.3 Humidity

Humidity is a fundamental measurement for the estimation of evaporative heat loss in direct calorimeters, except for the gradient-layer type where evaporative heat is measured as the heat of recondensation. Humidity measurement is also necessary in most forms of calorimetry for corrections to STPD.

Accurate measurement of humidity has been for many years a major problem. A great number of instruments have been devised based on measurements of changes in various physical properties of different hygroscopic materials. Nearly all have exhibited some degree of hysteresis and have lacked long-term stability. The most reliable humidity measurements have until recently been those provided by dewcells and wet- and dry-bulb psychrometers. However, a quite recent development is the Peltier cooled-mirror type of dewpoint apparatus. This definitive method is now commercially available and, although still relatively expensive, seems likely to transform the practice of humidity measurement in future. Meanwhile, the wet- and dry-bulb psychrometer will continue to be of use.

6.1.3a Wet- and dry-bulb psychrometers

The principle of the wet- and dry-bulb apparatus has already been discussed in relation to the psychrometric chart (see Chapter 5, Section 5.1.1). The two thermometer bulbs, one covered with a wetted wick, are placed in a moving airstream. The dry bulb must not be immediately downstream of the wet bulb, and if a reservoir is used to supply water to the wet-bulb wick, then the reservoir must be placed so that evaporation from it does not affect the humidity being measured. The wick is usually made of cotton cloth, and should fit snugly around the thermometer bulb. It is advisable to renew the wick every few days and to boil new wicks in distilled water to ensure removal of any soluble materials, such as starch.

Standard tables of vapour pressure against wet- and dry-bulb temperatures are based on an empirical relationship determined with mercury-in-glass thermometers ventilated at not less than 3 m/s. Errors in the estimation of humidity by this method arise mainly from two sources; namely error in the 'psychrometric constant' of the empirical relationship and errors due to radiative and conductive heat exchanges by the thermometers. Monteith (1954) analysed these errors and concluded that best results may be obtained by keeping the wet bulb as small as possible; a fine wire thermocouple is well suited for such work. Despite the many potential sources of error, well-maintained wet bulbs give surprisingly consistent results and are generally more reliable than instruments based on hygroscopic materials. But when wet bulbs are used as the primary measurement in the estimation of evaporative heat loss it is essential to calibrate them regularly against a known standard rather than rely on tabular values.

6.1.3b   *Dewcells*

The dewcell consists of a thermometer inside a cylindrical metal tube which is covered by a fibreglass wick, the whole being surrounded by two separate spirals of platinum wire. The wick is impregnated with a saturated solution of lithium chloride and an a.c. electrical supply is connected between the two platinum wires. Current then flows through the solution, the temperature of the whole element rises and water evaporates until the current is limited by the increasing electrical resistance of the solution. An equilibrium is reached close to the temperature at which the saturation vapour pressure of lithium chloride is equal to the surrounding vapour pressure. This method has the benefit of being a definitive one, in that humidity is measured in terms of a known physical relationship (i.e. between temperature and saturation vapour pressure over lithium chloride), but the element is somewhat cumbersome. However, it requires little maintenance and is particularly useful as a sensing element in environmental control systems.

6.1.3c   *Peltier cooled mirrors*

The Peltier cooled mirror dewpoint thermometer is a recent development of the classical Regnault dewpoint apparatus, in which air was bubbled through ether contained in an externally silvered test tube. The bubbling was maintained until dew was observed just starting to form on the mirror surface. The dewpoint could then be read from a thermometer immersed in the ether. The Peltier effect is the observation that when current is forced to pass through a thermocouple circuit there is a transfer of heat between the two junctions. Modern technology has made it possible to cool a small mirror using the Peltier effect by driving a current through a circuit formed from certain semi-conductor materials. Decreasing the temperature of the mirror cools the air in contact with it. When moisture begins to condense out of the air onto the mirror the temperature is equal to the dewpoint of the airstream. Detection of dew formation is by a photocell whose output is amplified and used to control the cooling current. Temperature is measured by a thermocouple or thermistor in contact with the mirror. Errors arising out of contamination of the mirror surface by deposits of dirt may be avoided by regular checks on the

Fig. 6.2. Peltier cooled-mirror dewpoint probe. (Photograph provided by Michell Instruments Ltd, Cambridge, UK.)

reflecting power of the mirror during periods when it is allowed to warm up to dispel the dew; this resets the optical reference point at which the photocell is deemed to detect dew. This is a truly definitive method of measuring humidity and can record dewpoint with an accuracy of $\pm 0.1\ °C$. Although Peltier sensors are quite bulky, it is only the small mirror that must be exposed to the air under test and the apparatus is suitable for measurements on airstreams of as low as 1 l/min. The measuring cell of an instrument working on this principle and produced by Michell Instruments Cambridge, England, is illustrated in Fig. 6.2.

Dewpoint temperature (T_d) is related to other measures of humidity as illustrated in psychrometric charts (see Fig. 5.1) and tables. Over limited ranges these relationships may be expressed in the form of equations which are readily incorporated into the software of computer programmes. For example, the equation

$$P_w = 4.666 + (0.2625 \times T_d) + (0.0190 \times T_d^2) \tag{6.1}$$

gives vapour pressure P_w accurate to ± 0.1 mm Hg over the dewpoint temperature range $T_d = 0\text{–}20\ °C$.

6.1.4   *Pressure*

Barometric pressure may be recorded continuously using a barometer with digital readout. Aneroid barometers with this facility are commercially available; they require to be checked regularly against a precision mercury barometer. Alternatively, use may be made of the stability of a high-quality oxygen analyser that is not compensated for changes in barometric pressure. Since dried fresh air has constant oxygen concentration (20.95 %), an oxygen analyser set to read 20.95 % at a time when the barometric pressure is P_i, will read $20.95 \times (P/P_i)$ % for fresh air at any subsequent time when the barometric pressure is P. If the measuring system of the calorimeter already contains such an oxygen analyser, it may be regularly switched to fresh air briefly and thus used as a recording barometer.

Other pressures likely to be measured in calorimetric procedures are all close to barometric pressure and may be measured relative to barometric pressure by means of a mercury or water manometer or by a sensitive differential pressure transducer.

6.2   Equipment for gas analysis

The purpose of this section is to describe in outline the methods by which the composition of respiratory gases is determined. We will avoid a detailed description of how to use the apparatus since this is adequately covered elsewhere in other publications and manuals.

6.2.1 Chemical analysis

Whilst the routine analysis of respiratory gases is now almost entirely carried out using electrical analysers, chemical analysis is still widely used for calibration purposes. Numerous methods are described by Lunge & Ambler (1934) but of these only the Haldane apparatus, and possibly the Orsat apparatus, are currently in common use. The Scholander method, introduced in 1947, is relatively new.

6.2.1a Haldane analysis

The Haldane apparatus for the determination of oxygen, carbon dioxide and methane was first described by J. S. Haldane in 1898. Since then it has been subjected to numerous modifications. A detailed description of its use, and some of the modifications, is given by several authors (Haldane, 1912; Cathcart, 1918; Henderson, 1918; Carpenter, 1923; Douglas & Priestley, 1924; Du Vigneaud, 1927; Carpenter, Fox & Sereque, 1929; Peters & Van Slyke, 1932; Swift, 1933; Bartels *et al.*, 1963; Consolazio *et al.*, 1963).

The apparatus (Fig. 6.3) comprises a burette on top of which is a two-way stopcock connecting the burette through glass capillary tubing with either the sample inlet, or the line to the two absorption pipettes. The bottom of the burette is connected by thick-walled rubber tubing through a stopcock to a mercury reservoir. The most accurate of the original Haldane burettes had a capacity of 21 ml but this was largely superseded by the smaller portable 10 ml burette. At Leeds we prefer to use a 21 ml burette for the extra accuracy it confers. Carpenter (Carpenter, 1923; Carpenter *et al.*, 1929) introduced a 40 ml burette for very accurate analysis. The bottom third of the Haldane burette is volumetrically graduated to 0.01 ml; the Carpenter burette is graduated to 0.004 ml. With a magnifying lens and very great care it is just possible to read to one part in ten of the graduation.

The burette must be calibrated by attaching a three-way tap, with a fine capillary tip, to the bottom. The burette, tap and capillary tip, all of which must be scrupulously clean, are filled with mercury, taking care to eliminate any trapped air. The mercury is run from the burette through the capillary tip, collected and weighed. It is delivered in aliquots that correspond first to the ungraduated portion of the burette, and thereafter to the 0.1 ml graduations. From the weight and density of the mercury the volume occupied by it can be calculated. The volumetric graduations are plotted against the measured volumes and a correction for any part of the burette determined. Because the column of mercury may cause a flexible joint between the burette and the tap to expand, the connection must be absolutely rigid, and preferably the burette and tap should be fused together. Peters & van Slyke (1932, p. 21) described an ingenious modification to the calibration apparatus that makes this unnecessary.

The burette is connected to the two absorption pipettes either through two individual stopcocks as in the original design, or through a single two-

Fig. 6.3. The Haldane apparatus: A, thermobarometer; B, measuring burette; C, mercury reservoir; D, E, absorption chambers; F, KOH reservoir; G, 1,2,4–benzenetriol reservoir; a, b, two-way stopcocks; c, d, one-way stopcocks; x, y, z, graduations for setting the levels of the absorbents at the beginning and end of analysis.

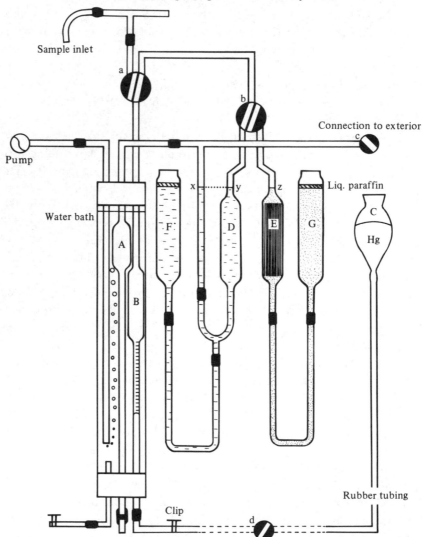

way stopcock introduced by Carpenter, Lee & Finnerty (1930). Of the two pipettes, one contains a solution of potassium hydroxide for the absorption of carbon dioxide, and the other a solution of alkaline pyrogallate (1,2,3-benzene triol) for the absorption of oxygen (it will also absorb carbon dioxide). There is some variation in the concentrations of the solutions, and each is claimed to provide some benefit (see reviews above). Whilst carbon dioxide is absorbed with ease, pyrogallol only slowly absorbs oxygen despite modifications to increase its contact with air (Carpenter, 1923; Carpenter *et al.*, 1929; Peters & van Slyke, 1932; Kilday, 1950; Colovos, Holter & Hayes, 1968). We have confirmed a dramatic improvement in the rate of oxygen absorption by using 1,2,4-benzene triol rather than pyrogallol (Margaria, 1933); we and others (Colovos *et al.*, 1968) have found that a commercial solution produced by Ruhrchemie Aktiengessellschaft, Oberhausen-Holton in W. Germany, is also better for oxygen absorption.

A third pipette or chamber can be added for the determination of methane. It contains a spiral of platinum which when connected to a four volt battery glows red hot and ignites any combustible gases such as methane. In methane analysis the volume of carbon dioxide is determined first by absorption in potassium hydroxide in the usual way and then the sample is introduced into the combustion chamber and ignited. It is passed back and forth into the combustion chamber until all the combustible gas is burnt and the volume is then read. One volume of methane requires two volumes of oxygen for complete combustion and in the process produces one volume of carbon dioxide; thus the volume of methane is equal to half the measured volume change. A further check is to repeat the absorption of carbon dioxide since its volume should equal that of the methane originally in the sample.

A thermobarometer is connected to the absorption pipettes to compensate for any temperature and pressure fluctuations that would affect the volumetric measurements during the analysis. Both the burette and thermobarometer are enclosed in a water bath, stirred by air bubbles, to minimise the effect of environmental temperature fluctuations. Lloyd (1958) produced a compact design in which the pipettes were also enclosed in the water jacket.

Water vapour will also affect the volume of gas, and since the sample inside the apparatus will come into contact with aqueous solutions during analysis it is important that it becomes saturated before any volumetric measurements are made. For this purpose a small amount of dilute sulphuric acid is introduced into the burette and into the thermobarometer.

Before the analysis begins, any trace of oxygen and carbon dioxide must be removed from the apparatus by raising and lowering the mercury reservoir to drive the gas into and out of the pyrogallol pipette. A gas sample is then drawn into the burette by lowering the reservoir, and after a short time to allow saturation of the gas with water vapour, the top stopcock is turned to connect the burette with the absorption pipettes and the volume is quickly noted. Carbon dioxide is then absorbed by mixing the gas with the potassium hydroxide; the reservoir needs to be raised and lowered about ten times. After adjustments to compensate for any pressure and temperature changes the new volume is read; intermediate checks are usually made to ensure that absorption is complete by the final reading. The procedure is repeated after connecting the burette to the oxygen-absorbing pipette, though with pyrogallol about 30 or more swings of the reservoir are needed to ensure total absorption of the oxygen. The composition of the gas can be calculated from the original volume and the volumes after the absorption procedures. There have been modifications to the reservoir system to provide greater control over the movement of the mercury to reduce the possibility of accidental contact between it and the absorption reagents (Blaxter, Graham & Rook, 1954). Automation of the movement of mercury can also provide greater control and remove much of the tedium associated with the analysis (Duffield, 1926; Cruickshank, 1930; Peters & van Slyke, 1932; Chamberlain, Dixon & Waters, 1954).

Most investigators claim that the Haldane apparatus is capable of achieving precision (measured in terms of the standard deviation) of better than $\pm 0.02\%$ for carbon dioxide and $\pm 0.03\%$ for oxygen (Haldane, 1898, 1912; Peters & van Slyke, 1932; Kleiber, 1933; Cormack, 1972) and with the improvements introduced by Carpenter (1923) the precision is about ten times better. Because the only current serious use of the Haldane apparatus is the calibration of gas analysers, we have discussed its precision and accuracy further in Chapter 7.

6.2.1b *Scholander gas analyser*

In 1947 Scholander described a compact portable apparatus which could analyse a 0.5 ml gas sample for carbon dioxide and oxygen with an accuracy and precision similar to that attainable with the Haldane apparatus. The analyser is shown in Fig. 6.4. A clearer description of the method than that given by Scholander exists in Bartels *et al.* (1963) and Consolazio *et al.* (1963).

The apparatus comprises a reaction chamber, the upper part of which is connected through a capillary to another chamber that functions as a thermobarometer. The bottom of the reaction chamber is connected by a three-way stopcock to a micrometer burette filled with mercury; the third arm of the stopcock leads to a mercury levelling burette. Two side chambers arising from the reaction chamber hold the two solutions with which carbon dioxide and oxygen are absorbed. The solutions are separated from the reaction chamber by a layer of mercury. All four chambers are immersed in a water bath that is supported on a pivoting rod.

Carbon dioxide absorption is by means of a potassium hydroxide–potassium dichromate solution, and oxygen is absorbed by a solution of potassium hydroxide, sodium hydrosulphite and sodium anthraquinone–sulphonate. The vapour pressure of the solutions must be equal since equilibration during the analysis may produce a volume change which is not due to the absorption of gas. Collins, Musasche & Howley

Fig. 6.4. The Scholander apparatus. A, compensating chamber (thermobarometer); B, reaction chamber; C, side-arm for carbon dioxide absorber; D, side-arm for oxygen absorber; E and F, solid vaccine bottle stoppers; G, receptacle for stopcock s-1; H, micrometer burette; I, levelling bulb; J, handle for tilting apparatus; K, tube for storing acid rinsing solution; L, pipette for rinsing acid; M, transfer pipette. (Reproduced from Consolazio *et al.*, 1963.)

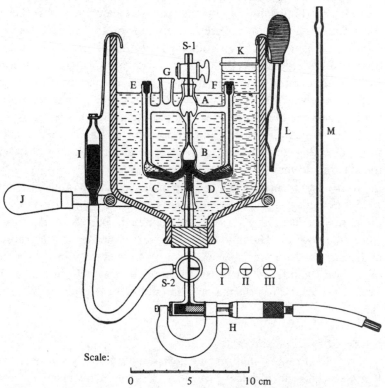

(1977) have described modifications to the solutions of Scholander which ensure stable vapour pressure.

Before beginning an analysis the reaction chamber is washed out by a rinsing solution which is introduced into the thermobarometer, and drawn down into the reaction chamber by lowering the levelling burette. When the reaction chamber has been well rinsed by raising and lowering the burette, the solution is returned to the thermobarometer by raising the level of mercury until it just reaches the top of the capillary. A pipette, holding the gas sample, is introduced into the top of the capillary and sealed by the rinsing solution. The sample is drawn from the pipette into the reaction chamber by lowering the mercury. The pipette is then withdrawn leaving the rinsing solution to cover the capillary. The solution is then removed except for a small drop in the capillary which is drawn down by adjusting the mercury micrometer until the bottom meniscus of the drop is level with an engraved ring about half-way down. After allowing the system to equilibrate for a short time, during which the drop is kept on the ring by adjusting the micrometer, the tap leading from the thermobarometer to atmosphere is closed and the reading on the micrometer noted. The apparatus is then tilted slightly to admit a small amount of the carbon dioxide absorbing solution into the reaction chamber, and is then gently rocked either by hand or mechanically to ensure contact between the sample and solution. The micrometer is adjusted periodically to keep the drop level with the ring as the volume in the reaction chamber is decreased with the absorption of carbon dioxide. Once the carbon dioxide is completely absorbed, the bottom miniscus of the drop is finally adjusted onto the ring and the micrometer reading is noted. The procedure is repeated with the oxygen absorber. The composition of the sample is calculated from the change in volume at each stage of absorption as a proportion of the starting volume.

The method is capable of achieving high accuracy (Gaudebout & Blayo, 1975) and good agreement with measurements made by the Haldane apparatus (Collins *et al.*, 1977). It is claimed (Bartels *et al.*, 1963) that, compared with the Haldane apparatus, it is easier to handle, more rapid and easier to clean for no greater cost. A simpler version for the analysis of carbon dioxide only has been described (Levy, Bernstein & Devor, 1958). A discussion of the major sources of error is given by Severinghaus (1960).

6.2.2   *Physical analysers*

Routine gas analysis in calorimetry is entirely by means of analysers working on physical principles. Compared to chemical analysers they are fast acting, less demanding to use and more precise – though considerably more expensive. Furthermore, their analogue output allows recording of data.

Many of the analysers measure the partial pressure of the gas rather than, as the scale calibration may suggest, its volume percentage contribution to the gas mixture. This assumes that the relationship between partial pressure and volume percentage is constant, which unfortunately it is not. Any change in the water vapour content of the gas mixture or the pressure across the cell will disturb the relationship, and the apparent volume percent displayed by the analyser may be considerably different from the true value. This leads to the requirement to condition the sample before it enters the analyser. Since this sample conditioning affects the validity of any calibrations that are undertaken, we have discussed it more fully in Chapter 7.

Although numerous different types of analyser have been used, the most common ones in calorimetry are paramagnetic oxygen analysers, oxygen electrodes, and infra-red gas analysers. We will deal with these in detail. The less often used analysers – mass spectrometers, diaferometers, fuel cells and gas chromatographs – will be only briefly discussed, though the references should allow the reader to obtain more information if necessary. Some of these gas analysers have been discussed in detail in Janac, Catsky & Jarvis (1971) and Verdin (1973).

6.2.2a *Oxygen electrodes or polarographic cells*

Oxygen electrodes are commonly used in commercial oxygen analysers, particularly those at the lower end of the cost range. The basis of the method is the technique of polarography which is the electrolysis of solutions of electro-oxidisable or electro-reducible substances between a noble metal electrode (e.g. mercury, silver or platinum) and a reference electrode. Since noble metal electrodes have a low solution pressure, they are effectively inert in the electrolyte. A continuously increasing potential is placed across the electrodes and the flow of current measured; the plot of current against voltage is called a polarogram.

Electrolysis involves two main types of activity, known as mass transfer and electrochemical processes. Mass transfer processes involve the movement of electro-active material up to the electrode surface by migration, diffusion and convection. In polarography the object is to have the oxidisable or reducible materials arrive at the electrode by natural diffusion alone. Convection can be largely eliminated by providing temperature regulation to prevent thermal gradients in the system. Migration, which depends on the transport number of the test substance (i.e. the fraction of the total current carried by it) is largely eliminated by the addition of large quantities of a supporting electrolyte. The large number of ions added decreases the transport number of the test material to a negligible figure; these added ions must not undergo electrode reactions over the range of potentials likely to be applied.

Electrochemical processes are those which occur at the electrode itself and in which there is a transfer of electrons. The test substances is measured by such an electrochemical process; it is important that the electrode does not itself contribute to the oxidation-reduction reaction and for this reason inert noble metal electrodes are preferred. The reaction occurs at the cathode for reducible materials such as oxygen and at the anode for oxidisable materials.

The relationship between the applied potential and the flow of current for reducible substances is shown in Fig. 6.5. Initially, as the potential is applied there may be little or no current flowing through the solution. At some point a reducible substance in the electrolyte will begin to be reduced at the cathode and the current will increase; the point of inflection is the decomposition potential. Current flow increases as the potential continues to increase until eventually a plateau is reached as current flow is limited by the rate at which the reducible substance can reach the cathode. In a solution of several such substances, a continuing increase in potential will cause them to be reduced in turn at the cathode, and a series of plateaus will be produced.

The substance causing the increase in current flow can be identified by its half-wave potential, which is the potential at which the current flow is half-way between the previous plateau and that due to the limiting concentration of the substance. The increase in current from one plateau to

Fig. 6.5. The main features of the relationship between current and voltage in a polarographic system.

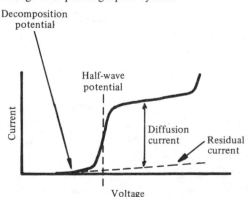

another is called the diffusion current and is proportional to the concentration of the reducible substance which gave rise to it.

Oxygen is reduced at the cathode when only a very low potential is applied to the electrodes and the polarographic cell therefore provides an ideal means for its detection and measurement.

The electrode used for the determination of oxygen is shown in Fig. 6.6. It has been described in detail elsewhere (for example Heitman, Buckles & Laver, 1967; Fatt, 1968; Severinghaus, 1968; Crampton Smith & Hahn, 1969). The cathode is a fine platinum wire which is embedded in an insulating material such as glass until only the tip of the platinum electrode is in contact with the electrolyte. The anode is a silver–silver chloride, silver–silver hydroxide or calomel half-cell. The electrolyte is either potassium chloride, potassium hydroxide or a potassium chloride–phosphate mixture; this should be periodically replaced. The tip of the cathode is covered with a thin membrane which is permeable only to oxygen; this makes the cell specific for oxygen and prevents contamination of the

Fig. 6.6. An oxygen electrode assembly. (Reproduced from Crampton Smith & Hahn, 1969.)

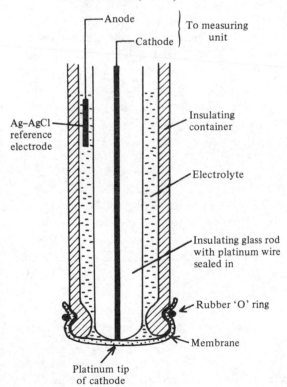

cathode. Various materials have been used for the membrane such as polyethylene and polypropylene but teflon seems most popular. There is only a thin layer of electrolyte between the cathode and membrane so that the diffusion distance is short.

The flow of current depends on the following reactions:

cathode $O_2 + 2H_2O + 4e^- \rightarrow 4OH^-$

anode $4Ag + 4Cl^- \rightarrow 4AgCl + 4e^-$

The electrode has been widely used for measurements of oxygen in respiratory gases and in fluids such as blood. Davies & Brink (1942) first used the principle to measure the oxygen content of animal tissues; however Clarke (1956) was responsible for the inclusion of a membrane that permitted oxygen, but not other contaminants, to enter the cell. This brought about a significant improvement to the electrode and led to its extensive use. Since then there have been numerous modifications to reduce the size and response time of the electrode (Kreuzer, Rogeness & Bornstein, 1960; Staub, 1961; Severinghaus, 1963; Friesen & McIlroy, 1970). The response time of the cell, which depends largely on the nature of the membrane and its thickness, is very fast: using a teflon membrane 3 μm thick, gives a 98% response time better than about 0.4 s.

The cell is inherently partial pressure and temperature sensitive. Most investigators, including ourselves, have found that it is linear over a wide range (Clarke, 1956; Kreuzer *et al.*, 1960; Beneken Kolmer & Kreuzer, 1968; Friesen & McIlroy, 1970) although Heitmann *et al.* (1967) have reported small deviations from linearity in some commercial electrodes. The accuracy and precision of the cell compares favourably in the short term with any other method of oxygen analysis. Although the cells must be replaced periodically, and the electrolyte quite frequently, they are cheap.

6.2.2b *Paramagnetic oxygen analysers*

It is our view that paramagnetic analysers provide the best means of determining the oxygen concentration of gases in indirect calorimetry. The analysers measure the partial pressure of oxygen and it is necessary to pay particular attention to sample conditioning. In order to understand the principle of paramagnetic analysers it is necessary to discuss briefly some aspects of atomic structure.

Each atom is made up of a nucleus, which contains neutrons and positively charged protons, surrounded by a planetary system of negatively charged electrons. These electrons orbit around the nucleus in a characteristic pattern. Each orbit can be occupied by up to two electrons and one or more orbits make up a shell of electrons around the nucleus.

In addition to orbiting the nucleus, each electron spins on its own axis. As the electron travels in its orbit it generates an electromagnetic field which is due mainly to the spin (each spinning electron behaves as if it were a tiny magnet). When two electrons share the same orbit they must have opposite spins (Pauli's principle) and the two electromagnetic effects cancel out. Inside a magnetic field atoms in which there are unpaired electrons will line up with the lines of force and tend to strengthen the field; atoms and molecules that show this behaviour are called paramagnetic. Molecules and atoms in which all the electrons are paired off do not align themselves with the lines of force in a magnetic field and are repelled by it; they are said to be diamagnetic.

The sensitivity of a molecule to a magnetic field is described by its magnetic susceptibility which is inversely proportional to the absolute temperature (Currie–Weiss Law); its values are usually expressed adjusted to 760 mmHg and 20 °C. Paramagnetic substances always have positive values for susceptibility whilst diamagnetic ones have negative values. Table 6.2 shows the magnetic susceptibility for several gases; only oxygen, chlorine dioxide, nitric oxide and nitrogen dioxide are paramagnetic, the rest are weakly diamagnetic. At first sight it might appear surprising that oxygen, with an even number of electrons, is strongly paramagnetic but two of the electrons are in unpaired orbits. Paramagnetism therefore offers an important means of measuring oxygen concentration particularly in air and respiratory samples where nitric oxide and nitrogen dioxide, the gases with magnetic susceptibilities closest to that of oxygen, are unlikely to be present.

There are three main types of analyser that make use of the paramagnetic principle and these are discussed below. Verdin (1973) has described individual instruments that were available at that time.

Table 6.2. *Magnetic susceptibility of several gases*

Gas	Magnetic susceptibility (cgs × 10^{-6})
Ammonia	−18.0
Carbon dioxide	−21.0
Carbon monoxide	−9.8
Hydrogen	−4.0
Methane	−18.0
Nitric oxide	+1461.0
Nitrogen	−12.0
Nitrogen dioxide	+961.0
Oxygen	+3449.0

Thermomagnetic or 'magnetic wind' instruments. The variation in the magnetic susceptibility of oxygen with temperature formed the basis for the first paramagnetic oxygen analysers (Senftleben, 1930; Klauer, Tur-owski & Wolff, 1941; Dyer, 1947; Lehrer & Ebbinghaus, 1950). The common feature of these instruments is that the measuring cell contains one or two heating elements within a strong magnetic field. If the gas is paramagnetic, it is attracted into the magnetic field causing an increase in flowrate and hence the rate of heat loss by the element. As the gas in contact with the element heats up, its magnetic susceptibility and attraction to the magnetic field decreases. Because of this, it is quickly displaced by colder paramagnetic gas following behind, which in turn is heated and then itself displaced; the effect is to produce a continuous airstream – the 'magnetic wind' – across the element. Since the rate of heat production is constant, the temperature and resistance of the element depends on the rate of flow and thermal conductivity of the gas.

With a magnetic field of constant strength, the flowrate of the gas depends on its magnetic susceptibility and thus its oxygen content. Assuming little variation in thermal conductivity, the resistance of the element is proportional to oxygen content and is a measure of it. The resistance of the wire element is measured by its incorporation into a Wheatstone-bridge circuit.

Two main forms of thermomagnetic instruments are available: the filament type (Fig. 6.7(*a*)); and the ring chamber type (Fig. 6.7(*b*)). The

Fig. 6.7. The main features of thermomagnetic or 'magnetic wind' oxygen analysers of the (*a*) filament type, and (*b*) ring chamber type. (Adapted from Verdin, 1972.)

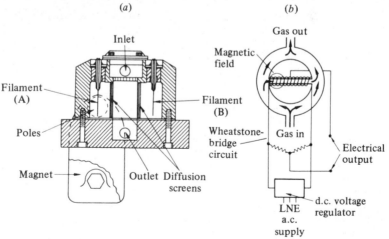

basic design of the filament type of analyser involves the separation of two detection chambers by diffusion screens from a central chamber through which the gas flows. A heated wire, typically of platinum, projects into each of the two detection chambers; one of the detecting chambers has a strong magnetic field applied to it. The two filaments form one arm of a Wheatstone bridge circuit. In the absence of a paramagnetic gas, or if the magnetic field is disabled, only natural convection will cause gas to circulate into and out of the detection chambers. The amount of convection is small and causes little cooling of the elements. Switching on the magnetic field or introducing oxygen into the gas stream causes an increase in the diffusion of gas into the magnetised detection chamber and a greater rate of heat loss from the wire therein. The other detection chamber continues to provide a reference signal. The change in resistance of the cooled wire in the magnetised chamber is a measure of the oxygen content of the gas.

This type of cell is very stable, simple and rugged and little affected by changes in its position. It is, however, sensitive to variation in the thermal conductivity of the sample gas (Verdin, 1973).

The second form of the 'magnetic wind' instrument is called the ring chamber analyser. The method was patented in 1942 by Lehrer and manufactured by Hartmann and Braun; it still forms the basis of their Magnos range of oxygen analysers. As its name implies it comprises an annular tube across the centre of which is a connecting hollow glass tube. A heating element of either nickel or platinum is wrapped around one end of the interconnecting tube and is surrounded by a strong magnetic field. A second element, free of the magnetic field, is wrapped around the other end of the interconnecting tube. The two heaters form one arm of a Wheatstone-bridge circuit. Air enters and leaves the annulus at opposite ends.

A stream of weak diamagnetic gas such as nitrogen divides itself evenly between the two limbs of the annulus. Providing that the connecting tube is placed accurately in a horizontal position no gas will enter it. When oxygen is introduced into the gas stream it is attracted into the connecting tube from the side of the annulus nearest the magnetic field. Once inside, the gas is heated and its magnetic susceptibility decreased; it is then displaced by colder gas. The procedure is continuous and a 'magnetic wind' is set up. The displaced air passes through the tube, over the second heater and back into the annulus. The air flow transfers heat from the first heating element to the second. The change in the rate of heat flow of the two elements alters their resistance and disturbs the balance of the bridge circuit. The output from the bridge is proportional to the concentration of oxygen in the gas.

Unlike the filament type of analyser, the ring cells are quite insensitive to fluctuations in thermal conductivity; they are also more sensitive to oxygen. Both types are temperature and pressure sensitive and some form of thermostating and barostating may be necessary. The output from the Wheatstone-bridge circuit is inevitably non-linear, although linearising circuits are usually employed to give a recorder output which is linear over at least the working range (Dyer, 1947).

Susceptibility pressure analysers. These are often referred to as Quincke analysers after the scientist who first described the principle on which they

Fig. 6.8. The mean features of the Quincke analyser: A, flow-channels; B, side-circuit; C, D, heaters forming one arm of a Wheatstone-bridge circuit; E, F, heaters forming the second half of a Wheatstone-bridge circuit. (After Verdin, 1972.)

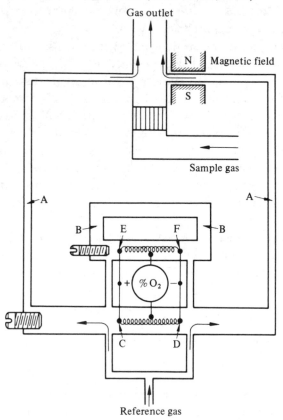

are based. Only a few oxygen analysers based on the principle have been produced.

A typical analyser layout is shown in Fig. 6.8. A reference gas is introduced through a common inlet into the two limbs of the circuit. Because the resistance to flow is equal in the limbs, the flow of gas separates evenly. Near the outlet of the circuit, it mixes with sample gas entering from a side arm and the mixture is exhausted to atmosphere.

At the end of one of the limbs is a magnetic field which, because of its strength and proximity to the sample line, attracts the sample gas if it contains oxygen. With a fixed magnetic field the strength of attraction is proportional to the magnetic susceptibility of the sample gas, i.e. its oxygen content. The attraction of sample gas into the limb increases its resistance to flow, causing an increase in pressure. This leads to diversion of some of the reference gas across a pair of heaters and into the other limb. The additional flow of reference gas increases the rate of heat loss from the heaters. All the changes are directly proportional to the oxygen content of the sample gas.

The heaters form one half of a Wheatstone-bridge circuit. A second pair of heaters in a side circuit forms the other half of the bridge, and acts as a compensator for any convection flow caused by changes in the position of the instrument. The flow of current across the unbalanced bridge is thus proportional to the oxygen content of the sample. The output signal from the bridge will be non-linear, but a linearising circuit can be used to provide a linear recorder output. Since the heaters are only exposed to a reference gas of fixed composition, thermal conductivity is constant and does not contribute to any variation in heat loss. The reference gas is typically nitrogen, air or, in the case of a differential analyser, 'inspired air'.

Magnetodynamic or dumb-bell analysers. In 1851 Faraday showed that a hollow oxygen-filled glass sphere mounted on the end of a horizontal rod, which in turn was supported by silk fibres, was attracted by a magnet. This observation led almost one hundred years later to the development by Linus Pauling and his colleagues of an oxygen analyser which became the forerunner of a range of highly successful and accurate instruments. The earliest instruments were however temperamental and somewhat unreliable. In the light of the success of the method it is interesting to recall Pauling's description of its development: 'On October 3, 1940, at a meeting in Washington called by Division B of the National Defense Research Committee, mention was made of the need for an instrument which could measure and indicate the partial pressure of oxygen in a gas. During the next few days we devised and constructed a simple and effective instrument for this purpose.' (Pauling, Wood & Sturdivant, 1946.)

The apparatus (Fig. 6.9) consisted of a dumbbell composed of a 4 mm length of glass rod with, at each end, a glass sphere of 3 mm diameter filled with a weakly diamagnetic gas such as nitrogen. The spheres were matched for volume. The dumbbell was suspended from its centre by means of a quartz fibre stretched between the arms of a quartz yolk. Finally a small length of glass rod and a small mirror were attached to the centre of the dumbbell which was then carefully balanced. The dumbbell was mounted in a chamber through which test gas could flow. The chamber was then placed between the poles of a large magnet which provided a non-uniform magnetic field. A light beam was played onto the small mirror and the reflected light focussed onto a scale; any displacement of the dumbbell caused a movement of the reflected light beam along the scale.

The position adopted by the dumbbell depends on the balance between the magnetic force acting upon it and the torsional force (the resistance to turning) of the supporting fibre. With nitrogen flowing through the chamber there will be no displacement of the dumbbell; if the magnetic susceptibility of the gas is increased by the introduction of oxygen, the dumbbell rotates to move the spheres into the weakest region of the

Fig. 6.9. The 'Pauling' dumbbell assembly: g, balancing glass rod; m, mirror. (Adapted from Pauling *et al.*, 1946.)

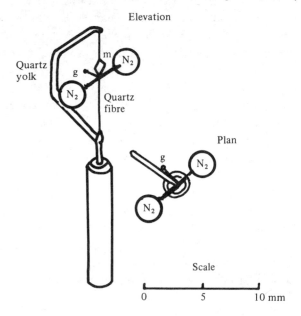

magnetic field. The rotation is opposed by the torsional force and the final balance position is proportional to the magnitude of the magnetic susceptibility (i.e. the oxygen content) of the surrounding gas.

The cell is thermostated to eliminate the affect of temperature variations on the magnetic susceptibility. The cells are also very sensitive to variations in pressure and flowrate (the gas stream can rotate the dumbbell mechanically) and suitable precautions are necessary to regulate these. The original Pauling cell produced a non-linear output because as the dumbbell progressively rotated with increasing oxygen concentration into the weaker region of the magnetic field the force acting on the nitrogen in the dumbbell per unit change in oxygen concentration decreased.

The problem of non-linearity was solved by the introduction of the Munday cell which is used in the range of Servomex analysers. In this system the quartz fibre is replaced by a stronger platinum iridium suspension. The main difference however is the introduction of a feedback circuit. As the dumbbell begins to rotate, the movement of the light beam produces the feedback signal, which in turn leads to the generation of a force which opposes the movement of the dumbbell and keeps it in the null displacement position. Servomex analysers rely on an electromagnetic force applied through a single turn of platinum wire wound around the dumbbell (Fig. 6.10). Beckman analysers use an electrostatic restoring force.

Fig. 6.10. The 'Munday' cell as used in the Servomex range of paramagnetic oxygen analysers. (Reproduced by permission of Servomex Ltd.)

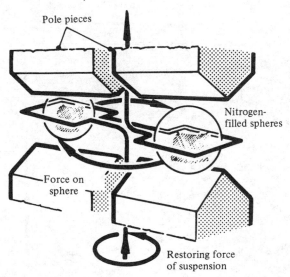

In the Servomex analysers the light beam is reflected to land evenly on two adjacent photocells (Fig. 6.11) and the current produced by the two individual photocells is supplied to a differential amplifier. In the null position the current from the two photocells is equal and there is no current output from the amplifier. Rotation of the dumbbell shifts the light beam so that it falls predominantly on one of the two photocells. This alters the electrical output from them and the difference is amplified. The output current from the differential amplifier is fed back to the platinum coil around the dumbbell and produces a restoring force that returns the dumbbell to the null position. The current required to restore the dumbbell is directly proportional to the magnetic susceptibility, and hence to the oxygen content, of the gas which initiated the displacement. A voltage output can be obtained by passing the current through a fixed resistance of appropriate value. Servomex claims the analyser to be inherently linear and our experience of several model 0A184 analysers confirms this (Chapter 7).

The stability of the Servomex analyser is exceptionally good and at the Hannah we have found that once the analyser is set up on dry fresh air, over a long period of time the voltage output with fresh air in the measuring cell varies in direct proportion to barometric pressure with accuracy of one part in 1000 (see Chapter 7). Nunn and his colleagues (Nunn *et al.*, 1964; Ellis & Nunn, 1968) have also reported favourably on these analysers, albeit in investigations of earlier models.

Fig. 6.11. The feedback circuit used in Servomex oxygen analysers using the 'Munday' cell. (Reproduced by permission of Servomex Ltd.)

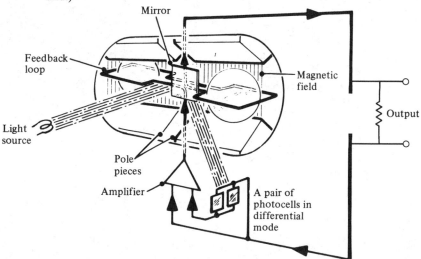

The potential weakness of the analysers is that the sensitive dumbbell mechanism is in contact with the airflow and is in danger of being damaged by the accidental introduction of contaminants such as liquids and particulate matter. Careful filtering of the air entering the analyser should prevent most problems and we have had no such difficulties in many years of use.

Differential analysers have been introduced, for example the Servomex 0A184 and its successor the 0A1184, which allow measurement of 'inspired' air at the same time as 'expired' air and give an output which is the difference between the measurements. Differential analysers can be regarded as two single channel instruments (and they can be switched to function as such) combined 'back-to-back'. This type of analyser is particularly useful in situations in which the composition of 'inspired' air is variable. The output sensitivity of differential analysers is not, as is often mistakenly claimed, independent of changes in barometric pressure.

6.2.2c *Infra-red gas analysers*

The concentrations of carbon dioxide and methane are measured most commonly by infra-red spectroscopy. To understand the manner by which infra-red analysers detect these two gases, it is necessary to describe briefly the way in which electromagnetic radiation is absorbed by molecules.

Electromagnetic radiation is a form of radiant energy and covers a wide range of frequencies (Fig. 6.12(*a*)). When atoms are exposed to electromagnetic radiation energy may be absorbed and the orbit of the electrons raised to a higher energy shell. The large differences in the energy levels of these shells limit the transitions to the visible or UV part of the spectrum. The atoms of different elements have different electron orbit energies and so absorb energy at different wavelengths. The emergent electromagnetic radiation is thus deficient in energy corresponding to a particular wavelength; and if this wavelength is in the visible band of the electromagnetic spectrum, the effect of the absorption is a sensation of colour. Molecules can also absorb energy in transitions between different electronic energy levels such as the molecular orbits; but because of the large differences in the electronic energy levels these transitions also only occur in the visible or UV part of the spectrum.

Whilst radiation is unable to bring about changes in the electronic configuration of atoms or molecules, it may contain sufficient energy to bring about other molecular changes. A molecule can change its energy level by two other ways that are not available to atoms: these are vibrational changes and changes in the rotational energy. It is by such means that molecules such as carbon dioxide and methane are measured by infra-red analysis.

One can envisage molecules as a collection of atoms bonded by electrical forces that behave like springs. Each molecule can vibrate in a manner analogous to the spring stretching and compressing, or bending from side to side; and the atoms can rotate about the bonds. Like a weight and spring system, the molecule will possess characteristic resonant frequencies. The molecule may vibrate at several different frequencies and the pattern of these will be related to the composition and structure of the molecule. The absorption of radiation to bring about specific changes in the energy level of a molecule will produce an electromagnetic spectrum deficient in energy at wavelengths characteristic of the molecule (Fig. 6.12 (*b* and *c*)). Transitions between vibrational levels produce absorption in the near infra-red ($< 20 \mu$m) and those between rotational states at the same vibrational level produce absorption in the far infra-red ($> 20 \mu$m).

Diatomic molecules will only absorb vibrational energy if they have a permanent dipole moment (i.e. if the two atoms have different electrical charges). Electromagnetic radiation alternating at the appropriate frequency will excite such dipolar molecules into resonant vibration as the atoms are periodically pulled to-and-fro; atoms with the same charge,

Fig. 6.12. The electromagnetic spectrum (*a*) and spectra for (*b*) carbon dioxide and (*c*) methane. The black bands show the position of the absorbtion of infra-red radiation by water vapour.

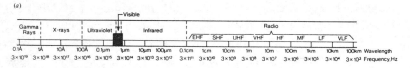

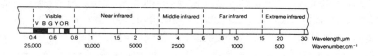

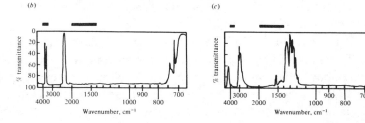

such as oxygen and hydrogen, would always be attracted in the same direction and no vibration would occur. There is no need for a polyatomic molecule to have a permanent dipole moment to have a vibrational spectrum in the infra-red, but they must be capable of undergoing a change from a symmetrical structure to an assymmetrical one with a dipole moment. Carbon dioxide, which is a symmetrical linear molecule, does not have a permanent dipole moment but this develops if the molecule is bent and the oxygen atoms are drawn together. When carbon dioxide, and methane, adopt an assymmetrical structure they absorb infra-red radiation strongly.

Whilst there are instruments which scan much of the electromagnetic spectrum for atoms and molecules which have absorbed radiation, it is more usual in gas analysis to have an instrument 'tuned' to one narrow band of wavelength. The wavelength selected is usually that over which the molecule most strongly absorbs radiation, although if there is competition from the absorption spectrum of another molecule a second weaker band may be chosen. Carbon dioxide (Fig. 6.12b)) absorbs mainly bands centered at 2.69, 2.77, 4.26 and 14.66 μm; the strongest band is that at 4.26 μm; those at 2.69 and 2.77 μm additionally suffer interference from water vapour if it is present. Methane (Fig. 6.12(c)) absorbs in bands centered at 3.29, 6.52 and 7.65 μm and that at 3.29 μm is usually selected for measurement.

Infra-red gas analysers operate on the same basic principle. Electromagnetic radiation is produced from a heat source and transmitted along an absorption cell, through which the sample gas passes, to an infra-red detector. The amount of radiation absorbed by the sample in a particular waveband is measured by the detector. The main type of analysers currently in use are double-beam non-dispersive instruments. Non-dispersive instruments use a broad band of radiation whilst the less well-used dispersive analysers use prisms or gratings to select out radiation of a wavelength appropriate to the gas being detected. Several types of analysers, and their components, are discussed in detail by Janac *et al.* (1971) and by Verdin (1973); the description here is limited to the most widely used instrument.

The non-dispersive double-beam analyser was developed by Luft in Germany before the Second World War (they are sometimes referred to as Luft-type analysers). It is shown in Fig. 6.13. Broad-band energy is supplied to two absorption tubes from either two nickel–chromium heat sources or a single source and a beam splitter or a parabolic mirror. A single source should in principle be most satisfactory since any variation in the output of radiation would affect both cells equally; in practice there are difficulties in achieving even reflection and absorption by mirrors and most analysers use a double heat source.

The two parallel absorption tubes, which are typically made of glass, brass or stainless steel, are internally plated and highly polished to allow efficient energy transmission. One tube contains a reference gas such as nitrogen which does not absorb infra-red radiation, whilst sample gas is passed through the second tube. The absorption of radiation follows Beer's Law:

$$I = I_0 \times e^{-laF}$$

where I = emergent energy; I_0 = incident energy; l = path length of the absorbing tube; a = absorption coefficient (a constant for a particular band of absorption); F = concentration of the absorbing gas.

The amount of radiation reaching the ends of the tubes is measured in the detector which is filled with gas which absorbs radiation in the same band as the gas being measured (it is usually the same gas). The absorption

Fig. 6.13. The main features of the Luft analyser. (After Legg & Parkinson, 1968.)

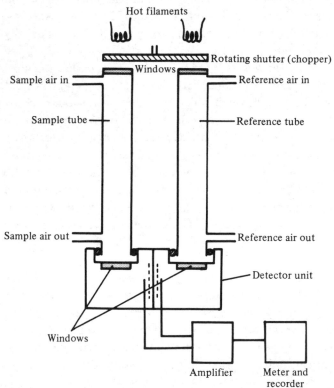

of radiation by the detector gas causes a proportional increase in its temperature and pressure. The detector cell is divided by a diaphragm which flexes as the pressure of the gas in one side of the cell varies. The diaphragm forms one side of an electrical capacitor and its movement changes the capacitance and the voltage, which is inversely proportional to capacitance.

A chopper between the radiation source and the absorption tubes rotates at high speed, typically between 5 and 20 Hz. It interrupts the passage of radiation, switching it alternatively between the sample and the reference tubes. As the amount of radiation arriving at the detector continually changes, it produces pulses of pressure which moves the diaphragm to-and-fro and generates an a.c. signal from the capacitor. The frequency of the signal depends on the frequency of the chopper motor and its amplitude on the concentration of the test gas. The voltage signal is amplified, rectified to produce a d.c. current and then supplied to a meter and recorder output. The concentration of the test gas is proportional to the logarithm of the difference between the voltages from the reference and sample tubes. The relationship to Beer's Law means that the output is nearest linearity when the proportion of the radiation absorbed is very low; however there is more danger of artefacts affecting the signal of these levels and a compromise is necessary. Typically the analysers are configured to produce between 5 and 20% absorption which produces a non-linearity of about 3–10% of full scale. Many analysers include a linearising circuit to transform the logarithmic output to a linear one.

Fig. 6.14. Transmission of radiation by several optical materials. (Reproduced from Verdin, 1972.)

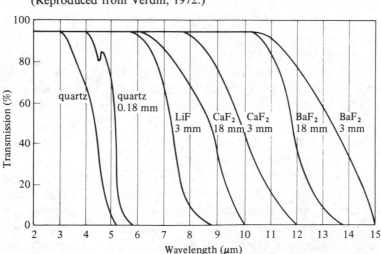

In some analysers the radiation to the absorption tubes is chopped simultaneously; the radiation from the reference tube is collected on one side of the cell and that from the sample tube on the other side. The movement of the diaphragm is then a measure of the difference in the concentration of the test gas in the two absorption tubes. A chopper is still used in this situation since many detectors do not respond to d.c. signals and a.c. amplifiers are more stable and simple than d.c. ones; it also eliminates detector responses to random temperature changes.

Care is necessary in the selection of materials used for windows through which the radiation must travel. Many such materials absorb infra-red radiation above particular frequencies (Fig. 6.14); calcium fluoride is often used in carbon dioxide and methane analysers.

Some analysers are fitted with interference filters to make them more specific to particular wavebands. The filters typically comprise alternate layers of materials with high and low refractive indices (for example, germanium and silicon monoxide respectively) deposited on a material which permits transmission of radiation. When radiation passes from one layer to the next some is reflected and some transmitted. If the layers are one-quarter wavelength thick, the reflected ray from the far face of the layer will be 180° phase delayed compared with the reflected ray from the front face, and the rays will cancel out. If many layers of different thicknesses are built up then all but one selected wavelength can be eliminated.

The analysers are very susceptible to the introduction of contaminants which absorb infra-red radiation and so must be kept scrupulously clean by filters. With these analysers considerable attention must be paid to sample conditioning. In our experience the major weakness of the analysers is the chopper motor which frequently must be replaced.

6.2.2d   *Mass spectrometers*

There are several reviews of the application of a mass spectrometer to the analysis of respiratory gases (Lilly, 1950; Fowler, 1969; Gothard *et al.*, 1980). There are also several useful chapters on the subject in the book edited by Payne, Bushman & Hill (1979). They may become of more importance in energy metabolism studies if the doubly-labelled water method becomes more widely used.

The main type of mass spectrometer used for analysis of respiratory gases is the quadrupole analyser. A vacuum pump draws the gas, either as a discrete sample or continuously, into an ionisation chamber where positive ions are produced from the molecules in the gas. The positive ions are accelerated into a quadrupole. This comprises four parallel circular rods placed as if to form four parallel edges of a cube and linked by diagonal electrical connections. Electrical current applied to the rods creates an electrical field at right-angles to them. At the end of the quadrupole is the detecting element, which is either a Faraday cup or an electron multiplier.

For any applied voltage only one ion with a specific mass-to-charge ratio will have a dynamically stable path through the electric field and reach the detector. Other ions with different mass-to-charge ratios will scatter and be dispersed elsewhere. If there is a steady change in the applied voltage, different ions will reach the detector in turn, and the components of the sample will be displayed as a series of peaks. By applying specific voltages to the rods in the quadrupole, the detection and quantification of components in the sample can be restricted to only those of immediate interest.

The mass spectrometer has a very fast response time and can detect the concentrations of several gases simultaneously. By the nature of its design, it is unaffected by pressure fluctuations, and the presence of water vapour does not influence the analysis and compensation for its diluting effect is easily achieved (Scheid, Slama & Piiper, 1971; Gothard *et al.*, 1980). Mass spectrometers are linear over a wide range of gas concentrations and are accurate and stable (Hallback, Karlsson & Ekblom, 1978; Gillbe, Heneghan & Branthwaite, 1981). By introducing a tracer such as argon, the mass spectrometer can also measure flow rate by the dilution principle (see Section 6.3).

Unfortunately mass spectrometers are extremely expensive and the cost considerably exceeds that of individual oxygen, carbon dioxide and methane analysers. Their use for metabolic measurements seems restricted to clinical studies (e.g. Patterson, 1979) presumably reflecting their presence in hospitals for breath-by-breath gas analysis in respiratory function studies. The major limitations to their use in calorimetry are their cost and their reliance on gas mixtures for calibration, which as we point out in Chapter 7 is a major restriction to accuracy.

6.2.2e *Thermal conductivity*

Perhaps the most well known of the thermal conductivity analysers is the Noyons diaferometer (Noyons, 1937); although some are still in use, it is now mainly of historical importance.

The principle of the method is that the resistance of a wire varies with its temperature. If a fixed electrical current is used to heat the wire in an airstream, any variation in resistance will reflect the rate of heat loss. If the flowrate of gas across the wire and the initial temperature of the gas are regulated, the variation in heat loss of the heated wire will be due to variations in the thermal conductivity of the gas stream, caused by differences in its composition. The resistance of the wire, which reflects the rate of heat loss and thus the thermal conductivity of the gas, is measured in a Wheatstone-bridge circuit.

In the diaferometer, two Wheatstone-bridge circuits are used, one to measure oxygen content and the other carbon dioxide, and their output is displayed on galvanometers. A dried sample of the gas to be analysed is depleted of carbon dioxide and passed over a pair of wires in each of the Wheatstone-bridge circuits. Carbon dioxide-free dry fresh air is passed over the other two wires in the oxygen measuring circuit; any difference in the ratio of oxygen and nitrogen in the two samples causes a difference in the resistance of the pairs of wires. In the carbon dioxide measuring circuit, dry sample gas (from which carbon dioxide has not been removed) is passed over the remaining two wires; the difference in resistance is caused by the difference in carbon dioxide content.

The output from the Wheatstone bridge is inevitably non-linear and although the manufacturer supplies a calibration chart to convert the galvanometer readings to gas concentrations, the apparatus should be calibrated by means of gas mixtures.

6.2.2f   *Fuel cells*

Weil, Sodal & Speck (1967) and Sodal, Bowman & Filley (1968) have described zirconium fuel cells which are suitable for measurement of oxygen in respiratory gases. There is some difference in the structure of the two cells but the theoretical basis for them is the same. The cell is composed of a tube of zirconium oxide (mixed with calcium or hafnium oxide) coated on both surfaces with porous platinum. The zirconium oxide, when heated to a temperature of about 800–1000 °C, acts as a semi-permeable membrane that is highly conductive to oxygen. The platinum surfaces act as electrodes. A difference in oxygen concentration between the inside and outside of the tube leads to a potential difference across it.

The potential difference is about 1 mV for a 1 % difference in oxygen concentration and can be directly recorded. The output is, however, logarithmic and some form of linearising circuit is usually added.

The sample gas is passed down the centre of the tube and a reference gas, usually dry fresh air, is passed along the outside. The fuel cell is therefore inherently a differential analyser. The cell is sensitive to sample flowrate which must thus be kept constant; a considerable increase in sensitivity occurs above 900 ml/min and even minor fluctuations lead to instability in output. Flowrate is typically set between 100 and 500 ml/

min. Although water vapour does not interfere with the analysis, it will dilute oxygen and so gases should be dried before analysis. The cell is also inherently sensitive to total pressure.

Because the cell must be maintained at a stable high temperature, it is contained in a temperature-regulated oven typically at 850 °C. Despite this, the instrument is small: it is about 15 cm long and, including an outer Bakelite insulating tube, is about 5 cm in diameter.

The fuel cell is accurate when checked against a Scholander analyser (Weil *et al.*, 1967; Sodal *et al.*, 1968) or a paramagnetic oxygen analyser (Torda & Grant, 1972). It has a very fast response with a time constant (i.e. the time taken to reach about 63 % of its full scale value) of 10–25 ms at a flowrate of 100–500 ml/min.

The output stability and response time of the analyser deteriorate after several weeks or months of intermittent use; however the cell can be restored by heating the cell to 1000 °C for a few hours. The need for a high operating temperature is however an important disadvantage; and if combustible gases are present, they will burn at this high temperature and reduce the oxygen concentration. The fuel cells are more fragile than other oxygen analysers.

Whilst the fuel cell has been used in indirect calorimetry (Poole & Maskill, 1975) and for monitoring respiratory gases, its main use has been industrial where its wide range and resistance to high temperature have made it particularly suitable for monitoring oxygen concentration in furnaces.

6.2.2g   Gas chromatography

Gas chromatography has gained popularity in recent years as a quite cheap, rapid and accurate method for analysis of respiratory gas samples (Consolazio *et al.*, 1963).

Its application to the measurement of respiratory gases is described in detail by Hamilton & Kory (1960). Helium, which is used as the carrier gas, is passed over two reference thermistors and through a sampling valve into which the sample is injected. The mixture is dried and then passed into the first separation column which is composed of an inert stationary phase coated with di-2-ethyl hexyl sebacate. Carbon dioxide, which is a polar compound, travels more slowly than oxygen and nitrogen and the first detecting thermistor measures two peaks, an early carbon dioxide-free composite and a later carbon dioxide peak. The gas stream continues into a second column containing a zeolite molecular sieve; the rate of flow of a molecule through this depends on its size and shape. Carbon dioxide is irreversibly retained on the column but oxygen is rapidly eluated followed by nitrogen. The gases flow over a second detecting thermistor and out to atmosphere. The four thermistors make up a thermal conductivity cell which detects changes in the thermal conductivity of the gas flowing over the detector thermistor relative to that of helium. The change in thermal conductivity alters the temperature of the detector thermistor and its electrical resistance. This affects the balance of a Wheatstone bridge and produces a change in voltage which can be recorded on a chart recorder.

The output is represented on the chart by a series of peaks appearing at various times after injection of the sample; a complete analysis can be achieved in five to ten minutes. The concentration of a component of the sample is determined by measuring its peak height or area and comparing it to a calibration chart produced by introducing several gas mixtures of known composition into the analyser. Argon is not separated from oxygen by the molecular sieve so that a correction factor must be applied.

Not all gas chromatography instruments have this dual column system and in some instruments the gases are usually analysed in two stages with a change of column between each. A single concentric column, which is now available (Alltech, USA), eliminates this problem since in a single run separation of oxygen, carbon dioxide, nitrogen and methane is achieved. The column consists of a central core to separate polar from non-polar molecules and an outer ring composed of a molecular sieve separating oxygen, nitrogen and methane.

The output from gas chromatography systems appears to be linear over the range 0–10% carbon dioxide and 0–21% oxygen (Dressler, Mastio & Allbritten, 1960; Hamilton & Kory, 1960). By using a zero suppression circuit, Hamilton & Kory (1960) were able to achieve high accuracy and precision of measurement as judged against Scholander analysis.

6.3   Instruments for the measurement of volume and flow

Measurement of energy expenditure inevitably requires measurement of flowrate and this can be achieved either by the measurement of the volume of gas expired over a period of time or by the integration of a continuous measurement of rate of flow. The measurement of total volume, such as in the Douglas-bag method, is the easiest method. Whilst it is used primarily for short studies it can be equally applied to 24 h measurements when the investigator is interested only in the total energy expenditure.

For long-term studies when measurement of variation in energy expenditure within the time period is required, continuous measurement of flowrate is unavoidable. This requires an instrument fitted with an electrical analogue output to supply a suitable flowrate signal to a computer or other recording device. This does not necessarily exclude those instruments such as wet and dry gasmeters and rotameters which are not usually supplied commercially with an electrical output of sufficient accuracy, since it is possible to fit electro-optical devices to count the number of revolutions or measure the height of the float as appropriate. Nevertheless the use of mass flowmeters with analogue outputs already fitted has obvious advantages. It is not always appreciated that if the meter includes a digital panel meter display this will be a source of a digital signal (a DPM is effectively an analogue to digital converter).

The effects of temperature, pressure and humidity on the measurement of gas volumes and flowrates must be considered from two standpoints. Firstly there is the effect on the volume of the gas being measured and secondly the effect on the measuring device itself. The output of volumetric flowmeters is not directly affected by the condition of the gas, but the observed volume or flow must be corrected to STPD (see Chapter 4, Section 4.5 and equation (4.1)). Measurement of mass flow is also normally independent of gas condition and the output of these instruments is often expressed in terms of the volume flow at some fixed temperature and pressure condition (typically 20 °C and 760 mm Hg). No further correction should be needed except that allowance may have to be made for any water vapour present, which affects the measurement not only by dilution of the dry gas volume but also through its effect on the specific heat capacity of the gas. The output of variable area flowmeters is also usually expressed as volume flow at some specific condition; but the calibration factor of the instrument itself is inversely proportional to gas density. For these instruments correction to STPD is more complex.

The following section will consider the main instruments used in calorimetry for volume and flowrate measurement. Measurement of flowrate is particularly difficult at rates greater than about 200 l/min, not through lack of instrumentation but because of the problem of calibration; this is something we will return to in Chapter 7.

There are several reviews which collectively give a detailed account of the main flowmeter designs (Nelson, 1971; Ower & Pankhurst, 1977; Pritchard, Guy & Connor, 1977; Brain & Scott, 1982; Baker & Pouchot, 1983a, b).

6.3.1 Measurement of volume
6.3.1a Wet gasmeters
The wet gasmeter has been described in detail by Hyde & Mills (1932); its principle features are shown in Fig. 6.15a. It comprises a

gastight cylindrical steel case containing a drum that is divided into four chambers of equal volume. The drum is concentrically mounted on a shaft. One end of the shaft is supported in a bearing in the back of the case whilst its other end passes through the case by means of a stuffing box and is connected to a pointer. The case is filled with water, or a light oil of low viscosity (to British Standard 148/1959), to a preset level slightly above that of the shaft, thus submerging a large part of the drum. A glass sight box is mounted on the side of the meter to show the water level (strictly speaking of that in the outer casing). The drum is free to rotate concentrically about the shaft with little friction, and resistance to its movement through the water is minimised by setting the vanes separating the chambers at an angle like the blades of a ship's propeller. Rotation of the shaft moves the pointer round a volumetric scale mounted on the front face of the meter.

A dome-shaped hollow hood, called the apron, covers one end of the drum; it has a circular opening surrounding the shaft and through this

Fig. 6.15. The main features of (*a*) the wet gasmeter and (*b*) the Hyde sight box. The dotted circles show the position of the Inlet (I) and Outlet (O) connections on the rear of the meter. The inner scale gives the flowrate over a timed 1-min period, and the outer scale gives the volume in litres; the actual scale values depend on the capacity of the meter.

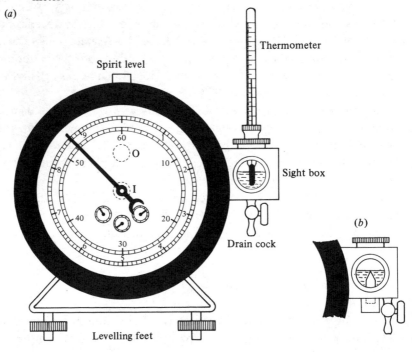

passes an inlet tube which emerges inside, above the water level. The operation of the meter is seen in Fig. 6.16; the pressure of the gas entering the partially submerged chamber A forces it clockwise until it reaches the position occupied by chamber B. In this position chamber A is completely full and yet is still sealed by the water. With further rotation it reaches the position occupied by chamber C in the diagram, and as it descends once more into the water its content of gas is forced into the space in the outer casing and exits through the outlet tube.

The volume of gas held by the chamber depends on its physical dimensions and on the height of the water level. Whilst the chambers are manufactured to specific dimensions, accurate volumetric calibration is carried out once the meter is assembled by adjusting the water level. Calibration of gas meters by the manufacturer is usually by means of a Harcourt meter proving bottle with a water jacket for the smaller gas meters (see Hyde & Mills, 1932), and for those of 5 dm³ capacity or greater a cubic foot bottle is employed. As their name implies both of these are glass bottles graduated with two marks to give a precisely known volume; air is displaced by means of water from the bottles into the meter. Once the water level in the meter has been adjusted so that the correct volume is registered, a graduation is made on the glass face of the sight box. Thereafter when the wet gasmeter is used it must always be filled to the graduation to give a correct reading.

Some wet gasmeters operate on the Hyde principle. In these the drum is reversed so that the apron backs on to the front of the meter. The gas inlets and outlets are also reversed so that gas enters through the 'outlet'

Fig. 6.16. The operation of the wet gasmeter; note the differences in liquid level in the meter.

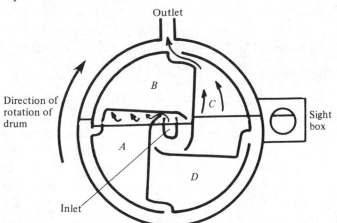

of the previous design and out through the central 'inlet'. Hyde meters are slightly more accurate than the standard gas meter and the water level in the sight box truly reflects the level of water in the chambers (in the standard meter the water levels in the chamber and gas space are not the same – see Fig. 6.15). The Hyde meter also uses a stainless steel indicating point in the sight box rather than an engraved line (Fig. 6.15(*b*)).

Water temperature is measured by a thermometer mounted in the sight box or the casing. Before any measurements are made, gas is allowed to pass through the meter for some time to ensure that the water temperature is in equilibrium with that of the gas and the water is saturated with the gas. The pressure of the gas in the chamber is equal to the driving pressure, which is conveniently measured by a water manometer on the inlet. If the meter is filled with water the gas is assumed to be saturated at the appropriate temperature; if dry gas is to be measured, it may be more convenient to use an oil-filled meter and eliminate any need to correct the volume for water vapour.

In the United Kingdom wet gasmeters manufactured by Alexander Wright are available in six models to cover the range of 1–200 l/min. The bulk of the meters is a significant limitation to their maximum practicable capacity. The measuring capacity of each model is restricted by the speed of rotation of the drum; typically there is a great loss of accuracy if the drum rotates at more than 1–4 revolutions per minute (the actual value depends on the model). In our experience the claim by the manufacturer of an accuracy of better than 0.5 % can be achieved provided that the meter is used in accordance with their instructions.

Whilst wet gasmeters can be used for the continuous measurement of flowrate, the lack of an analogue output is a major disadvantage. Continuous use will lead to evaporation of water from the meter and a gradual loss of accuracy and therefore oil-filled meters are better in these circumstances. In practice wet gasmeters are most frequently used as secondary calibration standards (see Chapter 7).

6.3.1*b*   *Dry gasmeters*

The dry gasmeter has been described in detail by Adams *et al.* (1967). Its typical construction is illustrated in Fig. 6.17. The meter is divided into two vertical parts by a horizontal valve plate: the upper section is called the attic; and the lower section is the body. An inlet delivers gas into the attic. The body is itself divided into two sections by a central division plate. In each of the two bottom sections is a sheepskin bellows, one end of which is attached to the division plate and the other to a metal disc. The valve plate holds two sets of valves, each corresponding to one of the lower chambers. In each set there are three valves arranged in line: the innermost valve communicates with the inside of the bellows; the outermost leads to the chamber itself; and the centre valve leads to an outlet duct. The outlet ducts from each side fuse to form a common outlet. Over each set of valves is a valve cover composed of a central dome and two flat lateral wings. The cover moves to-and-fro over the valves, connecting in turn the outlet port with either the inlet to the bellows or to the chamber; simultaneously the air enters from the attic through the uncovered valve into the opposite bellows or chamber. The operation of the two sets of valves is linked to allow a smooth flow of air through the meter.

A guide rod is attached to the metal bellows disc to ensure an even movement of the bellows; in addition a hinged plate, the flag, is connected to each disc and converts any to-and-fro movement to a rotation of a vertical flag rod. The rod passes from the body through a gastight stuffing

Fig. 6.17. The structure of the dry gasmeter. (After Adams *et al.*, 1967.)

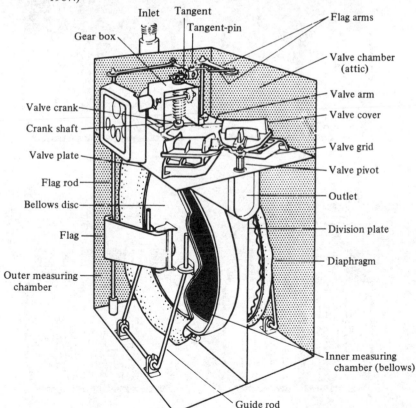

box or grease seal to the attic, where the rotation is transmitted to a swivelling metal bar called the flag arm. The flag arms from both lower chambers are connected to a lever (the tangent) fixed to a vertical crankshaft; the connection is through an adjustable tangent pin. The movement of the two flag arms rotates the tangent and this causes the crank to rotate and drive a pointer and dial counter. The pointer rotates around a volume scale mounted on the front of the meter. The crank is also

Fig. 6.18. The sequence of events in the operation of the dry gasmeter. The sequence of events is: (*a*) chamber 1 is emptying, 2 is filling, 3 is empty and 4 has just filled; (*b*) chamber 1 is now empty, 2 is full, 3 is filling and 4 is emptying; (*c*) chamber 1 is filling, 2 is emptying, 3 has filled and 4 has emptied; (*d*) chamber 1 is completely filled, 2 is empty, 3 is emptying and 4 is filling. (Reproduced from Pritchard *et al.*, 1977.)

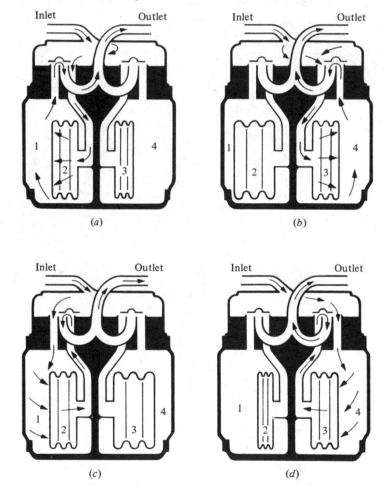

connected to swivelling valve arms which bring about the movement of the valve covers.

Air is pushed or drawn through the meter by external force and activates the sequence of events seen in Fig. 6.18. These events can be envisaged as occurring in four stages which are combined in such a way to provide continuous measurement of volume. Measurement of temperature and pressure of the gas passing through the meter can be made by placing a thermometer in the outlet and connecting a water manometer to the inlet.

Dry gasmeters have been widely used as stationary gasmeters (for example to measure the volume of gas in Douglas bags) and to continuously measure gas volumes in flow-through systems. They are much lighter than wet gasmeters but are less accurate (typically no better than $\pm 1\%$). The meters can be calibrated by the methods used for wet gasmeters but according to Adams *et al.* (1967) the accuracy achieved is low. The meters are much more susceptible to leaks than are wet gasmeters and the elasticity of the bellows may decrease with age; this results in an increase in the pressure of gas in the bellows so that the capacity of the bellows (at STPD) will be underestimated. Wear in the linkages and in the gears driving the volume registering mechanism also leads to inaccuracy.

The main source of dry gasmeters in the United Kingdom is now United Gas Industries Ltd (London), who make a range of meters operating at up to about 200 l/min or 20 l per revolution of the dial. A new dry gasmeter with a digital display has been introduced by Harvard; it must be used with some caution since the display can continue to cumulate in response to a to-and-fro movement of gas, with no net flow.

6.3.1c *Wright respirometer*

This is a miniature turbine meter designed by Wright (1955) and based on the rotary inferential anemometer. It has been described and evaluated by Byles (1960) and by Nunn & Ezi-Ashi (1962). The air enters the meter through a side arm and passes through a series of tangential slots into a cylindrical chamber. Before being expelled axially the flow of air rotates a flat Terylene rotor vane mounted in the chamber. The total internal volume of the system is only about 22 ml. The rotor is supported by jewelled bearings and its rotation is used to drive a gear train and a dial similar to that found in analogue watches. The dial directly indicates the cumulative volume flow.

Whilst it has had wide application in medicine, the meter is grossly inaccurate, being in error by up to -50% at low flowrates and up to $+10\%$ at high flowrates, and is unsuitable for calorimetry.

6.3.2   *Volume flowrate*

6.3.2a   *Turbine flowmeters*

The most common turbine flowmeter used in calorimetry is based on the Wright respirometer described above. It was developed by Cox *et al.* (1974) and has found use in the Oxylog and Miser portable systems described in Chapter 4. It consists of the basic Wright flow sensor of a two-bladed vane mounted on a rotor. The gear train and watch dial have been removed from the rotor and replaced with a disc divided into alternate clear and opaque sections. The disc is in the line of a beam of light arranged to fall on a photocell. As the disc rotates its opaque portion interrupts the light beam. The photocell transducer thus generates electrical signals at a rate directly proportional to the speed of rotation of the rotor which in turn is proportional to the flowrate of gas through it. Elimination of the friction of the gear train is claimed to overcome much of the inaccuracy of the original Wright respirometer.

The output from the respirometer is linear but its accuracy is somewhat variable. Cox *et al.* (1974) found that it constantly underread when compared with a volume measured by a Douglas bag and this could lead to a substantial error from about 30 % at 5 l/min to about 5 % at 30 l or more, whilst Eley *et al.* (1978) found that it overread by an average of about 14 %. Ballal & Macdonald (1982) reported agreement of about 99 % between the respirometer and Douglas-bag measurements over the range 15–60 l/min.

6.3.2b   *Pneumotachograph*

The pneumotachograph was first described by Fleisch in 1925. The instrument depends on Poiseuilles law which states that for a laminar flow of gas, the flow passing a given cross-section per unit time is inversely proportional to viscosity and directly proportional to the drop in pressure per unit length of the vessel and to the fourth power of the radius. Since the viscosity of the gas, the length and radius of the tube can be regarded as constant the relationship can be simplified to one analogous to Ohm's Law in which flow is equal to the pressure drop divided by resistance, which in this case is constant. Thus the pressure drop is directly proportional to flowrate.

Fleisch used a cluster of narrow parallel tubes packed within a cylinder to create the resistance and a laminar flow. More recently other resistances such as mesh screens or vertical plates have also been used successfully (Fry *et al.*, 1957). Resistance heads of different sizes are available to cover a wide range of flowrates. Exceeding the flowrate for which the meter head was designed usually introduces non-linearity. The pressure drop across the resistance is measured by means of a differential manometer or pressure transducer. The instrument is fast responding and can measure the instantaneous flowrate throughout the respiratory cycle (which was its original purpose).

The use of a differential pressure transducer allows the flowrate to be continuously recorded or integrated to provide a volume measurement (Hill, 1959; Pols, 1962; Lunn, Molyneux & Pask, 1965). It is thus suitable to measure either the volume of expired air or, in flow-through systems, flowrate.

Providing the flow is laminar then theoretically the output of the instrument should be linear. Whilst some investigators have confirmed the linear nature of the instrument (Fry *et al.*, 1957; Lunn *et al.*, 1965), others have noted some degree of non-linearity (Finucane, Egan & Dawson, 1972), particularly if the geometry of the instrument induces turbulent flow. The gradual deposition of respiratory moisture or saliva in the resistance may cause non-linearity through induction of turbulence, although the meter head is usually heated to prevent condensation (Grenvik, Hedstrand & Sjogren, 1966). Our own experience is that pneumotachographs require considerable attention from the operator if they are to be reliable. The instrument should always be carefully calibrated under the conditions of its intended use. Under ideal conditions the output is accurate to within 0.5 % (Smith, 1963).

6.3.2c   Orifice flowmeters

These are similar in principle to the pneumotachograph comprising a flow constriction such as a capillary tube and a means of measuring the pressure drop across the tube (Nelson, 1971). Since air flow through such a restriction is turbulent, the relationship between the pressure drop and flowrate is not linear. However, the use of sintered plates or other porous materials as the restriction can produce laminar flow and a linear relationship then exists (Grootenhuis, 1949; Chase, 1964). They are very susceptible to clogging and the gas samples must be scrupulously clean.

6.3.3   Mass flowrate

6.3.3a   Critical flowmeters

These are discussed more fully in Chapter 7 when we consider their application to the calibration of flowmeters. They have been used however as flow controllers in flow-through calorimeters by several investigators (Chasteau & Cronje, 1974; Gray & McCracken, 1976).

6.3.3b   *Thermal flowmeters*

These are the most widely used of the mass flowmeters in calorimetry. There are excellent accounts of the underlying principles of the different types of instruments by Bradshaw (1968) and by Ower & Pankhurst (1977).

Hot-wire anemometers. The principles of hot-wire anemometry have been reviewed by Comte-Bellot (1976). As their name suggests these meters comprise a heated wire which is placed in the air flow. The rate of cooling and thus the temperature of the wire will vary with the mass flow of air across it. The relationship between the rate of cooling and flowrate is expressed by King's Law (King, 1914) which states that the rate of cooling is inversely proportional to flowrate (according to Collis & Williams (1959), flowrate to the power 0.45 is more correct).

The temperature of the wire will affect its resistance which then becomes a measure of flowrate. The resistance is determined by incorporating the wire into a Wheatstone-bridge circuit and measuring the current across the unbalanced bridge. Alternatively, the current through the wire can be adjusted to maintain its temperature constant despite changes in air flow; the heating current is measured by determining the potential difference across a fixed resistance in the appropriate arm of the bridge by means of a potentiometer. A shielded cold reference wire is usually included in the bridge to provide compensation for any changes in the resistance of the sensor caused by fluctuations of air flow temperature. The output is inherently non-linear but can be linearised electronically (Bradshaw, 1968). The use of hot-wire anemometers in respiratory studies has been described by Godal *et al.* (1976) and by Yoshida, Shimada & Tanaka (1979).

The major disadvantage of the simple hot-wire system is that the sensor sits directly in the air flow and is subject to drift caused by the deposition of dust and other contaminants on its surface. In addition the wire is so thin that its resistance can alter due to mechanical strain caused by the air flow. Thus the stability of calibration is poor.

This disadvantage has been largely overcome by the Shielded Hot-wire anemometer (Simmons, 1949). The hot wire is placed inside one bore of a twin-bored silica tube. Because this reduces the sensitivity of the wire to air flow, it is combined with the greater temperature sensitivity of a thermocouple. The nichrome–constantan 'hot' junction of the thermocouple is placed in the other bore. The two free ends of the thermocouple form the 'cold' junction and are attached to a pair of copper wires which support the silica tube. The thermocouple produces a voltage which is a measure of the difference in the temperature of the hot junction, which is heated by the hot wire and cooled by the air flow, and the cold junction which is at the temperature of the air flow. Inevitably the instrument is slow in response.

Although the output of the thermocouple is linear with respect to temperature, it is inversely proportional to flowrate because of King's Law. Thermistors can also be used as temperature sensors. The instruments are suitable for use over a wide range of flowrates with an accuracy of $\pm 1\%$ and are independent of temperature and pressure. Although they measure mass flow their output sensitivity is usually expressed in terms of volume units adjusted to STP.

Heated-tube mass flowmeters. The earliest design was that of Thomas (1911). It comprised a central heating element placed in the airflow with temperature sensors placed equidistant either side. With zero flow the temperatures registered by both sensors are equal but when air flows through, the heat is transferred entirely towards the upstream sensor by convection. The basic design was improved by the introduction of the boundary layer flowmeter which comprised a heating element wrapped around the outside of the tube, again with temperature sensors upstream and downstream. The mass flow is given by:

$$\dot{m} = \frac{\dot{Q}}{c_p \Delta T}$$

where $\dot{m}$ = mass flow per second; $\dot{Q}$ = heat supplied per second; c_p = specific heat capacity of air at constant pressure; ΔT = difference in temperature of the two sensors.

We have had considerable experience of the Hastings–Raydist flowmeters which are based on this principle. They cover a range of flowrates up to 1500 l/min. The sensor tube is about 1 mm in diameter and has a maximum linear flowrate of about 100 ml/min; meters for flowrates greater than this incorporate a bypass which includes a restriction to produce a flow through the sensor which is a constant proportion of the main flow. The restriction is either a parallel series of tubes identical in length and diameter to the sensor tube or an orifice.

The meters have a more uniform temperature gradient along the sensor tube than the Thomas or boundary layer meters and this is achieved by heating two heat sinks, one at each end of the tube. Two thermocouples are attached to the walls of the tube, each an equal distance from a heat sink. As in the Thomas and boundary layer flowmeters, in the absence of flow the same temperature is registered by the thermocouples. With the introduction of a flow of air, heat is lost from the downstream wall and transferred upstream, and this is reflected in the output from the two thermocouples. A measurement is made of the difference in voltage from the two thermocouples and this is inversely proportional to mass flow. A converter gives a final output voltage which is the reciprocal of the difference in the voltages of the two thermocouples, and thus the output voltage is directly proportional to mass flow.

We have found the meters very reliable, accurate, linear and stable (see Fig. 7.14). The 0–5 V analogue output makes them ideal for continuous recording.

M. K. S. Instruments (Burlington, Mass.) have produced a range of fast responding meters to measure flowrates up to 50 l/min. They comprise three heaters, mounted in series, to produce a known temperature profile along the sensor tube. All three heaters are maintained at a constant temperature by means of individual auto-balancing bridge circuits. As a measure of mass flowrate the meters use the voltage or power which has to be applied to the heaters to maintain the temperature profile. From the equation above it can be seen that the amount of heat introduced to maintain a fixed temperature gradient will be directly proportional to mass flowrate and the meter will have a linear characteristic. The maker claims an accuracy of 0.5% of full scale and a repeatability of $\pm 0.2\%$ of full-scale.

Stream temperature rise flowmeters are in our view the best means of obtaining an accurate continuous measurement of flowrate for flow-through calorimeters and their cost is not excessive.

6.3.4   Variable area flowmeters

Variable area flowmeters or rotameters have been widely used for the measurement of flowrate. A rotameter consists of a vertically mounted graduated tube, usually of glass, with a bore which increases in diameter from the bottom to the top. Inside the tube is a free piston called the float which indicates flowrate; there are a variety of float designs (Nelson, 1971).

When gas flows into the bottom of the tube, it causes the float to rise to a point at which the weight of the float is just balanced by the upward force generated by the gas stream. For an equilibrium to be achieved the upward force on the float, which is proportional to the square of the velocity of gas passing through the annulus around the float, must remain constant. To keep this velocity constant, despite changes in the flow of gas into the tube, the area of the annulus must be altered appropriately by a vertical movement of the float. Thus the level of the float is proportional to the flowrate of gas entering the tube. The taper in the tube is usually such that the annulus area is proportional to height and the flow can be read directly from a vertical scale marked on the tube. No matter what design of float is used the reading is taken level with its widest part.

Rotameters can be made in varying sizes covering a range extending from a few millilitres to several thousand litres of air per minute, but individual tubes usually operate over a 1–10 range ratio. They are simple to install and operate. If the scale on the rotameter is used, accuracy is typically between $+2\%$ and $+5\%$ (Nelson, 1972; Ower & Pankhurst, 1977; Brain & Scott, 1982) but this can be substantially improved by careful calibration. Calibration charts produced by manufacturers however may not always be accurate (Barnes & Birdi, 1977).

A repeatability of $\pm0.5\%$ over a flow range limited to about 2% of full scale, combined with an analogue output, has been achieved by fitting an electro-optical sensor to the tube (McLean & Davidson, 1978). The device consists of a light shining through the rotameter tube onto a vertical bank of ten miniature photoelectric cells. The height of the float determines the number of cells illuminated and hence the output voltage. With such a sensitive detector the output voltage can be carefully calibrated against accurately known flowrates. Commercial systems of lower accuracy but providing analogue output over the full range of the tube are also available.

For any point on the rotameter scale the mass flowrate varies in proportion to the square root of gas density; hence the volumetric flow varies inversely in proportion to the square root of density. If the rotameter is calibrated volumetrically in terms of the flowrate ($\dot{V}_{cal}$) of dry air at STP (with density ρ_o), and it is used to measure the volumetric flow of moist air (V_m with density ρ_m), then

$$\dot{V}_m = \dot{V}_{cal} \sqrt{(\rho_o/\rho_m)} \tag{6.1}$$

But the measured flowrate will have to be corrected to the flow ($\dot{V}_o$) at STPD

$$\dot{V}_o = \dot{V}_m \times \frac{P-P_w}{760} \times \frac{273}{T+273} \tag{6.2}$$

The value of ρ_m may be calculated (Weast, 1979) from

$$\rho_m = \rho_o \times \frac{273}{T+273} \times \frac{P-0.38P_w}{760} \tag{6.3}$$

where P_w is vapour pressure of moisture in the air. Combining equations (6.1), (6.2) and (6.3) gives

$$\dot{V}_o = \dot{V}_{cal} \sqrt{\Bigg/\left(\frac{273}{T+273} \times \frac{(P-P_w)^2}{760(P-0.38P_w)}\right)} \tag{6.4}$$

It should be noted that when the flow measured is of a dry airstream then $P_w = 0$, and

$$\dot{V}_o = \dot{V}_{cal} \sqrt{\left/\left(\frac{273}{T+273} \times \frac{P}{760}\right)\right.}$$

i.e. the correction factor is the square root of the normal STP correction factor.

A further complication is that manufacturers' calibrations are often expressed in terms of 'standard cubic feet per minute' (a term used in fluid engineering to refer to the condition 14.7 p.s.i. and 70 °F) which is higher than flow at STP by the factor 1.038. Rotameters to be used for calorimetry should always be calibrated by the user, so this factor is of small account, but it may help to explain apparent discrepancies found when calibrating.

6.4　　Recorders and data loggers

Chart recorders, although now largely superseded by automated data-acquisition systems, provide a permanent visual record which serves as a useful 'see-at-a-glance' indication of the continuing satisfactory operation of a calorimeter. This facility is particularly desirable for measurements with uncooperative animals as subjects, who like to pass the time in eating or otherwise destroying anything available, and especially measuring sensors! However, most modern calorimeters are now equipped with systems that automatically log the data at predetermined intervals and frequently do processing calculations 'on-line'.

The heart of these instruments is one or more analogue-to-digital (A–D) converters which may or may not incorporate digital display units making them into digital voltmeters (DVM). The precision, that is the degree to which an A–D converter can discriminate between different levels of input signal, is limited by the maximum number of binary counts which it outputs at full scale of the measuring range. For example an A–D converter with '8-bit' discrimination effectively divides the operating range into 2^8 or 256 parts; it can never distinguish between input-signal levels differing by less than $\frac{1}{256}$ of the full-scale range. Bradshaw (1984) has described the various types of A–D converter currently in use and compared their performance capabilities in terms of accuracy, speed of response and price. The accuracy of recording any physical measurement by means of a transducer or other instrument coupled with a data-acquisition system is determined by

(1) the accuracy of the transducer or instrument in producing a voltage analogue of the quantity being measured;

(2) the accuracy and precision of the A–D converter, including its effectiveness in disregarding spurious a.c. voltages picked up by the signal leads.

(3) the accuracy of any amplifier used to match the transducer output-signal range to the input range of the A–D converter.

There are two possible approaches to the selection of a data-acquisition system. One is to employ a separate A–D converter for each analogue signal being measured; each converter and its associated amplifier being only of sufficient accuracy for that particular measurement. The other is to employ a single high-quality A–D converter and to switch the measuring devices to it sequentially by means of a multiplexing system. The choice is largely determined by whether the number of quantities to be measured is big enough to justify the cost of a multiplexing system.

Some inexpensive computers, even some of those sold primarily as home computers, have a limited number of A–D converters incorporated into them. These are not usually of high quality and tend to produce 'noisy' output signals; however there is often time available between measurement scans to accumulate the sum of many individual readings from each channel and for the computer to form an average which is relatively free from 'noise'. Provided that suitable operational amplifiers can be made to interface the transducers to A–D converters, either built-in or incorporated with the amplifier circuits, these 'home computers' can usually be adapted to monitor a small number of signals with at least 1 % accuracy. They also usually have some digital channels which can be used to 'read' the condition of on/off switch contacts or to operate external switches. But if greater accuracy is required or if very many analogue or digital channels are required then it is preferable to employ data-logging system based on a high quality A–D converter and multiplexing system.

Once the basic data have been fed into the computer in digital form, further calculations on the data can be made with great speed and pre-cision. The practical limit to the overall accuracy of the final results is set not so much by the capacity of the computer as by the accuracy with which the basic data are fed in. It is particularly important to make full use of the precision of the logging device by matching the amplitude of the input signals to it. Where the quantity of interest is a small difference between two relatively large analogue signals, as in oxygen concentration differences, every effort should be made to log the difference itself rather than the two individual quantities. This is discussed more fully in Chapter 7.

Computer-controlled data loggers may also be used to detect mal-functions. Any of the basic data read in or the calculated results can be checked routinely against preset limiting values and an audible or visual warning device operated when the limits are exceeded.

6.5   Other instrumentation

In long-term calorimetric observations it is desirable, so far as is possible, to avoid interference with the measurements by opening doors or entry of operatives. With the cooperation of humans as experimental subjects or patients there are no great problems in passing objects and materials in and out of the chamber through suitably designed airlock hatches. With animals, however, especially herbivores which consume large quantities of food and water and produce much excreta, considerable problems arise.

6.5.1   *Food supply*

We have found in our flow-through calorimeters for rats at Leeds that there are potentially substantial errors when food or energy intake is estimated from the weight difference between food offered and food refused. The moisture content of 'dry' diet is in equilibrium with atmospheric moisture. Since the humidity of the respiration chambers is usually higher than that of room air, the moisture content and thus weight of diet placed inside the chambers will increase somewhat. With rats at least, the amount of food placed in the chamber is typically several times what the animals will eat (because of the possibility of spillage), and so a small change in the moisture content will produce a weight change which is large in relation to the amount of food eaten. There is also a problem with moist diet which will lose moisture by evaporation in the air flow with consequent danger of overestimation of intake. Whilst we overcome these problems by measuring the change in moisture content, and by adding uneaten residues to the excreta and measuring the total energy in and the total energy out of the chamber (Armitage *et al.*, 1983) this may not always be appropriate particularly for large animals and special feeding equipment may be necessary.

A limited number of feeds may be stored in closed bins for making available to the animal at appropriate times. Such bins must be well sealed and of robust construction so that the animal is not tempted by delicious aromas to break into the food supply prematurely. It is preferable to store food high up in the chamber so that it can be released when required to fall into a feeding trough.

Water lying in open drinking bowls, apart from being a source of exchange between evaporative and non-evaporative heat, acts as an irresistible plaything for a bored animal. At the Hannah Institute we have found it necessary to make the supply of water available to drinking bowls only at regular times and for limited periods (usually for 30 min after each feed). Any water not drunk after the allowed period is pumped out of the bowl until the next time. Drinking water is always pumped into the chamber at calorimeter temperature to avoid heat exchange. The supply of both food and water and the removal of water not drunk are all automatically controlled by pumps and solenoids operated via the computer.

6.5.2   Collection of excreta

Collection of excreta from animals in respiration chambers can be effected by allowing the materials to fall into trays placed below a cage or stall with a mesh or slatted floor; but in direct calorimeters a large tray of water is unacceptable and the materials must be collected in closed containers. Many methods of achieving this involve collection bags hung from a harness worn by the animal. Collection bags often have to be tailored to suit individual animals and the optimum design, which defies written description, is usually found only after many days of messy experimentation. The only system of fairly general application to be described here is one which we have used regularly at the Hannah Institute for collection of urine from male farm animals (Watts, 1976) which is illustrated in Fig. 6.19(*a*).

The urine is collected initially in a rubber funnel (Wainman & Paterson, 1963) which is suspended from the harness. The presence of urine in the funnel triggers a transistorised switch by allowing conduction between either of two pairs of electrodes (A and B). The switch operates a pump to remove the urine, and then returns to a deactivated mode.

Electrodes (A) consist of bolts with large rounded heads (15 mm diameter) situated in the neck of the funnel placed so as to avoid direct contact when the animal lies down. Initial detection occurs as soon as urine reaches the level of the highest of this pair of electrodes. Electrodes (B) consist of two short lengths of copper piping inserted into the aspirating-tube, connected electrically in parallel to the first pair and situated at a high point on the animal's flank. This arrangement ensures that the switch remains on until all the urine has passed the highest point in the system. Without this capability urine, on deactivation of the switch, returns to the funnel reactivating the pump, sometimes resulting in oscillation of urine in the tubing.

The switch consists of a high-gain pnp transistor biased with a fixed and a variable resistance together with the electrodes for urine detection (Fig. 6.19(*b*)). The system contains three resistors (R_1, R_2 and R_3); R_2 and R_3 are normally smaller than R_1, ensuring that the transistor base current (through R_3) is maximally sensitive to changes in detector resistance $R_{A,B}$. High sensitivity must be maintained since the inside walls of the funnel remain 'wet' long after the urine has been pumped away. The variable resistance R_1 ensures some flexibility in the system so that although a value of 2.5 $k\Omega$ for R_1 is quite adequate for normal circumstances, unusual feeding or drinking procedures which alter the concentration of electrolytes in the urine can be catered for. For exceptionally high levels of conductivity (high levels of electrolytes) R_1 should be increased, and the

Fig. 6.19. A system for collecting urine from male farm animals: (*a*) Illustration showing location of electrical contacts A and B; (*b*) control circuit (C, relay coil, 12 V 110 Ω; D, diode, ISJ50; T, transistor, BC477; R_1, 5 $k\Omega$; R_2, 1 $k\Omega$; R_3, 2.2 $k\Omega$). (Reproduced from Watts, 1976.)

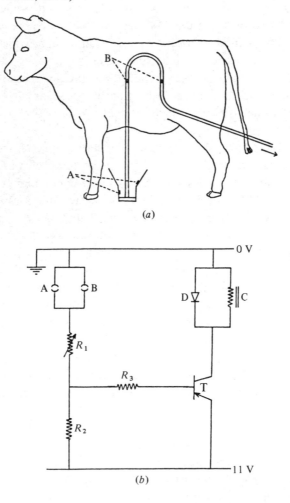

(*a*)

(*b*)

converse for low levels. The extension of low levels to detection of water is possible by reducing R_1 to zero.

For normal operation it has been found necessary to remove lengthy penile hair to prevent direct electrical contact with the animal. Loose hair and by-products of electrolysis must be cleaned from the system regularly (daily from the funnel). Water-repellent plastic tubing has been used to reduce the frequency of cleaning of the remaining apparatus; but it is still necessary to replace the electrodes (B) and the section of tubing between them after three or four weeks of continuous use.

6.5.3 *Safety aspects*

The two greatest hazards in a calorimeter are failure of the ventilation system and overheating. It is relatively simple to fit safety thermostats at appropriate points (e.g. water tanks, heater banks, air-supply duct, etc.) which trip the power supply to heaters if predetermined temperatures are reached.

Failure of the ventilation system can be very serious as animals may have the capability of consuming all the available oxygen within the volume of an unventilated chamber in a few hours. If animal chambers are to be left unattended overnight it is essential to install an emergency ventilation system or an emergency hatch in the roof of the chamber which is held closed only so long as the main ventilation system is operational. It is best to use a normally open micro-switch which is held closed only by the flow or pressure created by the main fan or blower; the current passed by this switch operates a solenoid latch which holds the hatch closed. If desired the emergency system may be operated through a delaying device so that it is not triggered by momentary power failures; but whatever is used, it must be a device which fails in the safe position.

The safety devices mentioned in this section should of course be in addition to the normal fuses or overload contactors of the electrical switchgear.

7

Calibrations and standards

7.1   Standardisation of basic measurements

Many instruments for basic physical measurements can be purchased with accuracy certified by the National Physical Laboratory in the UK, the National Bureau of Standards in the USA, or from equivalent authorities in some other countries. Instruments such as mercury-in-glass thermometers, mercury barometers, certified weights and resistance boxes should maintain their accuracy indefinitely; they may be used for regular calibrations of thermistors, thermocouple- or resistance-thermometers, manometers or pressure sensors, electronic balances, and resistors. Other electronic instruments require to be checked against a reliable standard at least once a year and often more frequently; this can be done by some suppliers who are approved by their national standards authority. A high-quality, multirange digital voltmeter, regularly serviced by an approved supplier, is now an essential piece of laboratory equipment for calorimetry, and has virtually replaced the old standard cell and potentiometer. It may be used as a secondary voltage standard for the calibration of potentiometric recorders as well as of a great many instruments which provide output in voltage analogue form. For direct calorimetry a high quality wattmeter is equally essential.

7.2   Calibration of gas analysers
7.2.1   *Introduction*

Accurate calibration of gas analysers provides one of the greatest challenges in calorimetry; it is an area whose importance is not always fully appreciated even amongst some workers engaged in calorimetry. The calibration methods used and the care with which calibration is carried out, are often inadequate to achieve the appropriate level of accuracy for their experiments.

238

Some workers question the need for high accuracy and careful calibration on the grounds that where there is high biological variability in heat production or loss, for example in many human studies, accurate measurement is pointless. This argument is tenable only for some applications, e.g. where the object is to measure the mean level of energy expenditure in a population (and calorimetry is unlikely to be the method of choice here). However, if an evaluation of the full energy balance of an individual subject, or the metabolic response to some applied experimental treatment is required, then heat production must be measured with the greatest possible accuracy and precision. Those who are tempted to save on the cost of equipment by accepting a lower precision of measurement should bear in mind that the maintenance of the same overall experimental precision needs an increase in the number of observations by the square of the decrease in precision of measurement: that is to say, if precision of measurement, expressed as the coefficient of variation, is decreased by half, the number of observations must be increased fourfold.

7.2.2 Sample conditioning

Several factors may lead to inaccuracy in the measurement of gas concentrations even when the analysers themselves have been carefully calibrated. It is important both in calibration and in subsequent measurements that gases entering the analysers are always conditioned to the same fixed levels of humidity, pressure, flowrate and temperature. Flowrate measurement will also be affected by moisture content, pressure and temperature but it is usually not difficult to adjust the flow measurements to STPD. We discuss below how these potential sources of error can be controlled.

Drying agents. A short list of some of the main drying agents used in calorimetry and their properties is given in Table 7.1; the efficiency of many of them has been given by Trusell & Diehl (1963). Some absorbents used for removing water vapour from respiratory gas samples can adsorb carbon dioxide to a limited degree. Silica gel is well known as having this property (Janac *et al.*, 1971) and should be avoided as a drying agent when the carbon dioxide concentration is likely to fluctuate. It can also occur with two other common desiccants, magnesium perchlorate (Samish, 1978) and Drierite – anhydrous calcium sulphate (Koller & Samish, 1964).

We have also experienced adsorption and desorption of carbon dioxide from newly obtained supplies of Drierite (anhydrous calcium sulphate). This even manifested itself as a change in the output of an oxygen analyser some minutes after a new gas sample was passed to it. When the gas was

Table 7.1. *Characteristics of several commonly used drying agents*

Desiccant	Initial Composition	Regeneration	Average efficiency (mg/l)	Relative capacity (l)
Drierite	$CaSO_4$	1–2 h at 200–225 °C	0.067	232
Silica gel		12 h at 118–127 °C	0.070	317
Anhydrous calcium chloride	$CaCl_2.O.18H_2O$	16 h at 127 °C	0.067	33
Anhydrous magnesium perchlorate	$Mg(ClO_4)_2.O.12H_2O$	48 h at 245 °C	0.0002	1168

Average efficiency – the amount of water remaining in a stream of humidified nitrogen, flowing at 225 ml/min, after drying to equilibrium by the stated desiccant.
Relative capacity – The average maximum volume of nitrogen which can be dried at the specified efficiency for a given volume of desiccant.

(After Nelson, 1971.)

first passed through the Drierite, carbon dioxide was either removed from, or added to the sample stream until the drying agent reached a new equilibrium with the carbon dioxide in the gas. During this period the change in the apparent oxygen concentration from that of the previously analysed gas sample appeared to overshoot its eventual new level. The effect appears to occur only with some samples of Drierite and it can to some extent be overcome by deliberately moistening a new supply of the material and oven-drying it again.

Permapure dryers eliminate this problem entirely; they have a central tube, with a wall made of a hygroscopic substance, enclosed within an outer tube. The 'wet' sample gas passes through the inner tube, whilst on the outside of the inner tube a dry gas, or a gas with a low water vapour pressure, flows in the opposite direction. Water vapour, but not the respiratory gases, diffuses through the wall out of the sample gas and into the drying gas. With appropriate rates of flow of sample and drying gases, the sample gas leaves the tube at the same water vapour pressure as the drying gas entered. The low water vapour of the recipient gas can be created by drying it (even previously dried sample gas can be used), or by reducing its total pressure by introducing it into the outer tube under partial vacuum. Although initially expensive the cost of these dryers may be recovered over a longer period by eliminating the need for frequent purchase of drying agents which are themselves costly. Such dryers have been used by Schoffelen *et al.* (1984) in their calorimeter.

Cooling the gas sample so that water condenses out is also used as a drying process. Although the gas may be saturated at a low temperature, the amount of water vapour it contains will be low; the sample is then rewarmed before it enters the analysers. Calibration gases will often be dryer than sample gas under these conditions so that they must also be saturated by passing over cold water, and then rewarmed and introduced into the analysers. Provided that water cannot condense out in the analysis line or in the analysers (and these are usually heated to prevent this), it is not essential that the samples be absolutely dry, only that they contain a constant level of water vapour and that this be the same for different calibration and sample gases. The use of temperatures below freezing point (0 °C) to cool the samples may lead to difficulties as condensing water freezes and blocks the gas lines. A temperature just above 0 °C is perfectly satisfactory; Auchincloss, Gilbert & Baule (1970) have described a system using a temperature of 15 °C. It is however important to remember that gases such as carbon dioxide will dissolve in the condensed water and that the solubility of gases increases as the water temperature decreases.

Regulation of gas analyser inlet pressure. Since many of the gas analysers are pressure sensitive it is essential that the gases enter them at a constant pressure relative to barometric pressure. This is most likely to be a problem when gases have different origins, such as sample gas and calibration gases. Some analysers such as the Servomex oxygen analysers have a device which gives a constant inlet pressure providing that the flow of gas is sufficient; however because of the cost of large quantities of calibration gases it might not always be possible to use this system. Whilst suitable barostats are available commercially or can be constructed (MacLeod *et al.*, 1985), they can be expensive. We have constructed simple water barostats (or more correctly constant pressure head devices) such as that described in Chapter 4 or by Janac *et al.* (1971); we emphasise that our devices only ensure fixed pressures relative to barometric pressure, which for most purposes is sufficient. If barometric pressure varies this too will affect the output of most gas analysers and will have to be taken into consideration.

Flowrate. Many analysers are sensitive to variations in flowrate. If the entry pressure is kept constant then a rotameter with a needle valve is usually sufficient to provide a constant flow. Servomex oxygen analysers are usually supplied with a constant-flow system.

Temperature control. Gas analysers are typically supplied with thermo-stated analysis cells which provide adequate temperature stability provided that the temperature of the incoming gas is within the manufacturers specification.

7.2.3   Calibration requirements

It is appropriate here to consider what is meant by *calibration* of gas analysers. In routine practice analysers are *set-up* rather than calibrated. The reading at some point towards each end of the analyser scale is adjusted to conform with the concentration of the gas in a known standard (or calibration-gas mixture). This procedure which sets the *calibration factor* of the analyser (the change in gas concentration F per unit change in gas analyser output reading R, i.e. dF/dR) is adequate only if it can be assumed that the standards are accurate and that the analyser is linear over its operating range. True calibration as opposed to setting up, involves checking on the accuracy of the standards and on the linearity of the instrument. For a given standard-gas mixture or analyser this basic calibration may be necessary only once, or at least infrequently.

As shown in Chapter 4, heat production is estimated (from equation

(4.21)) with terms involving $\dot{V}\Delta F$, where $\dot{V}$ is the ventilation rate of an open-circuit system and ΔF are gas concentration differences (for oxygen, carbon dioxide and methane) between outlet and inlet air. It follows that for 1 % accuracy in estimating heat production, we require at least 1 % accuracy in estimating ΔF and hence at least 1 % accuracy in setting up the calibration factor of the analyser. This estimate assumes that an accuracy of gas exchange measurement of about 1 % is acceptable and ignores any contribution from errors in the measurement of flowrate. It is unlikely that flowrate is measured with no error, so that in practice one has to achieve an accuracy of gas concentration measurement somewhat better than 1 %. This will only be obtained by the utmost attention to calibration, and Blaxter (1956*b*) has shown that substantial errors in estimates of energy balance can occur when gas exchange is measured with even this high order of accuracy.

The requirement for this accuracy is especially important for oxygen analysis because the oxygen term makes by far the greatest contribution to the overall estimate of heat production.

The following discussion will concentrate mainly on ventilated flow-through calorimetry which is the most demanding situation for the calibration of gas analysers, since the analysers are required to work over an absolute range of 1 %. The same approach to calibration can be taken for those methods that involve estimation of the gas concentrations of undiluted expired air; but here the changes in gas concentration from fresh air are about 4–5 % and the requirement for absolute accuracy is less demanding.

For carbon dioxide and methane, 1 % accuracy in setting the calibration factors can be achieved by setting up the two ends of the analyser scales by use of a zero gas which contains neither carbon dioxide nor methane, and a span gas which contains 1 ± 0.01 % carbon dioxide and 0.1 ± 0.001 % methane, i.e. the span gas concentrations must be known with an accuracy of one part in 100.

For oxygen the requirement is considerably more stringent. Fig. 7.1 shows some possible relationships between oxygen concentration F_{O_2} and analyser reading R. The required range of operation is from $F_{O_2} = 0.2095$ (fresh air) to $F_{O_2} = 0.2000$. If calibration gases corresponding to these two points (A and B) were to be used to set up the analyser calibration factor with 1 % accuracy, each would have to be accurate to ± 0.00005 or 1 part in 4000 of the stated composition. For oxygen analysers which operate over such a limited range it is not possible with current standards to achieve anything approaching this level of accuracy.

If, however, the analyser range extends to 0% and is linear over the whole concentration range (0–21%) then the calibration factor may be set with great precision using a zero gas and fresh air (points A and C). If the relationship between analyser output and F_{O_2} is non-linear, as illustrated by the broken curve, then the same high level of accuracy is still attainable providing that the slope of the graph in the region 20–21% (between points A and B') can be related to the mean slope over 0–21%. Fortunately, this comparison may be readily made in some analysers by means of varying the pressure of the gas in the analyser cell (see Section 7.2.7).

The foregoing discussion has established the requirements for calibration gases, namely:

(1) a zero gas containing neither oxygen, carbon dioxide or methane.

(2) span gases for carbon dioxide and methane accurate to better than 1 part in 100.

(3) for oxygen either a span gas containing close to 21% with an accuracy of better than 0.01%, provided that the degree of linearity is known, or two standard mixtures of approximately 20% and 21% oxygen accurate to 1 part in 4000.

We will now go on to consider what calibration gases are available in practice.

Fig. 7.1. Diagram illustrating linear and non-linear output responses from an oxygen analyser.

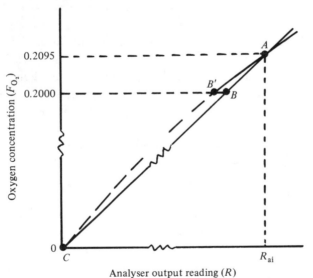

Analyser output reading (R)

7.2.4   Standard gases
7.2.4a   *Zero gas*

Nitrogen, which can be readily purchased with 99.999 % purity and less than 1 part per million of carbon dioxide, is an ideal gas for setting the zero point of gas analysers.

7.2.4b   *Dry fresh air*

Fresh air contains 20.93–20.96 % oxygen according to different authorities. Carpenter (1937) reported results of very many routine analyses of fresh air samples taken at three different laboratories over a number of years and found a mean level of 20.939 % ($\pm$0.004 % SD) oxygen and 0.031 % ($\pm$0.002 % SD) carbon dioxide, with no evidence of variations related to season or to proximity of large consumers of fuel. Machta & Hughes (1970) measured a large number of air samples around the world and found a remarkably constant level of oxygen at 20.946 % ($\pm$0.002 % SD); they also reported that the extreme range of reliable values for oxygen reported since 1910 was 0.007 %. Our routine continuous measurements of oxygen concentration over ten day periods using a paramagnetic oxygen analyser at the Hannah Institute also indicates a stable level to better than $\pm$0.01 %. The stability of oxygen concentration in dry fresh air thus seems adequately proven. Systematic differences between the actual value determined by users of different chemical analysers will, however, depend on the accuracy of the volumetric calibration of each apparatus. In this book we have adopted the value of 20.95 % for oxygen which is the most widely quoted in modern reference books.

7.2.5   Calibration-gas mixtures

These may be either purchased or made up in the laboratory by various methods. Manufacturers of gas mixtures usually state a filling tolerance (typically about 5 % of the stated composition of each minor gas constituent) and an analytical tolerance (typically 1 %). These tolerances mean that if a cylinder of 20 % oxygen is ordered, the product may contain anything between 19 and 21 % oxygen but its composition will be known within 0.2 %. This is of little practical use. However, it is possible to obtain special gravimetric mixtures to closer tolerances. In our experience analysis certificates even of special mixtures can be unreliable and we strongly recommend that the composition of purchased gas mixtures should always be checked by chemical analysis. We have even found cylinders of 100 % carbon dioxide gas to contain considerable levels of contaminants such as air and sulphur dioxide.

There are several good reviews describing the production and use of standard gas mixtures (British Standards Institution, 1979–83; Janac *et al.*, 1971; Nelson, 1971; Verdin, 1973; Barratt, 1981). The description here is limited to those methods which are most likely to be useful in the calibration of gas analysers in calorimetry.

7.2.5a Static methods

In these methods the pure gases required to form the mixture are pumped into a receiver one after another. The quantity of each gas fed in is measured either volumetrically (e.g. from a syringe, or by displacement with a weighed quantity or volume of a liquid), manometrically (by measurement of the pressure increase in the previously evacuated container) or gravimetrically (for a review see Nelson, 1971). The gravimetric method is probably the most accurate and precise method of producing gas mixtures and is widely used commercially; typical absolute accuracy is about 0.1 % (Nelson, 1971) though up to 0.002 % is claimed to be possible (Daines, 1969). The procedure requires a very accurate balance of high capacity and weighing resolution (to 1 part in 100 000 or better). A gas cylinder is evacuated by means of a vacuum pump, placed on the balance and tared. The individual high purity gases are introduced in turn at high pressure until the cylinder and its contents reaches a predetermined weight. For particularly accurate gas mixtures, allowance must be made for effects such as buoyancy changes between weighing, and changes of humidity, temperature, and barometric pressure which alter the density of the air surrounding the balance (Miller, Carroll & Emerson, 1965).

Whilst the method is primarily one used commercially because of the sophistication of the necessary weighing and pressurising apparatus, Daines (1969) has described a system suitable for the laboratory.

7.2.5b Dynamic methods

In dynamic methods, streams of the individual gases are continuously blended in proportions which give the required gas concentrations. The essential features of the method therefore are accurate and precise control of the flow rates of the component gases followed by thorough mixing. Some devices display the flowrate of each component gas and allow some adjustment of the flowrate, whilst in others the flowrate is rigorously regulated to some previously set value. In general dynamic methods are preferable to the static ones for laboratory use. They eliminate several problems: the need for storage when large volumes of calibrating gases are required; and the change of gas concentrations during storage caused by adsorption and diffusion onto or through the walls of the container. The main drawback is that the best dynamic systems are usually costly.

Restriction systems. The simplest dynamic method involves mixing the component gases after they have passed through gas lines each containing a restriction such as an orifice plate or a nozzle. The pressure drop across the restriction, measured by means of an accurate manometer, gives the rate of flow of the constituent gases. The contribution of a single gas to the mixture is determined as the ratio of its flowrate to the combined rates of flow. The typical accuracy of the method is approximately one to three percent of the range (Angely *et al.*, 1969; Janac *et al.*, 1971). A commercial system (G. A. Platon) uses needle valves to give a variable resistance and rotameters to measure the individual gas flows; however the gas mixtures so produced are inevitably inaccurate and must be further calibrated by chemical means.

Mass flowmeter systems. The principle of mass flowmeters has been described in Chapter 6. The electrical output of the flowmeter can be used to drive a motorised needle valve to regulate the rate at which a gas passes through the flow transducer. Initially the flowrate, and thus the electrical output from the meter, are preset by the operator. Thereafter any deviation from this preset value is automatically compensated by an appropriate adjustment of the motorised needle valve and the flowrate is then maintained at its preset level. Several flow transducers can be combined to provide control of the flowrate of several gases entering a mixing chamber. Flow transducers can be calibrated with an accuracy and precision of about ± 0.1–0.2% of their range (see Sections 7.3.1 and 7.3.2), but the overall accuracy of the gas mixture will depend on the number of component gases and the ranges of the flow transducers selected.

Signal gas blenders. These work on the critical orifice flowmeter principle described later in this chapter. Their use for calibration of gas analysers in calorimetry has been described (Hervey & Tobin, 1982; Tobin & Hervey, 1983). The system blends a diluate gas with a diluent by means of calibrated orifices and differential regulators (Fig. 7.2) with a precision of better than 1% of final concentration. Several models are available permitting fixed or variable blends of two or more gases in either one or several steps. Fig. 7.3 shows the results of using a Signal model 820B with three outputs providing nominally 1%, 2% and 4% of the diluate (in this case 21% CO_2 in N_2) in the diluent (dry fresh air) to calibrate differential oxygen and carbon dioxide analysers on each of four days. The orifices can be used singly or in combination to produce blends of diluate from zero to 7% in 1% steps – effectively this produces gas mixtures giving differentials of zero to 1.4% in 0.2% steps. In practice, as can be seen in Fig. 7.3, the nominal orifice calibration was in error and, when checked

Fig. 7.2. Diagram showing the gas mixing components of a Signal gas blender. The amount of diluate gas introduced into the stream of diluent gas is balanced by a proportional reduction in the flow of diluent so that the total output remains constant. The component gases are supplied to provide a pressure drop across the blender of at least 30 lb/in²; if the gas density and temperature remain constant, the diameter of the orifices determines the flowrate. The diluate can be introduced through a single variable or preset orifice, or through one or more of a set of three preset orifices of different size each with an on–off valve.

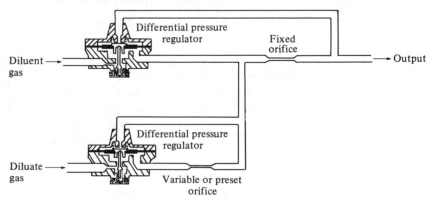

Fig. 7.3. The reproducibility of gas mixtures produced by a Signal gas blender, to calibrate oxygen and carbon dioxide analysers. The analysers were calibrated once per day for four days. For oxygen the coefficient of determination (linear regression), $r^2 = 0.999$; the 95% confidence limits of b, in the units of the graph, are ± 0.002. For CO_2, the respective values (second order regression) were the same.

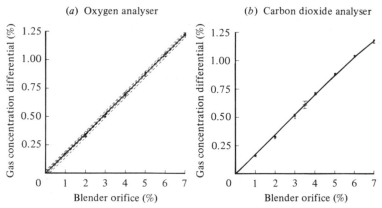

against Haldane analysis, the range of the blender gave differential gas mixtures from zero to 1.26 % in seven equal steps. Provided the pressure drop across the orifice is maintained constant the gas blenders produce highly repeatable gas mixtures, but their composition is not necessarily what the manufacturers claim, and we have found blenders occasionally to be in error by up to 20 % of the expected value. It is extremely unwise to use such devices without calibrating them by chemical analysis of their product, but once this is done the blenders are reliable and an asset to calibration.

Wosthoff gas mixing pumps. Wosthoff pumps are widely used gas mixing devices for the calibration of gas analysers and have earned a good reputation for reliability. They are, however, very expensive. They comprise bronze piston and cylinder assemblies to which gases are supplied at equal pressure. The gas is dispensed from the cylinder by the movement of the piston whose rate of travel is determined by preselectable gears. Each pump comprises a pair of cylinder assemblies that can deliver two gases at different rates. The relative rate is determined by the diameters of the cylinders and the difference in gearing applied to the two pistons. The gears and cylinder assemblies are housed in an oil bath and are driven by a synchronous motor. Multicomponent gases can be produced by cascading a series of pump assemblies. The accuracy of the pump is in the range ± 0.3–1 % of the selected ratio. A check of the output of a single Wosthoff pump by Lloyd, using the Lloyd–Haldane apparatus, showed that there was good agreement between the predicted carbon dioxide concentration delivered by the pump and the determined value. The greatest discrepancy was 0.02 % absolute.

Other methods. Gas mixtures may also be prepared by collecting, drying and compressing expired gas from a subject (Brockway *et al.*, 1977). This produces mixtures of convenient composition for calibration purposes, the precise values of which must be determined by chemical analysis.

7.2.5c   *Checking the composition of calibration mixtures*
The various methods used to produce the calibration gas mixtures fall below the accuracy required for gas analysers operating over a 1 % range. Thus whatever the method, the composition of the gas which is produced must be checked by chemical analysis using methods such as those of Haldane and Scholander (these methods and their potential errors are discussed in more detail in Chapter 6). Even in experienced and careful hands, consistently achieving 0.01 % absolute accuracy for each

gas is stretching these chemical methods to the limit. In the standard Haldane apparatus, the reading error is about 0.005 ml which alone, for the usual 10 ml burette, represents a potential error of 0.05% units. For a gas analyser operating over a 1% range and calibrated with zero and span gases, whose composition has been determined in this way, there could be an error of up to 10% in its calibration factor (0.1% units in a range of 1%). Fortunately, with care, the magnitude of such errors can be substantially reduced by using a larger burette, attempting to estimate volumes with a precision better than 0.005 ml and by making multiple analyses of a gas (Carpenter, 1923).

To reduce the possibility of additional random and systematic errors, it is important that the operator observes rigorously the procedures recommended by Haldane himself (1898) or by later workers (Carpenter *et al.*, 1929; Peters & van Slyke, 1932; Carpenter, 1933*b*). It is difficult to define an absolute test for the absence of systematic errors in chemical gas analysis since this requires analysis of a gas whose composition is exactly known. Fresh air provides the best standard. Repetitive analysis of fresh air should provide a mean value of 20.93–20.95% for oxygen and 0.03% for carbon dioxide, each with 95% confidence limits of about ± 0.01–0.02% (Peters & van Slyke, 1932; Carpenter, 1933*b*, 1937; Machta & Hughes, 1970). It is important to ensure that the sample is uncontaminated fresh air; room air usually contains stagnant expired air in sufficient quantities to be detectably different from fresh air. A check analysis on fresh air should always be applied before attempting to measure the composition of the standard mixtures. As Consolazio *et al.* (1963) state: 'If one cannot achieve this accuracy (20.93 $\pm$ 0.02% oxygen and 0.03 $\pm$ 0.01% carbon dioxide) there is no use in analysing the unknown samples. More practice or else cleaning of the apparatus is indicated.' For an important span gas the effect of random errors can be very much reduced, and a more precise estimate of the true composition obtained by carrying out a large number of replicate analyses (typically 9 or 16). Many of the comments above can equally be applied to analysis by means of the Scholander apparatus.

Whilst we are not enthusiastic in recommending the determination of gas concentrations by methods such as those of Haldane and Scholander, we see no alternative where gas mixtures form a necessary part of calibration. However we hope it is clear that the introduction of even apparently small measurement errors can lead to substantial errors in the determination of respiratory exchange over the 1% oxygen range (20–21%), and that chemical measurement of gas content must be to the highest practicable standard.

7.2.6 Linearity checks of oxygen analysers

Whilst many oxygen analysers are claimed by their manufacturers to be inherently linear it is still important to check that this is true in practice; this may be done in one of two ways:

7.2.6a By means of gas mixtures

Gas mixtures can only be regarded as secondary standards, but for some gas analysers there is no alternative to their use. Although we have employed them extensively at Leeds to determine the linearity of gas analysers, their use presents several difficulties. The first problem is the need to make up a large number of such mixtures, usually to a preselected approximate concentration. Many laboratory methods for this purpose are tedious, slow and unreliable. Whilst gas blenders are most suitable, they are expensive and the range of gas mixtures available to provide points for the calibration may be limited. The second problem is to

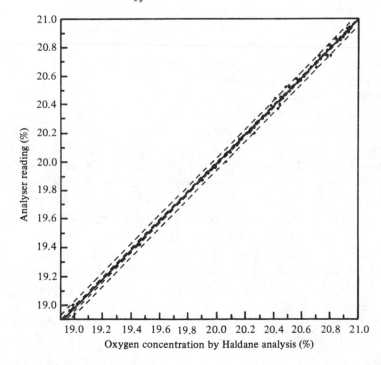

Fig. 7.4. Calibration of a paramagnetic oxygen analyser by means of gas mixtures. The parameters of the regression of analyser reading on oxygen concentration determined by chemical analysis were: n, 53; A, -0.1670%; B, $1.0075 - 95\%$ confidence limits of ± 0.0138; r^2, 0.9976; SD_{xy}, 0.0225%.

determine the actual gas concentration of the mixture, however made, by chemical analysis (see above).

To determine its linearity by this method, the analyser is first set up with a zero and a span gas. For oxygen the zero gas is oxygen-free nitrogen and the span gas is dry fresh air (it is important to emphasise that the air must be dry). Known gas mixtures are then passed through the analyser and the electrical output measured. To define accurately the relationship between gas concentration and the output of the electrical analysers, we have found it necessary to use a large number of gas mixtures, all analysed in duplicate by an accurate Haldane apparatus, to try to eliminate random analytical errors. The output from the electrical analysers, expressed as units of partial pressure or volume percent is regressed against the chemically determined gas concentration in the various gas mixtures. The coefficient of determination (r^2), the slope and intercept all give a measure of the linearity of the analyser. Whilst the method has been highly successful, it has proved laborious; Fig. 7.4 shows the response of a Servomex paramagnetic oxygen analyser to calibration over a 2% range by means of gas mixtures. The slope and the intercept appear to confirm the linear nature of the analyser over the range 19–21% oxygen, and by extrapolation over the range 0–21%.

7.2.6b *By variation of pressure in the analytical cell*

Most electrical oxygen analysers actually measure the partial pressure of the oxygen so that in principle using a single gas of known oxygen content and introducing it into the measuring cell at a range of pressures provides a means of calibration.

The pressure system is ideal for determining the linearity of paramagnetic oxygen analysers, which are the most common type of oxygen analysers used for flow-through calorimetry where the greatest precision and accuracy is needed. The technique is, however, unsuitable for use with methods of oxygen analysis such as mass spectrometry and gas chromatography in which the measuring cell is unaffected by the total inlet pressure. The method used at the Hannah has been described by McLean and Watts (1976), and it has been used at Leeds with slight modification. In our experience at both the Hannah and at Leeds, the pressure method of calibration is more accurate and involves considerably less effort than using gas mixtures. With the pressure method an oxygen cell can be calibrated in a matter of a few hours with an accuracy and precision better than that seen in Fig. 7.4 which was achieved by using about 60 gas mixtures (and analysis alone of these took more than 40 h). A detailed description of the method is included in the next section.

7.2.7 *Calibration of pressure-sensitive oxygen analysers by the variable-pressure method*

This offers the nearest approximation to a primary calibration of gas analysers that one can obtain. It requires a voltmeter or digital panel meter accurate to 1 part in 10000, a water manometer 250 cm long and a barometer. The arrangement of the calibration equipment is shown in Fig. 7.5. Two gases are used: high purity nitrogen; and dry fresh air because of its constant known composition (any error in using a value in the range 20.93–20.96% creates at most an error in calibration of 0.14% of full-scale).

The nitrogen is delivered to the cell through a needle valve directly from a cylinder; the dry fresh air is pumped through by means of pump (A) of 5 l/min capacity. It is important to ensure that there is constant flow through the cell. The Servomex analyser is fitted with a constant flow device: provided that a minimum flow (about 2 l/min) is maintained through the bypass valve and flowmeter, there will be a constant flow through the sample cell. The sample flowrate is adjusted by means of a needle valve to about 80% of maximum. The pressure in the cell is assumed to be equal to barometric pressure plus the change in pressure as measured by the water manometer (in our experience the pressure drop across the cell is no more than about 1–2 mm of water). All the pressures determined by the water manometer are adjusted to millimeters of mercury (1 cm H_2O equals 0.736 mm Hg). A second pump (B) and needle valve are

Fig. 7.5. Apparatus used to calibrate an oxygen analyser by the variable-pressure method.

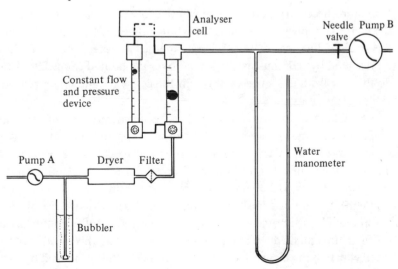

attached to the outlet of the analyser beyond the manometer. By adjust-ment of this needle valve and that on the bypass the pressure in the cell can be varied about barometric pressure (the pressure in the cell is de-termined by the balance of pressures created by the two pumps) and yet constant flow maintained through the cell.

With the cell held at the prevailing barometric pressure (P_i) the output of the analyser is adjusted to zero using oxygen-free nitrogen and to the required initial full-scale reading (R_{ai}) using dry fresh air. This procedure sets the calibration factor of the analyser to a mean value for the range 0–21%. It may be expressed either in terms of concentration

$$\bar{G}_i = \frac{dF_{O_2}}{dR} = \frac{0.2095}{R_{ai}} \text{ concentration units/V} \tag{7.1}$$

or in terms of partial pressure

$$\bar{k} = \frac{dP_{O_2}}{dR} = \frac{0.2095 P_i}{R_{ai}} = \bar{G}_i P_i \text{ mm Hg/V} \tag{7.2}$$

Because the instrument is inherently sensitive to partial pressure of oxygen rather than to concentration, the former factor $\bar{G}_i$ is applicable only at the barometric pressure (P_i) when the analyser was initially set up. If barometric pressure subsequently alters to any new level (P), the ana-lyser output for fresh air will change proportionately to

$$R_a = R_{ai} \times P/P_i \tag{7.3}$$

and the new calibration factor in terms of concentration becomes

$$\bar{G} = 0.2095/R_a$$
$$= \bar{G}_i \times (R_{ai}/R_a) = \bar{G}_i \times (P_i/P) \tag{7.4}$$

It remains necessary to determine whether the mean slope represented by the factor $\bar{G}$ applies throughout the whole range 0–21% as represented by the straight line AC in Fig. 7.1, or whether another slope such as that represented by AB' must be employed over the limited range 20–21%. To do this dry fresh air is passed through the cell while the pressure is varied as described above.

A typical result is plotted in Fig. 7.6: the linear response of the analyser to changes in total pressure (and therefore in partial pressure of oxygen) over the range of the test is clearly apparent. The equation of the straight-line fit to the points is found by linear regression; in the example given the slope $dR/dP = S_a$ is 2.304 mV/mm Hg and the intercept is $+48.1$ mV. At first sight it might be expected that for an analyser with linear response the extrapolated line should pass through the origin. But account has to be taken of the small diamagnetic effect of nitrogen; this produces a very small effect on the measuring cell which opposes that produced by oxygen.

The point represented by extrapolation to zero total pressure (and therefore to zero partial pressure of both oxygen and nitrogen) is thus slightly different from the setting up conditions where the cell contained nitrogen at atmospheric pressure. The slope (and intercept) requires adjustment to allow for this effect.

The magnitude of the correction can be calculated theoretically from the magnetic susceptibilities (3449 and -12×10^{-6} cgs units) and the relative proportions (0.21 and 0.79) of oxygen and nitrogen respectively. But it is simpler to determine it practically by repeating the variable pressure test on the analyser when its cell is filled with pure nitrogen (it is important to ensure that there is no leak or insufficiency of nitrogen which might lead to contamination with air of the nitrogen flowing through the cell – see below). Fig. 7.7 shows the result of such a test conducted immediately after that depicted by Fig. 7.6. Linear regression of this plot yields a slope (S_n) of -0.061 mV/mm Hg. The corrected slope of the graph of Fig. 7.6, which for a linear analyser would pass through the origin, is thus

$$\frac{\mathrm{d}R}{\mathrm{d}P} = S_a(1 - S_n/S_a) \tag{7.5}$$

Fig. 7.6. The relationship between the output of a Servomex paramagnetic oxygen analyser and variation in the pressure of its analytical cell when filled with dry fresh air. The parameters of the regression were: n, 20; A, 48.077 mV; B, 2.3046 – 95% confidence limits of ± 0.0129; r^2, 0.9999; SD_{xy}, 1.1622 mV.

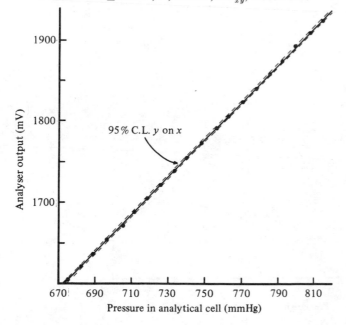

(In the original paper describing this method (McLean & Watts 1976) the factor 0.7905 was incorrectly included as a multiplier of S_n/S_a in this equation.)

The inverse of this slope multiplied by 0.2095 is the calibration factor of the analyser expressed in terms of partial pressure of oxygen over the range of the variable-pressure test.

$$k = 0.2095 \frac{dP}{dR} = \frac{0.2095}{S_a}(1 + S_n/S_a) \text{ mm Hg/volt} \qquad (7.6)$$

The constant k is the slope of the line AB' in Fig. 7.1, whereas $\bar{k}$ is the slope of the straight line ABC. The ratio $\mu = k/\bar{k} = G/\bar{G}$ is a fixed property describing the non-linearity of the analyser which needs to be determined only once (though it is sufficiently straightforward that periodic checking is easy).

If the analyser is linear over the whole range then $\mu = 1$. In fact Servomex instruments are normally linear over the whole range (their claim is to $\pm 2\%$). In our example in Fig. 7.6 the value of μ is 1.0017.

Fig. 7.7. The relationship between the output of a Servomex paramagnetic oxygen analyser and variation in the pressure of its analytical cell when filled with nitrogen. The parameters of the regression were: n, 12; A, 46.674 mV; B, $-0.0606 - 95\%$ confidence limits of ±0.0008; r^2, 0.9997; SD_{xy}, 0.0296 mV.

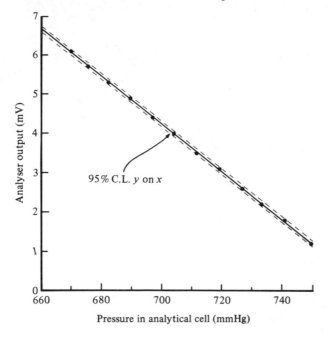

95% C.L. y on x

Analyser output (mV)

Pressure in analytical cell (mmHg)

Substitution from equations (7.1) and (7.4) now gives

$$G = \mu\bar{G} = \mu\bar{G}_i \times \frac{R_{ai}}{R_a} = \frac{0.2095 \times \mu}{R_a} \text{ concentration units/V} \qquad (7.7)$$

G is the calibration factor of the analyser expressed in terms of concentration units, its dependence on barometric pressure is automatically included by incorporation of the value R_a (the analyser output reading for fresh air at any time subsequent to the original setting up).

If there is a leak in the gas line before the analytical cell, then the dry fresh air or dry nitrogen will be contaminated with room air. The effect of this is most obviously seen in the calibration with nitrogen when a leak produces a clearly different relationship between voltage output and cell pressure above and below barometric pressure. When such a leak is found it is important to use only the values for the slopes for dry fresh air and nitrogen obtained at positive pressures to determine μ.

A high quality pressure-dependent oxygen analyser can also serve as a recording barometer. At the Hannah Institute our practice has been to set up the Servomex analyser at the commencement of each 10 day experimental run, and to enter the setting-up values of the full-scale reading (R_{ai}) and barometric pressure (P_i) into the computer. Subsequently whenever the analyser is switched to measure fresh air, the computer programme calculates the current barometric pressure from equation (7.3).

$$P = P_i R_a / R_{ai}$$

The value is displayed at regular intervals for comparison with a mercury barometer. The two values normally agree to better than 1 part in 1000 which provides evidence both of the stability of the analyser and of the constant composition of fresh air. The rare occasions when the comparison is not so close serve as a warning of some fault condition (e.g. the need for renewal of absorbent in the drying tower).

The ease and accuracy of the variable-pressure method for checking the linearity and calibration of pressure-dependent oxygen analysers and the possibility of continuous monitoring of satisfactory operation by checks against barometric pressure are very powerful assets. We regard the addition of pressure-compensation devices or facilities for proportional operation to oxygen analysers as a positive disadvantage, and they are something else that can go wrong.

The Servomex OA 184 analyser, as used at the Hannah Institute and at Leeds, is considerably more precise and accurate over the whole range of operation, and very much faster, than the Haldane method for the measurement of oxygen concentration (although its basic calibration factor does depend on a Haldane-determined constant – the oxygen content of dry fresh air). In fact we use the Servomex analyser as a standard against which we check the compositions of standard-gas mixtures used for other applications.

All that we have written so far emphasises the importance of measuring the *difference* between outlet and inlet air concentrations and of ensuring identical measuring conditions for both samples. In particular flowrate and dryness must be the same for both. Consequently there is much to be said for using a single channel analyser with periodic switching between the two gas streams and *employing the same final drying tower for each.* The disadvantage of this procedure is the loss of measurement of the exhaust stream during periods when the inlet air is being measured. Where the reference or inlet air may fluctuate in oxygen concentration and the use of a dual-channel instrument is essential, inlet air should still be passed periodically through the sample side of the analyser as a check – in the Leeds multi-chamber systems (see Chapter 8), one animal chamber is always kept empty for such purposes. A dual-channel instrument inevitably has the problem of inter-channel drift arising either from within the instrument, or through sample conditioning (for example, outlet air is normally more humid than inlet air, and the moisture absorbent in the drying tower in the gas analysis line will deteriorate the more quickly, particularly if there was no drying in the main line prior to measurement of flowrate). For dual-channel instruments used as two independent analysers, regular switching of the gas between the two channels (taking care to leave each channel associated with its own drying tower) is necessary to check on drift. Similarly, in differential mode regular checking of the zero differential condition (i.e. fresh air passed through the two channels) must be done.

On the Leeds animal calorimeters, which are used in differential mode, the analyser output with zero and span gases is determined hourly for 6 min each. The values over the last minute in each case (i.e. allowing five minutes wash-out time) are used to calculate the calibration factor G in units of per cent differential per volt output reading. The experimental measurements are always bracketed by periods of calibration and the effects of any drift, and of pressure variations, are eliminated by interpolation.

By adoption of the precautions listed in this section we have found that the accuracy and precision (judged from continuous monitoring of fresh air) of the Servomex OA 184 and the OA 540 can be improved considerably over the levels quoted by the manufacturer. We have found the built-in thermostat adequate in our (air-conditioned) laboratories. A successor to the OA 184, the OA 1184, is now in the final stages of testing and our experience of it, albeit limited, would confirm the manufacturers optimism that it will be a very satisfactory replacement.

7.2.8   Calibration of carbon dioxide and methane analysers

There are many similarities in the approach to calibrating oxygen, carbon dioxide and methane analysers (and that is particularly true of those methods such as gas chromatography and mass spectroscopy in which the principles of measurement are identical). In this section, we will concentrate on those aspects of calibration which are particular to the determination of carbon dioxide and methane concentrations. In practice, the most common means of determining their concentrations is by means of infra-red analysers. The method is perfectly suitable for the measurement of carbon dioxide in the 0–1 % range that is likely to be encountered in flow-through calorimetry. Methane is typically encountered in concentrations of about 0–0.1 % which is not much greater than the highest sensitivity of the analyser (Brockway, McDonald & Pullar, 1977) and creates particular calibration problems. Fortunately the methane term, like the carbon dioxide term, contributes little to the calculation of energy expenditure (Section 4.5) and accuracy is of less importance than for the measurement of oxygen. The outline sequence of steps to calibrate these analysers is similar to that described for oxygen analysers, beginning first with a check of linearity.

7.2.8a   Linearity checks

The electrical output of an infra-red gas analyser is inherently non-linear; however most analysers are provided with adjustable linearising circuits. Whilst it is possible to fit mathematically a response curve to a non-linear output from a carbon dioxide analyser (Tamas, Urquhart & Ozbun, 1971), it is usually sufficient to make use of the linearising circuit to produce a linear output over the working range.

Whilst the pressure method has proved very satisfactory in such a role with both paramagnetic oxygen analysers and oxygen electrodes, and has given accurate and precise calibration, it has proved less suitable for use with infra-red carbon dioxide and methane analysers.

The infra-red absorption bands for both carbon dioxide and for methane broaden as the pressure in the cell increases (Burch, Singleton & Williams, 1962; Burch, Gryvnak & Williams, 1962; Burch & Williams, 1962; Legg & Parkinson, 1968) and the analysers are typically fitted with narrow pass filters. The effect of this is that:

$$dP_{CO_2}/dR \neq P \cdot dF_{CO_2}/dR$$

If $dP_{CO_2}/dR = S_t$ and $P \cdot dF_{CO_2}/dR = S_p$, then S_t/S_p is a calibration factor that can be used to adjust the pressure calibration to a true calibration.

Theoretically if the carbon dioxide detector were a black body and absorbed all wavelengths equally the correction factor would be 1.75 (Legg & Parkinson, 1968). However, as the filling pressure of the detector is decreased, it deviates from a black body absorber and becomes less sensitive to pressure broadening in the sample tube. Experimentally determined factors have been somewhat different (Legg & Parkinson, 1968; Bate, D'Aoust & Canvin, 1969) partly because of this but also because the factor is different for different detectors – and it may change as the detector ages (Legg & Parkinson, 1968; Parkinson & Legg, 1978). Nevertheless, if one ignores the limitations of the method in providing an absolute calibration standard, it can still be used to check and, if necessary, provide a basis for adjusting the linearity of the analyser (Bate *et al.*, 1969; Sinclair, Buzzard & Knoerr, 1976).

Gas mixtures can be used to determine linearity but the method is laborious if the mixtures must be individually made up, and then analysed by a chemical method. The calibration shown in Fig. 7.8 was done in this way and in total took about 40 h of work to complete. The method is much improved by the use of a gas blender known to give consistent

Fig. 7.8. Calibration of an infra-red carbon dioxide analyser by means of gas mixtures. The parameters of the regression of analyser reading on carbon dioxide concentration determined by chemical analysis were: n, 60; A, -0.0205%; B, $1.0210 - 95\%$ confidence limits of ± 0.0281; r^2, 0.9891; SD_{xy}, 0.0295.

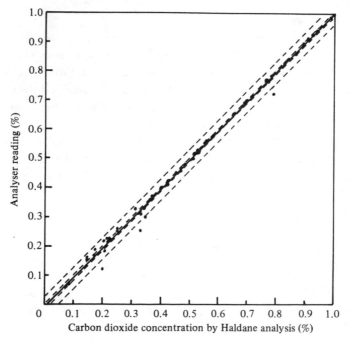

mixtures of gases. In this respect Wosthoff pumps have been most widely used (Bate *et al.*, 1969; Lundy *et al.*, 1978; Enoch & Cohen, 1981). We have used a Signal gas blender to produce suitable gas mixtures for calibration; Fig. 7.3 shows the calibration of a carbon dioxide analyser by this means. These devices permit regular checks of linearity, which is important with an inherently non-linear instrument.

7.2.8b   Routine calibration

The pressure method can be used to provide regular calibration of a differential carbon dioxide analyser (Sinclair *et al.*, 1976; Parkinson & Legg, 1978). The major drawback to the use of the pressure method is the continuing need for the analyser to be calibrated, at least initially, by gas mixtures. Any errors in the analysis of these mixtures are inevitably included in the determination of the pressure correction factor.

The use of gas mixtures prepared gravimetrically or by means of a mixing device as described above, and analysed by means of the Haldane apparatus, is perfectly satisfactory to achieve the level of accuracy needed in indirect calorimetry for carbon dioxide and methane. Typically these analysers are set up with CO_2-free air or nitrogen as the zero gas, and provided they have been linearised, a single span gas.

It is also important to be aware that if the background concentration of oxygen is substantially different in the calibration gases from that in the sample gases an error in the carbon dioxide reading will ensue. When nitrogen is replaced by oxygen (i.e. $F_{O_2} > 0.2095$) the carbon dioxide absorption band narrows, there is less absorption and the true concentration of carbon dioxide is underestimated. In addition to oxygen, several gases which also do not absorb infrared radiation (for example helium, hydrogen, argon, nitrous oxide) can alter carbon dioxide band width and thus introduce an error (Burch *et al.*, 1962; Hill & Powell, 1967.)

Where extremely high accuracy of calibration is required, as for example in some areas of plant physiology, the method described by Parkinson & Legg (1971) in which the analyser tubes are divided into two cells of different but known pathlength may be used (the absorbance of infra-red radiation depends on the concentration of the test gas and the path length, thus if a single gas is used the output of the analyser is proportional to pathlength which is easy to measure).

7.3   Calibration of flowmeters

Flowmeters should be calibrated against a primary standard that itself can be directly checked against a measurement of length, mass and time. The selection of the appropriate specific method of calibration depends largely on the range of flows to be used. Nelson (1971) has reviewed several of the available methods. Unless a single flowrate is to be used, it is best to calibrate the flowmeter at several points over the working range and then to carry out a linear or higher order regression (as appropriate) on the data. Regression analysis is a powerful method for assessing the calibration: it provides information on the linearity of the flowmeter and the 95% confidence limits of Y on X gives the magnitude of the possible error in the measured flowrate at any point over the working range.

7.3.1   *Soap-bubble flowmeters*

Calibration of flowmeters operating below about 1 l/min is best and most easily achieved with the soap bubble flowmeter. The method is described by Barr (1934) and is considered in detail by Levy (1964); it is simple and yet very accurate. Levy states that it can be used for flows in the range 6 ml–6 l/min with an accuracy of $\pm 0.25\%$.

The apparatus is shown in Fig. 7.9; it comprises a graduated tube or

Fig. 7.9. A soap-bubble flowmeter.

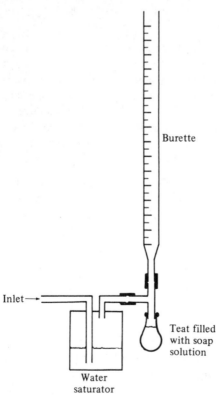

Burette

Inlet →

Teat filled
with soap
solution

Water
saturator

burette connected through a tee-piece to a rubber bulb or teat. The volume between the graduations must be determined accurately. For this the teat is replaced by a burette tip and the burette filled with water to the upper graduation. The water is slowly run out through the burette tip and collected until the water level reaches the bottom graduation. The volume between the graduations is determined from the weight of water collected and its density.

For calibration the rubber bulb or teat is filled with dilute detergent (Teepol is ideal) and the walls of the burette wetted with the solution. The gas, which must be saturated with water vapour, is introduced into the burette through the side arm of the tee-piece. The teat is squeezed briefly to introduce the detergent into the throat of the burette, and then quickly released. With practice this produces a soap bubble that forms a visible airtight seal and this is carried upwards with negligible friction by the flow of gas. The pressure drop across the bubble is less than 0.01 mm Hg and providing that there is no resistance at the outlet of the burette, the pressure behind the soap bubble can be assumed to be equal to barometric pressure. The passage of the bubble between the graduations is timed by means of a stopwatch. The temperature of the system is determined to ±0.05 °C by lowering a sensitive mercury thermometer down the burette and into the gas stream; pressure is measured to ±0.15 mm Hg by means of a barometer. Repeated estimations of flowrate are made.

Levy has estimated that with care the fixed measurement errors, which are independent of the design and size of the bubble flowmeter, will amount to no more than ±0.08%. The errors are made up of ±0.02% from the measurement of temperature, ±0.04% from the measurement of pressure and ±0.02% from the measurement of the volume of water. If the overall error is to be no greater than ±0.25%, then remaining errors should be ±0.17% or less. These residual errors arise from the measurements of the time taken for the bubble to travel between the graduations, and the volume between the graduations; the size of the errors will depend on the design and size of the flowmeter.

The volume of tube recommended for a particular flowrate to meet a specified level of accuracy can be determined by the following formula:

$$V = \frac{22.5 \times \dot{V}}{e - 0.08}$$

where V = volume in ml; $\dot{V}$ = flowrate in ml per s; e = overall specified percentage error of the flowmeter, typically 0.25%.

Table 7.2 shows the relationships between flowrate, intended accuracy, and the diameter and minimum volume of the tube recommended by

Levy. If tubes with a constant bore are used, the length of tube between graduations is given by $90/(e-0.08)$, which for $\pm0.25\%$ overall accuracy would be about 530 cm. To reduce this length, the central region of the tube can be gradually expanded to about five times the diameter of the region where the graduation marks are found. The length is decreased

Table 7.2. *The recommended relationships between flowrate, intended accuracy, and the diameter and minimum volume of the tube (assumes that temperature is measured to $\pm0.05\,°C$)*

Flowrate (ml)	Diameter (cm)	Minimum volume (ml) $e = 1.00$	$e = 0.50$	$e = 0.25$
0.1	0.18	2.5	5.4	13.0
1.0	0.57	24.5	53.5	132.0
10.0	1.78	245.0	535.0	1320.0
100.0	5.67	2450.0	5350.0	13 200.0

$e =$ overall specified percentage error of the flowmeter
(Adapted from Levy, 1964.)

Fig. 7.10. Calibration of a mass flowmeter by means of a soap-bubble flowmeter. The parameters of the regression of mass flowmeter reading on the actual flowrate determined by means of a soap-bubble flowmeter were: n, 19; A, -0.132 ml/min; B, $0.9998 - 95\%$ confidence limits of ±0.0027; r^2, 1.000; SD_{yx}, 0.1728 ml/min. (Courtesy of Dr K. Sivapalan.)

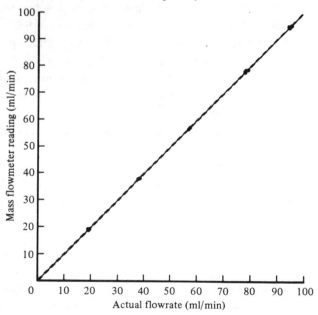

about n^2 times for each n increase in diameter. The diameter at the graduated portions of the tubes should remain narrow for at least 10 cm on the proximal side of each of the graduations.

The flowrate, adjusted to 0 °C and 760 mm Hg, is calculated as:

$$\dot{V} = \frac{V \times 60 \times (P - P_w) \times (273 + T)}{t \times 760 \times 273}$$

where $\dot{V}$ = flowrate in ml/min; V = volume between the two graduations in ml; t = time of travel of the bubble between the graduations in seconds; P = barometric pressure; P_w = saturated water vapour pressure at T °C; T = temperature in the burette.

We have used a standard 50 ml burette as the basis for a soap bubble flowmeter to calibrate a mass flowmeter operating over the range 0–100 ml/min. Even with this simple device a theoretical accuracy in the volume measured by the bubble flowmeter of better than $\pm 0.5\%$ is possible; an example of a typical calibration is shown in Fig. 7.10.

7.3.2   *Wet gasmeters*

Wet gasmeters covering the range of 1–200 l/min with an ac-curacy of $\pm 0.25\%$ are available in the UK from Alexander Wright. Al-though there is some overlap with the bubble flowmeters in the coverage at the bottom end of the range, above 1 l/min the wet gasmeters are preferable, despite their higher cost. The mode of action of the wet gasmeters is described in the previous chapter. Because introduction of water vapour from the meter to the gas stream causes difficulties in the measurement of flowrate, we prefer to fill the wet gasmeter with oil rather than water. Any light oil of a non-viscous nature is satisfactory providing that it meets the British Standard BS 148/1959; we use Shell Diala B oil.

Despite the manufacturers claims to accuracy, the meters should be checked. This can be achieved by connecting the wet gasmeter to a large aspirator or some other suitable large container (Fig. 7.11). Before cali-bration, the liquid level in the gasmeter must be adjusted to correspond with the manufacturers indicating point or engraved calibration line. Air is displaced from the aspirator or container by adding successive accu-rately known volumes or weights of water. It is important to ensure that no trapped air is introduced into the aspirator with the water. This is prevented by introducing the water slowly into the aspirator through a wide bore tube. The water added to the aspirator, and the oil in the wet gasmeter, should be at ambient temperature.

Wet gasmeter readings are taken between additions of water once the

Fig. 7.11. Apparatus for the calibration of a wet gasmeter.

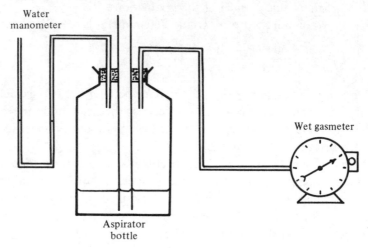

Fig. 7.12. Calibration of a wet gasmeter by displacement of air from an aspirator by water. The parameters of the regression of wet gasmeter reading on the volume of air displaced were: n, 21; A, 0.002 l; B, 1.0019 – 95 % confidence limits of ± 0.0033; r^2, 1.0000; SD_{yx}, 0.0197 l.

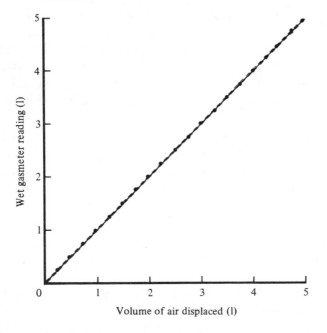

air is stationary. A water manometer allows the pressure inside the aspirator to be measured. Usually this shows a small positive pressure of a few millimeters of water; this probably reflects 'stiction' in the gasmeter but because it is so small it can be ignored. Larger pressure changes may need to be included in the calculation of the volume of air actually

Fig. 7.13. Apparatus for the calibration of a mass flowmeter by means of a wet gasmeter.

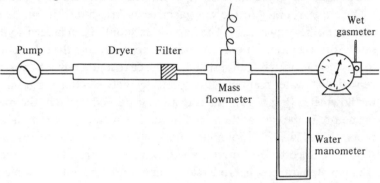

Fig. 7.14. Calibration of a mass flowmeter by means of a wet gasmeter. The parameters of the regression of mass flowmeter reading on the actual flowrate determined by means of a wet gasmeter were: n, 15; A, -0.0077; B, $1.0048 - 95\%$ confidence limits of ± 0.0047; r^2, 0.9999; SD_{yx}, 0.0119 l.

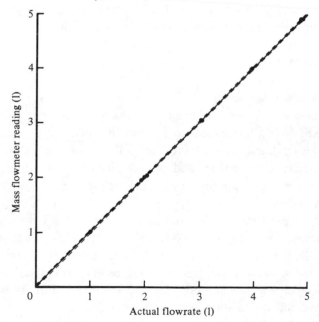

displaced. Otherwise no corrections for temperature and pressure are needed. A small portion of the water added to the aspirator remains in the wide bore inlet tube and does not displace air into the gasmeter; this volume is calculated from the cross-section of the tube and the length of tube involved and is allowed for. Fig. 7.12 shows the result of the calibration of a wet gasmeter operating in the range of 0–5 l/min. The meter was linear over the range and the slope (1.0019 litres displaced per litre added) showed a systematic error within the manufacturer's specification.

Once the accuracy of the wet gasmeter is ascertained it can be used to calibrate other flowmeters. The apparatus should be arranged so that one side of the wet gasmeter is open to the atmosphere and the other connected to the flowmeter to be calibrated and thence to a pump whose flowrate can be varied. A water manometer is connected between the wet gasmeter and the flowmeter (Fig. 7.13). Air is drawn or pushed through the meters by the pump. In the case of thermal dilution flowmeters, in which the thermal properties of water vapour may have an effect on the reading, the air should be dried before entering the flowmeters. When the temperature of the oil is stable and equal to the air temperature, the volume of air flowing through the wet gasmeter is measured over a period of time, typically 5–10 min and flowrate calculated. It is necessary to adjust the flowrate to standard conditions.

If possible the flowmeter reading should be adjusted to correspond to that of the wet gasmeter. The procedure should be then repeated at several different flowrates and the reading from the flowmeter regressed against that from the wet gasmeter. Fig. 7.14 shows the results of the calibration of a mass flowmeter calibrated in this way over the range of 0–5 l; the systematic error was less than 0.5%.

7.3.3   *Critical-orifice flowmeters*

In the previous chapter we mentioned the use in indirect calorimeters of choked-nozzle flowmeters to produce fixed flowrates. These flowmeters, variously described as choked-nozzle, sonic-nozzle and critical-orifice flowmeters, can also be used for the absolute calibration of flowmeters. They provide an important means of calibration at flowrates greater then 200 l/min.

The theory of the pressure/gas flow relationships in sonic nozzles used as calibration devices has been described in detail by Arnberg (1962) and Sparkes (1968). In the case of air passing through a suitably designed nozzle to discharge to atmosphere, the relationships may be reduced to a simple formula involving only the pressure and temperature at the upstream side of the nozzle and its throat dimensions.

The design of the nozzle derived from Sparkes, is shown in Fig. 7.15. It includes an entry package to eliminate flow distortion and swirl, followed by the nozzle itself. The theoretical pressure ratio that will establish critical flow in a gas stream is given by

$$P/P_1 \leqslant \left(\frac{2}{\gamma+1}\right)^{\gamma/(\gamma-1)}$$

where P is the downstream pressure, P_1 is the upstream pressure, and γ is the specific heat ratio (c_p/c_v) of the gas. For air, $\gamma = 1.4$ and thus for critical flow the value of the upstream pressure (P_1) must be 1.9 times that of the downstream pressure.

Fig. 7.15. The design of the sonic nozzle: (*a*) the nozzle installation; (*b*) the nozzle. *d*, nozzle throat section diameter; *X*, distance downstream of the measurement section *A-A*. (After Sparkes, 1968.)

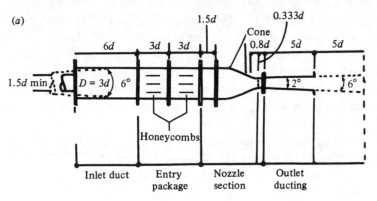

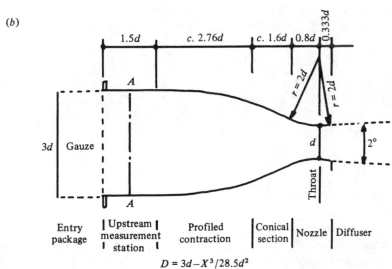

$$D = 3d - X^3/28.5d^2$$

If this can be achieved then the mass flow of gas per second (ṁ) through the throat is given by:

$$\dot{m} = kA\frac{P_1}{\sqrt{T_1}} \cdot \sqrt{\left(\frac{g\,\gamma}{Z\,R}\right)\left(\frac{2}{\gamma+1}\right)^{(\gamma+1)/2(\gamma-1)}} \quad \text{kg/s}$$

where k = the coefficient of discharge, a dimensionless constant describing the boundary layer flow effects within the nozzle (see below); A = the cross-sectional area at the nozzle throat (m^2); g = gravitational constant, 1 kg m^{-1}/N s^2; T_1 = upstream temperature (°C); Z = compressibility factor; R = gas constant, 8314 J/(k mol), and pressure is measured in pascals.

In practice there is a small discrepancy between the measured and calculated flow; the ratio between them is k. The discrepancy occurs largely because of boundary layer effects, but these are very much reduced by curving the wall profile continuously through the throat of the nozzle as suggested by Stratford (1964) and Sparkes (1968). The discrepancy will be at a minimum (i.e. $k \to 1$) when the wall profile at the throat has a radius of curvature of about two throat diameters (see Fig. 7.15) (Stratford, 1964). According to Stratford if the recommended conditions are met, then the theoretical discharge coefficient is 0.995 ± 0.0025 over a wide range of Reynolds numbers, and the method is then suitable for absolute flow calibration.

The final outstanding problem is that the real gas compressibility factor, Z, is unity only for an ideal gas. The introduction of a non-ideal gas such as carbon dioxide will decrease the value of Z at positive pressures. In practice unless very high pressures are used, and this is unlikely in indirect calorimetry, then the effect will be small and can safely be ignored. Thus the basic equation can be simplified by inserting values for the constants to give

$$\dot{m} = \frac{0.0405\,k\,A\,P_1}{\sqrt{T_1}}\,\text{kg/s}$$

(adapted from Perry, Green & Maloney, 1984).

The mass unit of flow is converted to a volume flowrate at 0 °C and 760 mm Hg by:

$$\dot{V} = \frac{\dot{m} \times 1000 \times 22.4 \times 60}{M \times 273}$$

where $\dot{V}$ = flowrate in l/min; M = molecular weight of the gas (for air 28.96).

For a fuller treatment of the theory and practice of calibration using sonic nozzles, the reader is advised to consult Arnberg (1962), Stratford (1964) and Sparkes (1968).

Despite its apparent attractiveness as a method for calibrating flow-meters operating at high flowrates in indirect calorimeters, we are aware of only one reference to its use in this way: Patterson (1979) reported that he recalibrated a 1000 l/min mass flowmeter by means of sonic flow nozzles that were accurate to better than 99.5%.

7.3.4 *Spirometers and gasometers*

The construction, testing and operation of spirometers and gas-ometers have been described in Chapter 4. When used as a calibration source for flowmeters, spirometers must themselves be calibrated carefully. Krogh (1920) showed that the volume of a gasometer could be determined by making numerous measurements of the internal diameter of the bell. Nelson (1971) describes a method of calibration called 'strapping' in which a metal measuring tape is used to determine the outside radius of the bell. After correcting for the thickness of the wall of the bell, the internal radius and cross-sectional area (r^2) of the bell can be calculated. The volume of air displaced by lowering the bell can be then determined by measuring the bell's vertical movement. As the bell moves deeper into the water, its walls displace an equal volume of water; the water level rises and displaces further air from the bell. The additional amount of air displaced is calculated as:

$$V = V_1 \times (a_i/a_t)$$

where V_1 = the additional wall-volume submerged; a_i = surface area of water inside the bell; a_t = surface area of water inside and outside the bell.

Nelson (1971) suggests that the method of 'strapping' can achieve accuracies of $\pm0.2\%$ in the volume calibration.

The major source of error in the method lies in the need for very accurate measurement of the vertical displacement of the bell; a small error in this will produce a large error in the volume measurement because of the large contribution of the surface area term. The fitting of electromechanical 'stops' to give fixed start and end points for the calibration can help to improve accuracy (Krogh, 1920). Turney & Blumenfeld (1973) connected the pulley wheel of a Tissot gasometer directly to a precision potentiometer so that the movement of the bell was accurately measured in the form of a voltage.

Calibration of a flowmeter is achieved by passing air drawn from, or delivered to, the spirometer through the flowmeter for a known period of time. The spirometer measures the volume of air which has passed through the flowmeter. Where it is necessary to adjust the volumes for temperature and pressure (for example when calibrating a mass flowmeter) these measurements can be easily made by a thermometer mounted in the spirometer and a water manometer between the spirometer and flowmeter. Any adjustment for water vapour should be made on the assumption that the air is saturated at the appropriate temperature.

7.3.5   *Gravimetric calibration*

Strømme and Hammel (1968) have described a very useful gravimetric calibration for flowmeters. The apparatus is shown in Fig. 7.16. Crushed dry ice is placed in ethyl alcohol in a thermos flask that contains an adjustable heating element. The thermos flask is further insulated. As the dry ice is heated a flow of carbon dioxide is produced. The ethyl alcohol ensures a stable flow of gas but since its vapour pressure at $-80\ °C$ (the point at which dry ice sublimes) is small, its contribution to the flow is negligible.

The flow of carbon dioxide is passed through a heat exchanger placed in a water bath held at room temperature, and through the flowmeter where the volume, temperature and pressure of the gas are measured. Ideally the thermos flask should be mounted on a high resolution balance throughout the calibration so that the rate or amount of weight loss from the supply of dry ice can be determined accurately. Alternatively the flask can be weighed at the beginning and end of a timed period, and the volume of carbon dioxide produced calculated as the weight loss of dry ice in grams divided by the density of carbon dioxide at STPD (1.977 g/l).

By varying the heat output of the element, the rate of carbon dioxide production can be varied and, by means of linear regression, a calibration line can be produced for the flowmeter. If the problem of buoyancy errors

Fig. 7.16. Apparatus for the gravimetric calibration of flowmeters. (Reproduced from Strømme & Hammel, 1968.)

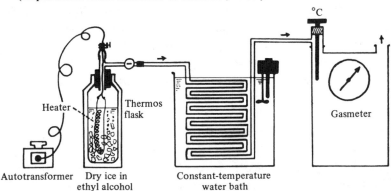

in weighing a cold object can be overcome, the accuracy of the method should be very high.

7.3.6 Dilution method

The dilution method, whilst not a primary calibration method, is a very useful means of calibrating flowmeters operating at high flowrates when methods described above cannot be used. In the method a known quantity of a tracer gas is introduced into the tubing leading to the flowmeter and well mixed with the main air flow. The concentration of the tracer gas in the air is measured downstream and from this, and the rate at which the gas was introduced into the airstream, the extent of its dilution can be determined. If the tracer gas is normally present in air then its contribution to the total concentration must be taken into account. From the dilution the flowrate can be calculated.

Carbon dioxide or nitrogen are suitable tracer gases for most purposes. Carbon dioxide is relatively easy to determine by chemical means or by a carefully calibrated analyser; if nitrogen is used its effect on the concentration of oxygen can be used to determine the extent of the dilution. When a mass spectrometer is available for gas analysis, argon or nitrogen are commonly used as a tracer. The principle of the method and its use in conjunction with a mass spectrometer has been described by Davies & Denison (1979) and Heneghan, Gillbe & Branthwaite (1981).

The rate at which the tracer is introduced should be sufficient to produce a final tracer concentration which is sufficiently high to be determined accurately by whatever method is chosen. The amount of tracer admitted must be determined accurately and this can be achieved by the use of a calibrated flowmeter of appropriate size. Alternatively, if carbon dioxide is used this can be added gravimetrically as in the method of Strømme & Hammel (1968) described above.

7.3.7 Calibration of flowmeters that measure pulsatile flow

It has been suggested that a steady flow of gas may not provide a satisfactory way of calibrating a flowmeter, such as that in the Kofranyi–Michaelis meter, which is used to measure pulsatile flow.

Where a pulsatile flow calibration is necessary delivery of known volumes of air from a reciprocating piston pump or syringe assembly is suitable. Each of the cylinders or syringes is calibrated gravimetrically.

Riendeau & Consolazio (1959) have developed a calibration technique for the K–M meter which comprises a large hand-operated respiration pump, obtained from an iron lung, connected to a Tissot gasometer. An airtight piston divides the pump cylinder into two opposing chambers with a combined capacity of up to five litres. By to-and-fro movement of the piston, air is delivered from the Tissot to the flowmeter in a pulsatile flow. The volume displaced from the chambers is adjusted by limiting the travel of the piston by means of a clamp on the piston rod, and the frequency of pumping is controlled by using a metronome. The total volume delivered is measured by means of the Tissot.

Using this method, Consolazio and colleagues (Riendeau & Consolazio, 1959; Consolazio *et al.*, 1963; and Consolazio, 1971) found that the manufacturers calibration based on continuous flow was somewhat different to their own determined one: of six meters tested, four underestimated, and two overestimated the true volume of air by an average of 4% and the overall 95% confidence limits were ± 0.9 l/min.

7.4   Whole system checks

Although it is both desirable and necessary to calibrate individual measuring instruments used in calorimetry, the ultimate calibration of the system as a whole must be in terms of its overall performance. It is necessary to verify the response of the system when a known quantity of heat is generated within it, or when a known quantity of gas is injected into (or in the case of oxygen removed from) the system. These are known as recovery trials or checks. Gas recovery trials, in addition to providing a combined check on the measurements of gas concentration and gas flow, also serve to test that the system is free from leaks.

7.4.1   *Recovery of heat*

Recovery trials for sensible heat are straightforward. A high-precision wattmeter is used to measure the electrical power delivered to a heating element inside the chamber. The element should preferably produce a similar quantity of heat to that usually measured from the subject, and ideally should have a similar shape to the subject under study. A resistance wire spread out across a light wooden frame makes a good heating element for calibration purposes. Heating elements that glow red-hot and electric light bulbs should not be employed as calibration heat-sources unguarded, although they are acceptable if placed inside protective radiation screens.

Recovery trials for evaporative heat are more complex and cannot generally be conducted in the absence of sensible heat exchanges. When there is no heat source in the calorimeter and water is run in at chamber temperature and allowed to evaporate completely, the total quantity of evaporative heat (Q_E) recorded should be equal to the mass of water (m) multiplied by the latent heat of evaporation at chamber temperature

(L_{T_a}). Since there is no alternative heat source, all the heat for vaporisation must have come from within the chamber and it will also be recorded as negative sensible heat, i.e.

$$Q_E = mL_{T_a} = -Q_N \qquad (7.8)$$

(It should be remembered that for gradient-layer calorimeters, the plate-meters also record the heat of cooling the recondensed water from calorimeter to platemeter temperature, which must be subtracted from the recorded level of Q_E – see Chapter 5, Section 5.3.4b.)

At the start of the test the latent heat of vaporisation is provided by cooling the pool of water and any tray or cloth used to contain it, but eventually all return to chamber temperature. Sufficient time must there-fore be allowed during the test to ensure re-equilibration of temperature. Because of the transient temperature changes no attempt should be made to apply equation (7.8) dynamically to relate instantaneous rates of heat exchange and weight loss.

The test just described *employing no additional heat source* provides the only sure way of checking the sensitivity of the evaporative measurement of a calorimeter. If a known quantity of electrical heat (M) is produced in the chamber while the water is being evaporated then the recovery of total heat is given by

$$M = Q_E + Q_N$$

but the partition of M between Q_E and Q_N now depends on whether the heat of vaporising the water comes from the heat source, the air or the water itself and it is no longer justifiable to relate Q_E to m (see Chapter 5, Section 5.7).

7.4.2 *Recovery of carbon dioxide*

Recovery trials for the estimation of carbon dioxide may be made by injecting a gravimetrically determined quantity of gas from a cylinder into the chamber over a convenient period. Another alternative is to enclose a piece of solid carbon dioxide ('dry ice') in a plastic box with a tightly fitting push-type lid. There is then just sufficient time available to weigh the box and place it in the chamber before the subliming gas blows the lid off the box. The volume of carbon dioxide (l at STP) which should be recovered is the weight change of the cylinder or plastic box divided by the density of carbon dioxide (1.977 g/l). The cylinder method requires a very precise balance to measure the relatively small weight decrease. The main problem with the other method is to ensure that there is no con-densation of atmospheric moisture onto the 'dry ice' before it is weighed.

7.4.3   *Recovery of oxygen*

For oxygen a recovery trial involves removing the gas from the chamber rather than adding it. This can be achieved by replacing air by nitrogen or by combusting oxygen in the air.

7.4.3a   *Nitrogen injection*

When normal inlet air is fresh air, it contains 20.95% oxygen. Hence every litre of nitrogen injected is equivalent to the removal of 0.2095 l of oxygen. The method depends on the accuracy with which the oxygen content of inlet air is known; it cannot be applied to closed-circuit systems, nor to fixed-volume confinement systems.

Fig. 7.17(*a*) illustrates the effect of injecting nitrogen on the oxygen concentration difference developed across an open-circuit. The calorimetric estimation of the volume of nitrogen injected is

$$V_{N_2} = -\frac{\dot{V}_E}{F_{I_{O_2}}} \int \Delta F_{O_2} \cdot dt$$

where $\dot{V}_E$ is exhaust ventilation rate of the system (l/min STPD); $F_{I_{O_2}}$ is inlet oxygen concentration; ΔF_{O_2} is the increment in oxygen concentration (a negative quantity).

If the quantity of gas injected is measured as the weight loss of a nitrogen cylinder (W_{N_2}) the recovery of nitrogen is given by

$$\text{Recovery} = \frac{\dot{V}_E}{F_{I_{O_2}}} \cdot \frac{1.2506 \, A_{N_2}}{k W_{N_2}} \%$$

where A_{N_2} is the area under the curve of Fig. 7.17(*a*) and k is a constant representing an area equivalent to a 1% change in oxygen concentration for 1 min. The constant 1.2506 is the density of nitrogen. The accuracy of gravimetric recovery trials using injected nitrogen is greatly improved if a lightweight (aluminium) cylinder is used as the nitrogen container.

At Leeds we make recovery checks on both carbon dioxide production and oxygen consumption simultaneously by injecting hourly a mixture of 21–79% carbon dioxide–nitrogen into the outlet airstream of an empty chamber of our multichamber open-circuit rat calorimeter (Tobin & Hervey, 1983; see Fig. 8.7). This simulates the gas exchange of animals having a respiratory quotient of one. The method could be theoretically extended to ruminant animals by including methane in a three-gas mixture. The test-gas mixture is made up from cylinders of the pure gases using a gas blender (Signal Instruments, Camberley) and it is then compressed to 250–500 kPa (2.5–5 atmospheres) for storage in a 10 l reservoir. We have found the composition of the mixtures delivered by the blenders constant to better than 1% over periods of years, though different from

Fig. 7.17. Typical output traces showing the decrement in oxygen concentration of the outlet airstream produced by (*a*) injection of nitrogen, and (*b*) oxidation of 'steel wool'.

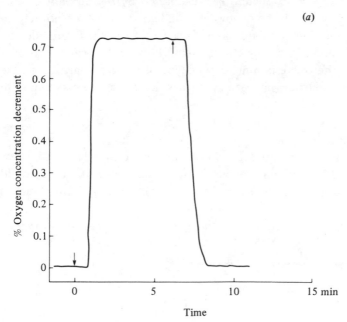

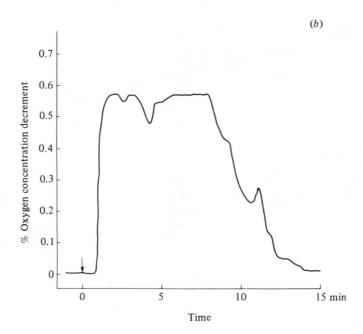

the manufacturer's nominal figures (Hervey & Tobin, 1982). The ultimate standard used for determining the composition of these gas mixtures is a 21 ml Haldane apparatus. The injection rate of the test-gas mixture is measured by a mass flowmeter, itself calibrated by a soap-bubble flowmeter. Providing that the main flowmeter and gas analysers do not rely on the same calibration standards as the recovery system, it provides an independent check. In our system the main flowmeter is calibrated by a wet gasmeter, but there is some dependence of the gas analysers on the primary carbon dioxide–nitrogen mixture which is used to make up span gases.

7.4.3b   *Steel-wool burners*

Young, Fenton & McLean (1984) have recently described a new gravimetric method of consuming a known quantity of oxygen by rapid oxidation of steel wool (the Fe-burner method). The apparatus, which is

Fig. 7.18. Apparatus for consuming a measurable quantity of oxygen by 'burning steel wool'.

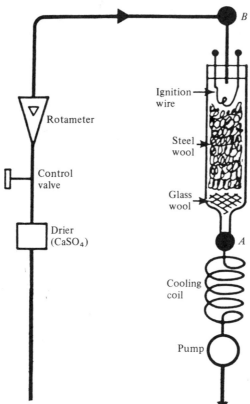

shown in Fig. 7.18, consists of a sealed air pump which draws fresh air through a drying tube containing calcium sulphate, a flowmeter (0.5–30 l/min) with a needle control valve, a combustion chamber, and a cooling coil. The combustion chamber, mounted vertically, is made of Pyrex glass, 1.8–5.0 cm ID, 0.2 cm wall, and about 45 cm long, narrowing at the base to a neck 1.25 cm OD and *c.* 10 cm long. The neck is joined by an airtight seal at point *A* to the cooling coil by use of a ground-glass fitting. At its upper end the combustion chamber is closed by a rubber stopper traversed by a plastic air-inlet tube and the two heavy copper electrode wires of the ignition system. The cooling coil is made of a 75 cm length of aluminium tubing of 0.3 cm ID.

The combustion chamber is packed with a plug (1–2 g) of glass wool at the bottom and then with degreased steel wool, compressed to a density of 0.1–0.2 g/ml, to approximately 8 cm from the top. The ignition system consists of a length of fuse wire attached to the copper wire electrodes to form a short loop (about 7 cm) which comes into contact with the steel wool when the stopper is placed in the top of the combustion chamber. The combustion chamber complete with stopper is weighed and connected into the apparatus at points *A* and *B*. The ignition system electrodes are wired to a power source which provides 2 A via a contact switch.

Once the Fe burner is set up and joined to the calorimetry system, the pump is started and air flow adjusted to produce a velocity of between 15 and 50 cm/s in the combustion chamber. After at least 30 s of steady baseline output from the calorimeter recording system, the ignition system is momentarily activated, and the steel wool starts to oxidise immediately. By control of the airflow rate through the combustion chamber, the exothermic oxidation reaction maintains a steady self-sustaining orange-red glow that travels slowly down the chamber until all the steel is oxidised. At the conclusion of the burn the airflow is maintained for a further 5 min to allow the apparatus to cool. The combustion chamber is then removed and reweighed. Combustion chambers can be made in various sizes to calibrate calorimeters varying in size from those designed for rats up to those for cattle.

A typical trace of the output from an oxygen analyser is shown in Fig. 7.17(b). The calorimetric estimate of the volume of oxygen absorbed (V_{O_2} litres) is given by

$$V_{O_2} = \frac{-\dot{V}_E}{(1 - FI_{O_2})} \int \Delta F_{O_2} \cdot dt$$

Recovery of oxygen may be expressed as

$$\text{Recovery} = \frac{\dot{V}_E}{(1 - FI_{O_2})} \cdot \frac{1.429 \, A_{O_2}}{kW_{Fe}} \%$$

where A_{O_2} is the area under the curve of Fig. 7.17(b), W_{Fe} is the weight gain of the combustion chamber, and 1.429 is the density of oxygen. Young *et al.* (1984), from comparisons using two calorimetry systems, reported that the observed recovery by the Fe-burner method was 1.3% higher than that by the nitrogen method. However, this discrepancy was subsequently found (McLean & Young, 1985) to be due to carbon impurity in the steel wool used, with the consequent formation of carbon monoxide and dioxide gases. It may be shown theoretically (McLean & Young, 1985) that the expected recovery of oxygen by the Fe-burner method is

$$\text{Recovery} = 100 + (114 \times R_{CO}) + (164 \times R_{CO_2}) \%$$

where R_{CO} and R_{CO_2} are the ratios of the volumes of gases produced to the volume of oxygen consumed in the burner. The values of R_{CO} and R_{CO_2} may be simply determined by passing samples of exhaust gas from the burner through Draeger tubes.

7.4.4   Combined methods

7.2.4a   *Alcohol burners*

The classical calibration method in animal calorimetry is combustion of alcohol (Carpenter & Fox, 1923). The chemical formula for the combustion of ethyl alcohol (M.W. 46.0869) is:

$$2C_2H_5OH + 7O_2 \rightarrow 4CO_2 + 6H_2O + 1371 \text{ kJ/mol}$$

Hence for every gram of alcohol burned, the oxygen consumed is 1.703 l (STP), the carbon dioxide produced is 0.973 l and the heat evolved is 29.75 kJ.

The method has been used very extensively (see Chapter 4, 4.2.4e) and there are countless reports of oxygen and carbon dioxide recovery to $\pm 5\%$ and better. The method must certainly be accurate to at least this level. However the method is not definitive because there is no means by which the observer can be sure that the alcohol is all burnt and that there has been no loss through vaporisation. For this reason it is not our method of choice. Brockway *et al.* (1977) after experimenting with many burner designs estimated that their best design resulted in 98% combustion of alcohol. They pointed out however that despite incomplete combustion and additional errors due to inaccuracy of air flow measurement, the burning of alcohol should always produce a respiratory quotient of 0.667. This provides a cross check on the accuracy of oxygen and carbon dioxide analysers.

Various hydrocarbons have also been combusted as a check on calorimeters, for example Young *et al.* (1975) used propane in their open-circuit calorimeter for cattle and Jequier & Schutz (1983) have reported recoveries of both oxygen and carbon dioxide accurate to $\pm 2\%$ from combustion of butane gas in their respiration chamber for humans. Commercial sources of butane, particularly those sold for camping stoves should be treated with caution since they are often, in fact, mixtures of variable composition containing isobutane, butane and propane.

7.4.4b   *Agreement between direct and indirect calorimetry*

In calorimeters that record heat both directly and indirectly, there should be perfect agreement between the two measurements from an alcohol burner regardless of whether or not the combustion is complete. Similar agreement cannot necessarily be expected when there is a living subject in the chamber because the body-heat content of the subject may change.

The rate at which heat is produced in large animals would be sufficient, if it were all retained within the body, to raise body temperature by 1 °C in about an hour; in small animals the equivalent time is much less. Consequently, in order to maintain a reasonably steady body temperature, the rate of heat production must be very closely matched by its rate of loss. However, in the short term the matching is seldom perfect and part of the adjustment to changes both in metabolic activity and environmental temperature involve changes in mean body temperature, i.e. in heat stored in the body. The change in body temperature can be sufficiently large to result in considerable imbalance between simultaneous observations by direct and indirect calorimetry over short periods. In homeotherms body temperature is regulated to within a few degrees (any failure of this quickly leads to death) so that over an extended period of time the agreement between direct and indirect calorimetry should become closer. But conversely estimation of body-heat storage over a prolonged period of time from the difference between direct and indirect calorimetry would demand almost impossible accuracy. A systematic error of 1% in either would appear for man to imply a change in mean body temperature of about 1 °C per day.

It has often been claimed that agreement between direct and indirect calorimetry proves the validity of applying the first law of thermodynamics to living organisms. Garby (1985) has criticised this statement on the grounds that proof of the first law requires observations of the flow of energy in and out of a system as well as energy change within the system. We can and indeed must assume that the laws of thermodynamics apply to all animate and inanimate systems. Agreement between direct and indirect calorimetry of living subjects merely confirms that our measurements are accurate and that the assumptions we make in the course of our computations are valid.

7.4.5   *Calibration: the golden rules*

Whole-system checks on calorimeters represent the final validation of the method. Very many individual measurements and computations are made and the end-result also depends on correct functioning of
the whole apparatus, freedom from leaks, thermal stability, etc. No instrument should be relied upon to provide a certain degree of precision just
because the manufacturer's specification says so. It always should, but it
often does not. Sometimes, of course, this is because it is not being used
properly or within the conditions stated in the specification. But whatever
the cause, the golden rule is to check, so far as is possible, the accuracy of
all instruments in the system; and then check the accuracy of the entire
system by a recovery trial. If the recovery is significantly different from
100 % then there must be a reason, and the error must be found. The use
of 'fiddle factors' to adjust the result to compensate for the observed over-
and under-recovery is bad practice and should be avoided. It can lead to
totally incorrect conclusions in the interpretation of results.

It is perhaps worth quoting here the words of Du Bois (1927):
'The man who conducts metabolism experiments should be perfectly
frank with himself regarding the accuracy of his results. He should estimate the possible percentage error and never conceal from himself the
fact that the technic of a certain experiment was rather poor. This seems
only common honesty but it is rather difficult to practice. Having made an
experiment it becomes your own, and you are fond of it. It is only after
bitter experience that you learn to discard all questionable results. They
lead you to false conclusions which bring endless trouble. They destroy
your self confidence and the confidence of others.'

7.5   Some additional problems encountered

Many hours can be spent in searching for the cause of some
discrepancy or malfunction in a calorimetry system. It would be impossible
to draw up a comprehensive list of all possible faults and the means of
curing them. Here, in no particular order, we list some of the more elusive
ones which we have encountered, and which have not so far been mentioned in the text. They are mostly concerned with respiratory gas sampling and analysis. Some additional problems are covered in the section in
this chapter on sample conditioning.

Gas piping. Rubber and rubber-based materials can exhibit a similar
adsorption–desorption capability for oxygen to that described in the section on sample conditioning for silica gel and Drierite in respect of carbon
dioxide. This tends to be a long-term effect on gas concentrations and can

be detectable for up to an hour after changing the sample. We have no proof as to whether this effect we have observed is a true adsorption of oxygen or a temporary evolution of some interfering gas from the rubber after a period when it has not been ventilated. Diffusion of carbon dioxide is possible through a variety of plastic materials including polythene and PVC. Whilst this may not be a problem when gas is flowing continuously along short lengths of tubing made of such materials, particularly when the gas concentrations are similar to those of fresh air, it is a problem if long lengths are used or gases are allowed to stand in them for long periods (Janac *et al.* (1971)). In such cases we use stainless steel tubing. It is also a major problem if gases are stored in bags made of rubber or rubber-based materials for later analysis (see Chapter 4); metalised bags are preferable.

Leaks. Contaminating gases which affect the output of gas analysers can enter calorimeters through leaks or can arise from equipment within the system, e.g. leaking refrigerant gas. It is always desirable to check for equality of composition between air entering and leaving the system in the absence of any subject. In our multichamber calorimeters one cage is always left empty for such a check. Loss of expired air is always a potential danger and great care is necessary in design and use of mouthpieces, facemasks, headboxes, hoods and chambers. In human studies a beard or stubble creates particular problems if facemasks are used, and it is important to anticipate this, since these may be grown during the course of a long field study (I. F. G. Hampton, personal communication). Direct calorimeters should also be checked to detect any leaks or unexpected sources of heat; if work is done in a direct calorimeter it is important that any energy transformation is restricted to the calorimeter or accounted for (Chapter 5).

Refrigerant capacity. The accumulation of dust between the fins of the evaporators can markedly reduce the capacity of air-cooled refrigerators and they should be blown clean regularly.

8

Some complete open-circuit systems

Closed-circuit or confinement chambers for measuring oxygen consumption of small animals can be constructed mainly from standard laboratory ware. The system described by Smothers (1966) is a good example. Alternatively some degree of automation may be added (e.g. Heusner *et al.*, 1971; Stock, 1975). The classical gravimetric method of Haldane (1892) combined with a modern electronic balance has still much to commend it. All of these chambers may be easily assembled without resort to expensive equipment, but the animal is confined and cannot readily be handled; and measurements are restricted to collection periods of short duration. Whilst these methods are valuable in some circumstances, they should be used with an awareness of their limitations and the results obtained with them interpreted with caution. For example, our experience is that scaling up a few short measurement periods does not give an accurate estimation of 24 h energy expenditure; and in addition the animals may take some time to adapt to their new housing and measurements over this period may be misleading.

For fast response measurements, particularly on humans and large animals, and for long-term studies, an open-circuit system employing flowmeters, electronic gas analysers and recording equipment is necessary. Various combinations of equipment are possible depending on the degree of accuracy required and the finance available. In this chapter we describe a few complete systems that have been found to operate satisfactorily, and which could be adopted in part or in entirety by the reader.

8.1 Using a simple voltage recorder

In this basic system (Fig. 8.1) fresh air is drawn by a pump through an animal chamber, headbox, hood or facemask and the flowrate measured. Either all, or for large animals part of, the exhaust air is passed via a drying tower containing $CaCl_2$ or $CaSO_4$ to a linear single-channel oxygen analyser (e.g. Servomex OA137 or 540). Provision is included for measuring the temperature, humidity (e.g. wet-bulb temperature) and pressure of the air at point A for adjustment of the observed flowrate to STPD.

The simple circuit of Fig. 8.2, which can be easily made up from a dry battery and resistances, is used to interface the analyser output to a single-channel potentiometric chart recorder. The values of the various resistors are chosen to suit the voltage output of the oxygen analyser and the range of the recorder employed. In the example given, it is assumed that the

Fig. 8.1. A simple open-circuit system for measurement of oxygen only and using a chart recorder. D, drying tower; P, pumps.

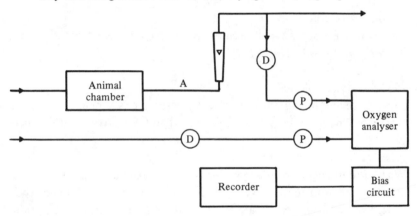

Fig. 8.2. Detail of the bias circuit used in the simple system depicted in Fig. 8.1.

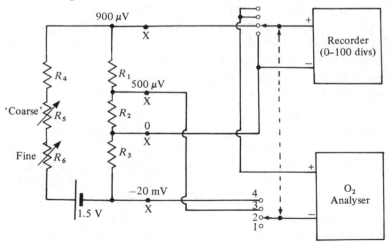

analyser produces approximately 1 mV per 1 % oxygen content of dry air, and that this is being fed to a recorder with a fullscale of 100 divisions and a span of 1 mV.

The critical part of the circuit is the resistance chain comprising R_1: R_2: R_3 which must have the resistance values accurately in the ratio 0.4:0.5:20.0. These resistors are wound from manganin or some other resistance wire that is insensitive to temperature changes, and short copper connections are included between the manganin and terminal posts at points X. This construction prevents accidental alteration to the resistance values when external wiring is connected to the terminal posts.

The switch is a four-position wafer-switch and is operated to set up the system according to the following procedure:

Position 1 (recorder zero).
The recorder input is short-circuited for adjustment of its electrical zero to 0 divisions.

Position 2 (set voltage).
The coarse and fine resistors R_5 and R_6, respectively, are adjusted to make the recorder read 90 divisions (equivalent to 0.9 % oxygen differential).

Position 3 (analyser zero).
With pure nitrogen from a cylinder passing through the analyser, its zero controls are adjusted to give 50 divisions on the recorder. (The recorder is here biased to its mid-scale so as to allow adjustment from both sides of the analyser zero point).

Position 4 (analyser span and run).
Dry fresh air (containing 20.95 % oxygen) is fed to the analyser and its sensitivity adjusted so that the recorder reads 95 divisions. On this position the recorder is biased by a voltage equivalent to -20%.

The results of this four-part procedure has been to adjust the analyser and recorder so that the latter is now spanned to operate over the range 20–21 % oxygen. It will be noted that the system does not depend on the absolute accuracy of either the analyser or recorder, but only on the accuracy of the ratio $R_3/(R_1 + R_2) = 20/0.9$. Provided that this ratio is accurate, the action of setting the voltage supplied by the battery to give the equivalent of 0.90 % oxygen at switch position 2 automatically sets the bias on position 4 to -20%. It is also essential that the oxygen analyser output is linear over the whole range 0–21 % (see Chapter 7, Section 7.2.6) and that the span gases, nitrogen and fresh air, are pure and dry. These are not difficult conditions to fulfil. This simple system is susceptible to change in barometric pressure and will require setting up again with nitrogen and fresh air should barometric pressure change.

8.2   Switched-range recorder and twin-channel oxygen analyser

This system is similar to that just described but employs slightly more sophisticated equipment; namely a twin-channel oxygen analyser (Servomex OA184 or 1184) and a multi-range recorder, again scaled from 0 to 100 divisions. The resistance network for providing voltage bias is no longer needed. The procedure is as follows:

(1) With zero input to the recorder, its zero control is adjusted to give 10 divisions on the 10 mV range.

(2) Each analyser channel in turn is supplied with dry nitrogen and the zero set using the panelmeter of the instrument.

(3) The analyser is now set to differential mode and 1 % oxygen span, and its output fed to the recorder. If necessary the zero of the reference channel may be trimmed to ensure that the recorder reading is exactly ten divisions.

(4) The recorder range switch is set to a low sensitivity position to protect it against excess voltages. The analyser is then returned to independent mode and 25 % span, and dry fresh air is supplied to each channel. The recorder can now be switched back to the 10 mV range and the span of each channel set to read 93.8 divisions (note that 93.8 minus a zero setting of 10 divisions gives 83.8 % of the 25 % span which is equivalent to 20.95 % oxygen).

(5) The analyser is returned to differential mode and 1 % span and the span control of the reference channel trimmed to give zero difference (10 divisions on the recorder).

(6) The reference and sample channels can now be fed with dried gas samples from the inlet and outlet airstreams of the calorimeter.

The system is thus set up to give a differential oxygen span covering the range -0.10 to $+0.90$ % with the zero point at ten divisions. If barometric pressure alters, the zero will not drift but there will be a change in sensitivity, and the setting up procedure must be repeated.

8.3   Systems using divided reservoirs

In systems such as those described in the last two sections, the output trace usually exhibits continuous fluctuations caused by frequent changes in the respiratory activity of the subject; these can make it difficult to estimate the mean level of oxygen concentration from the chart record. Also it is necessary to interrupt measurements of exhaust gas concentration from time to time in order to check inlet gas concentration or zero drift. A method overcoming these problems was described by McLean (1974) and is illustrated in Fig. 8.3. Exhaust air from the chamber, headbox or mask is passed into one side of a divided reservoir, made up from two stainless steel basins which are clamped face to face with a loosely-folded rubber diaphragm in between. As one side of the reservoir fills, air is expelled from the other side and passed to the gas analysers. When one side is completely filled and the other side empty, the pressure rise in the filled side is detected by a pressure switch which operates solenoid valves reversing the flow through the reservoir.

The time taken for filling each side may be adjusted over a wide range depending on the flowrate and the size of the reservoir. In some applications it is possible to use the reservoir as the air flowmeter by recording the time taken to fill the two halves using the switching operations as markers. For large animals this may become impracticable because of the large size of reservoir required; in this event only a sample of the airstream can be passed through the reservoir.

Because the gas sample being analysed between switching operations is of fixed composition, the output trace of the recording instrument reaches a steady level as soon as all the residual gas from the previous sample has

Fig. 8.3. Collection of exhaust air using a divided reservoir. D, drying towers; P, pumps; R, reservoir; PS, pressure switches; SV, solenoid valves.

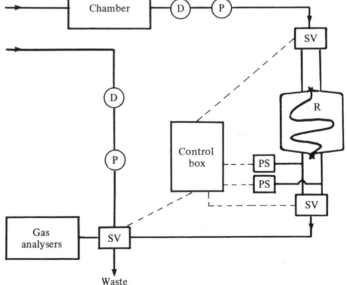

been washed out of the analysers; hence the time for each measurement can be reduced to 2 min or less. Provided that the reservoir's switching time is at least twice this duration, a timer can be included which switches the gas analyser to chamber inlet air for part of each reservoir cycle. By this means it is possible using a single-channel analyser to monitor both inlet and outlet airstreams without any loss of sampling time. We have also used a divided reservoir at the Hannah Institute to monitor the outlet gas from a bank of six rat-chambers. A series of solenoid valves operated sequentially by the switching action of the reservoir diverts the gas outflow from each chamber in turn into the reservoir for a full switching cycle; meanwhile gas from the remaining chambers is exhausted to waste.

The divided-reservoir system of gas collection lends itself well to being monitored by a low-cost computer. The six-chamber system just described employs a BBC micro-computer. This device has four analogue input channels and eight digital input or output channels. The four-analogue channels are used to monitor oxygen concentration, carbon dioxide concentration, barometric pressure and reservoir temperature. Two digital channels are used to detect filling of the two sides of the reservoir and the remaining digital channels to sequence the sampling between the six individual cages. Flowrate is estimated from the time between switching operations as measured by the internal clock of the computer. Humidity does not require measurement because the gasstream is dried before it is passed to the reservoir. Once set up this system can be left on automatic operation for 24 h. The system is quite new and could still be improved. One problem we have experienced is a settling-down time required after first switching on, and which appears to be associated with gas adsorption or release by the rubber diaphragm (see Chapter 7, Section 7.5). A polythene diaphragm has been found to overcome this problem, but this material becomes brittle after a few days of continuous flexing and developes leaks.

8.4 The Leeds indirect flow-through calorimeters

The systems described below show successive increments in their level of sophistication of control and data handling. Most of the differences between the systems have largely been determined by the availability of funds and ideally most of the components of the computer-controlled system described in Section 8.4.3 would have been incorporated into the other two systems. Two major criteria have affected their design:

(1) Accuracy and precision: as we discussed in Chapter 7 the essential measurements in flow-through calorimetry are of flowrate and the dif-

ference between outlet and inlet oxygen concentration. In all our systems we have incorporated what we think are the best flowmeters and oxygen analysers – mass flowmeters and Servomex paramagnetic oxygen analysers. If insufficient funds are available for these (or good alternatives such as wet gasmeters or calibrated electronic rotameters), there is little point in attempting accurate indirect flow-through calorimetry. If funds are limited it is better to restrict measurement to that of energy expenditure by means of a good flowmeter and oxygen analyser, and to avoid attempting a more ambitious system of poorer quality.

(2) Flexibility: we have built our calorimeters on the basis that as more funds became available the calorimeters could be upgraded by the addition of further components without any need for replacement of anything but the cheaper simple items. The order of priority for our calorimeters has been accurate measurement of energy expenditure, automated calibration of gas analysers, computer control and data handling, and automated recovery checks; the importance placed on additional gas analysers for carbon dioxide or methane or both depends on the application for which the calorimeter is designed.

8.4.1   The semi-automated human calorimeter

We have built a semi-automated flow-through calorimeter for studies on humans (Hampton *et al.*, 1982). It comprises (Fig. 8.4) a perspex headbox with a polythene skirt that can be fixed around the neck. Fresh air is drawn through a wide-bore plastic tube into the headbox and then through the outlet by means of a pump placed at the end of the sample line; flowrate is adjusted by means of a needle valve. The relative humidity of the outlet air is measured by means of wet- and dry-bulb thermistors; because of the high rate of air flow there is no attempt at this stage to dry it. The rate of air flow is measured by a 0–150 l/min Hastings–Raydist mass flowmeter and its volumetric output (see Section 6.3.3) is adjusted to dry conditions by using the humidity data.

A sample is taken at a rate of about 1 l/min from the main airstream by a small pump, passed through a solenoid valve and introduced into the gas analysis line. There are two other lines, each with a solenoid valve, leading to the gas analysis line; one supplies fresh air (the zero gas for differential measurements) and the other supplies a range of calibration gases from 19.55 to 20.95 % oxygen in seven 0.2 % steps. The calibration gases are produced by a Signal gas blender; in normal use one setting is selected to provide a suitable span gas. The blender is also used for occasional linearity checks.

A small d.c. pump passes the gas in the analysis line to a Permapure drier, through a constant pressure head device (the 'barostat' – see section 7.2.2), and then into a Servomex OA 540 oxygen analyser. The bypass system on the analyser which normally provides a constant pressure head is not used because of the large amount of calibration gas that would be necessary. The analyser is fitted with a zero suppression module; a back-off voltage of 8 V is applied to the usual 10 V output for 25%, and this gives 0 V for 20% oxygen and 0.4 V for 21% oxygen. The output from the mass flowmeter and oxygen analyser are displayed directly on a twin-channel chart recorder.

A simple control system determines which of the three gases i.e. sample gas, fresh air, or calibration gas is allowed to enter the analysis line. It was designed, like the rest of the control circuits described below by E. S. Stainthorpe in the Department of Physiology at Leeds. Control is achieved through the electrically operated Danfoss solenoid valves. They can be operated from a control box by individual three-position toggle switches giving on, off and auto. For automatic control, the control box also contains three thumbwheel switches which allow the operator to dial in the number of minutes (from 1 to 99) each of the three solenoid valves is open in turn; they operate in the sequence fresh air (zero gas), span gas, and then sample gas. The cycle is repeated until the apparatus is switched off or manual control selected. For automatic control all three toggle switches are set to auto.

Fig. 8.4. Diagram of the flow-through calorimeter used at Leeds for measurement of energy expenditure of man.

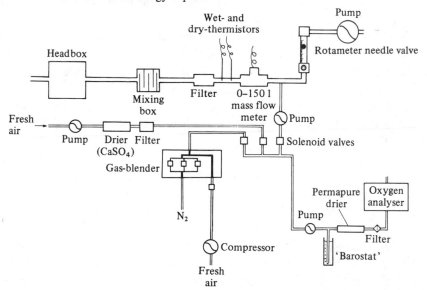

The thumbwheel switches are connected to Intersil ICM 7250 timers which convert the time setting to a binary code decimal format; each timer produces sufficient output to drive a LED indicating the state of the circuit and a solid state relay that switches mains current to and from a solenoid valve. To start the automatic sequence, a button (START) is pressed and all three solenoid valves opened. A second press button (RUN) starts the first cycle by triggering a monostable circuit. This provides a current pulse of the correct width to give a clean trigger pulse that switches on the first timer, which continues to activate the solenoid allowing entry of fresh air, and the appropriate LED is lit; the other two solenoid valves are closed. At the end of the set time the first timer, solenoid and LED switch off, and timer two and its solenoid and LED are activated; this allows entry of the span gas to the analysis line. At the end of the preset time, timer two switches off, timer three switches on and sample gas is admitted to the analysis line. Finally, at the end of its preset time, timer three activates the monostable circuit which begins the sequence once more. The control system is convenient and yet quite cheap and easily constructed from commercially available components.

8.4.2   The semi-automated, multichamber small animal calorimeter

The circuit of our simple multichamber system is shown in Fig. 8.5. It has five perspex chambers that act as animal cages, four contain rats and one is empty. Each chamber contains a food hopper and water bottle and has a mesh floor, beneath which is a tray to collect scattered food and excreta. The chambers are kept in a room which provides a controlled environment; air is continuously circulated through the room from atmosphere (analysis shows that the composition of the room air is not measurably different from fresh air). Fresh air enters the chambers from the room through small holes around the base; internal baffles prevent a rat placing its nose over a hole and expiring directly through it. Air leaves the chamber by two tubes, each pierced with numerous 1 mm diam. holes, which cross the chamber below its roof and join to form a single outlet line. Each chamber is provided with a tower 30 cm high × 15 cm diameter containing a wire-mesh cylinder of anhydrous calcium sulphate (Drierite) or silica gel (adsorption of carbon dioxide is not critical at this point since we are not monitoring rapid changes in gas concentration); the air in the chamber is continuously circulated through this, by an adjustable speed fan, to control cage humidity.

The air lines from the five cages pass through the wall of the controlled-environment room to the measuring system in an adjacent room. Visual flowmeters (G. A. Platon, range 0–5 l/min) indicate the flow through each

Fig. 8.5. Diagram of the semi-automated multichannel calorimeter used at Leeds for measurement of energy expenditure of rats.

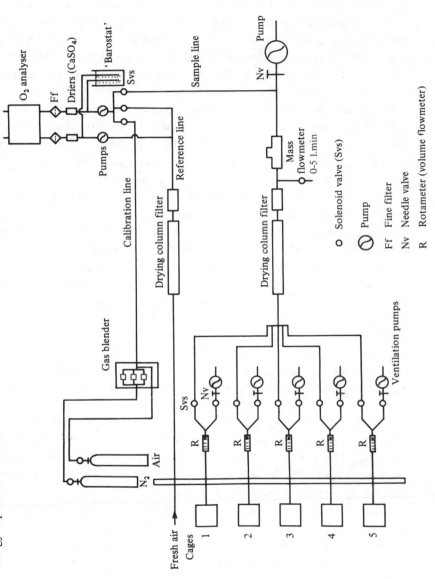

cage; the lines then bifurcate to lead either to the measurement line or to pumps that ventilate the cages to air. Switching is performed by a pair of Danfoss solenoid valves for each cage; cage ventilation is by means of diaphragm pumps, and flowrates are adjusted, typically to 1 l per rat, by needle valves attached to each pump. The measurement line has a more powerful centrifugal pump since a drying tower and filters cause a greater resistance in that line to flow; control of flowrate is by means of a needle valve. The flowrates through all cages, on-line and off-line, are made as equal as possible by adjusting pump speeds.

In the measurement line the air passes through a drying column 100 cm long × 5 cm internal diameter containing Drierite, a short column of glass wool which acts as a filter, and then through the 0–5 l/min Hastings Raydist mass flowmeter. The needle valve and pump are the last components on the line.

The oxygen concentration of the gas samples is measured by a Servomex OA 184 differential gas analyser; the sample enters the analyser through the gas analysis line whilst dry fresh air, identical to that entering the cages, enters through the reference line. Entry to the gas analysis line is controlled by a three Danfoss solenoid valves. These admit either air from the measurement line, dry fresh air (the 'zero' gas in differential measurement) from the reference line or a calibration gas produced from a Signal gas blender identical to that described above. Air is drawn off the measurement line or the reference line for analysis at approximately one litre per minute by a small pump that is electrically coupled to the appropriate solenoid valve. The calibration gases enter directly from the Signal gas blender that is electrically coupled to the third solenoid valve.

The gases introduced into the analysis line are pumped through a constant pressure head device – the water 'barostat', a small column containing Drierite, and a Microflow LF32 disc filter and into the oxygen analyser at a rate of about 600 ml/min. The reference gas is also pumped through an identical conditioning pathway.

The cycle of operations normally lasts one hour. Each chamber is in turn 'on-line', i.e. connected to the measurement line for 9 min. The oxygen analyser is then calibrated for nine minutes with zero gas and six minutes with a span gas. The empty cage provides a 'blank' check every hour. As in the previous design control is achieved through the action of the solenoid valves and additionally by switching the pumps on and off.

The control circuit operates through a master clock which is a RS 2240 programmable timer chip. Although the chip has a variable time constant, in this system it is set to 9 min. The timer is connected to a ring counter made up of seven *J–K* flip flops or bistables (for details of such components the reader should refer to Williams, 1984). Each flip flop has two outputs Q or $\bar{Q}$ (i.e. logic levels of 1 or 0), and four inputs, a clock pulse line, a $+5$ V power supply line and *J* and *K* inputs which are connected to the Q and $\bar{Q}$ outputs of the previous flip flop (see Fig. 8.6). A clock pulse travels to all the flip flops every 9 min; this reverses the state of only the flip flop which is currently 'on-line' and the next 'off-line' flip flop in the sequence. The current output of the flip flops is inadequate to drive solid state relays and so each flip flop is connected to a pair of Darlington drivers. One driver in each pair is connected to the Q output of the flip flop, and the other to the $\bar{Q}$ output. The driver is a logic device which provides either zero current output or one which is sufficiently large to drive one or more solid state relays and LEDs (it also acts as a buffer and prevents spurious signals activating the relay).

The flip flops, through the drivers and relays, control the solenoid valves and pumps and bring the five cages 'on-line' in turn. Finally they admit the zero and span gases to the analyser. To obtain the 6 min of span gas the clock switches on a subsidiary timer (RS 308–850 Precision timer) and itself switches off. At the end of the six minutes this timer switches the main clock back on and then itself switches off. The main timer once more changes the state of the first flip flop, cage one comes on-line and the sequence begins again.

As in the previous system the automatic control can be over-ridden.

Fig. 8.6. The organisation of the timer mechanism used in the semi-automated rat calorimeter at Leeds.

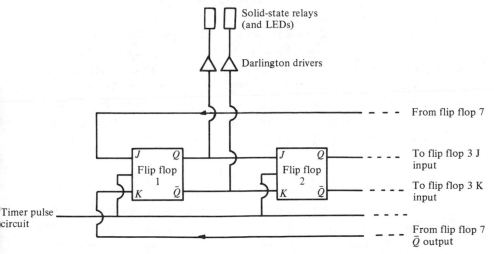

Three toggle switches on the panel allow the three solenoid valves that control entry to the gas analyser to be individually switched on, off or placed under auto control through the timer system. In the off position the appropriate flip flop is electrically disconnected from its Darlington drivers whilst in the manual on mode the solid state relay and LED are connected to a potential which is sufficient to energise them.

The control box also conditions the output from the oxygen analyser. The signals from the sample and reference sides of the analyser are fed to a differential amplifier and the differential signal (10 mV for each per cent differential) amplified 100-fold. The output is displayed on a 1.999 V digital panelmeter and supplied to a twin channel chart recorder which also records the flowrate.

Despite the additional complexity of this over the previous system the control mechanism is still quite cheap and easy to implement.

8.4.3 *The computer-controlled multichamber small animal calorimeter*

Our most advanced flow-through calorimeter (Fig. 8.7) is similar in basic design to that just described and is briefly mentioned in Chapter 4. It has been described in detail elsewhere (Tobin & Hervey, 1984). The major differences are that it includes a Grubb Parsons IRGA 20 carbon dioxide analyser, an additional Signal gas blender to provide a 21 % carbon dioxide in nitrogen mixture, apparatus for checks on the accuracy of the gas exchange measurements, and computer control and data handling.

The 21 % carbon dioxide in nitrogen gas mixture is used as the diluate in the main Signal gas blender rather than just nitrogen as in the previous designs; thus it is possible to calibrate the oxygen and carbon dioxide analysers simultaneously. In addition the gas mixture is injected through a 0–100 ml/min Hastings–Raydist mass flowmeter into the outlet from the blank cage once in each hour. Introducing a known amount of carbon dioxide and replacing a known amount of oxygen in the gas stream forms the basis of a recovery check for the whole system (see Chapter 7).

The cages are individually ventilated by 12 V d.c. centrifugal pumps; because these remain on throughout, an additional solenoid valve is needed so that when a cage goes on-line the pump is able to draw air from the atmosphere.

The sequence of operation of the calorimeter is also slightly different from that of the calorimeter described in Section 8.4.2 above. The analysers are calibrated for twelve minutes in each hour by introducing 'zero' and 'span' gases for six minutes each into the sample line. The animal chambers come on-line in turn for 8 min; cage three comes on-line twice,

Fig. 8.7. Diagram of the computer-controlled multichannel calorimeter used at Leeds for measurement of energy expenditure of rats.

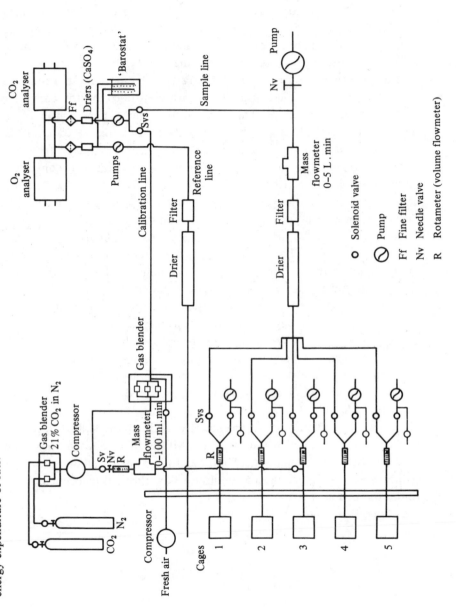

first for the recovery check when the carbon dioxide–nitrogen mixture is added to the air leaving the cage, and secondly as the blank cage (effectively as if it were a sixth cage). The output from the analyser calibrations over the last minute of each 6 min cycle is recorded by the computer; the output from the analysers during measurement of gas concentrations in the air from the cages is recorded over the last 4 min.

A minicomputer controls the operating cycle by means of the solenoid valves on the various gas lines. These are controlled by the computer through solid state relays. The state of the solenoids is continuously monitored, and the state of the system as a whole displayed on LED indicators. After appropriate amplification, data are collected from the analysers and mass flowmeters by the computer via 10 bit A–D converters (giving a resolution of 1 part in 1024 or 0.1 % of full scale); the data are also displayed on digital panel meters. Each datum is measured repeatedly during the measuring phases of the cycle and the readings in each 4 min reading (or 1 min calibration) phase averaged. At the end of each hour, the mean readings for what are effectively six cages are calculated from interpolation of the calibrations at the beginning and end of the hour. The results are printed out hourly. The present computer program is written in an extended Fortran developed by Professor G. R. Hervey for DEC computers (Hervey, 1973), and implemented here on a PDP–8A minicomputer. This is currently being replaced by a MOCOM 6809-based computer, programmed in a combination of Pascal and Basic. An alternative version of the program reads continuously from one cage and computes results at 5 min intervals; other schedules could easily be implemented.

Such a calorimeter is inevitably expensive and requires skilled electronic engineers to help build it (alternatively MOCOM in Southampton can now supply the electronic control and recording systems); it does however reduce much of the tedium of running a flow-through calorimeter continuously for many weeks.

Appendix I

Calculation of calorific factors from elemental composition

Suppose that 1 g of a substance contains f_C, f_H and f_O g of carbon, hydrogen and oxygen respectively, with atomic weights $a_C = 12.011$, $a_H = 1.008$ and $a_O = 15.999$.

Oxygen required for CO_2 form- $= 22.41 \times f_C/a_C = 1.8658 f_C$ l
ation

Oxygen required for H_2O form- $= 22.41 \times f_H/4a_H = 5.5580 f_H$ l
ation

Internal oxygen available $\quad = 22.41 \times f_O/2a_O = 0.7004 f_O$ l

Net oxygen consumed $\quad = (1.8658 f_C + 5.5580 f_H$
$\quad -0.7004 f_O)$ l

CO_2 produced $\quad = 22.41 \times f_C/a_C$
$\quad = 1.8658 f_C$ l

Respiratory quotient $\quad = 1.8658 f_C/(1.8658 f_C$
$\quad +5.5580 f_H -0.7004 f_O)$
$\quad = 1/(1 + 2.9789 f_H/f_C$
$\quad -0.3754 f_O/f_C)$

Water produced $\quad = f_H \times (2a_H + a_O)/a_H$
$\quad = 8.936 f_H$ g

Appendix II

Solution of equations for heat production, and carbohydrate and fat metabolised

The equations to be solved (see Section 3.4) are:

$$V_{O_2} = Ka_K + Fa_F + Na_N - V_G a_G \tag{A2.1}$$
$$V_{CO_2} = Ka_K r_K + Fa_F r_F + Na_N r_N - V_G a_G r_G \tag{A2.2}$$
$$M = Ka_K q_K + Fa_F q_F + Na_N q_N - V_G a_G q_G \tag{A2.3}$$

Multiplying equation (A2.1) by r_K and subtracting equation (A2.2):

$$V_{O_2} r_K - V_{CO_2} = Fa_F(r_K - r_F) + Na_N(r_K - r_N) - V_G a_G(r_K - r_G)$$

$$\therefore F = \frac{[V_{O_2} r_K - V_{CO_2} - Na_N(r_K - r_N) + V_G a_G(r_K - r_G)]}{a_F(r_K - r_F)} \tag{A2.4}$$

Multiplying equation (A2.1) by r_F and subtracting it from equation (A2.2):

$$V_{CO_2} - V_{O_2} r_F = Ka_K(r_K - r_F) + Na_N(r_N - r_F) - V_G a_G(r_G - r_F)$$

$$\therefore K = \frac{[-V_{O_2} r_F + V_{CO_2} - Na_N(r_N - r_F) + V_G a_G(r_G - r_F)]}{a_K(r_K - r_F)} \tag{A2.5}$$

Substituting K and F from equations (A2.4) and (A2.5) into (A2.3):

$$M = q_K[-V_{O_2} r_F + V_{CO_2} - Na_N(r_N - r_F) + V_G a_G(r_G - r_F)]/(r_K - r_F)$$
$$+ q_F[V_{O_2} r_K - V_{CO_2} - Na_N(r_K - r_N) + V_G a_G(r_K - r_G)]/(r_K - r_F)$$
$$+ Na_N q_N - V_G a_G q_G$$

Rearranging gives:

$$M =$$

$$\frac{\left\{ \begin{array}{l} V_{O_2}(q_F r_K - q_K r_F) + Na_N[q_N(r_K - r_F) - q_F(r_K - r_N) - q_K(r_N - r_F)] \\ + V_{CO_2}(q_K - q_F) - V_G a_G[q_G(r_K - r_F) - q_F(r_K - r_G) - q_K(r_G - r_F)] \end{array} \right\}}{(r_K - r_F)}$$

$$= \alpha V_{O_2} + \beta V_{CO_2} + \gamma N + \delta V_G \tag{A2.6}$$

Using the calorific factors of Brouwer (1965 – see Appendix III) the solutions are:

$$F = 1.719 \ V_{O_2} - 1.719 \ V_{CO_2} - 1.963 \ N + 1.719 \ V_{CH_4} \ \text{g}$$
$$K = -2.968 \ V_{O_2} + 4.174 \ V_{CO_2} - 2.446 \ N - 1.761 \ V_{CH_4} \ \text{g}$$
$$M = 16.18 \ V_{O_2} + 5.02 \ V_{CO_2} - 5.99 \ N - 2.17 \ V_{CH_4} \ \text{kJ}$$

Appendix III

Report of sub-committee on constants and factors

Prepared by Professor E. Brouwer of the Agricultural University, Wageningen, The Netherlands
(Reproduced from Brouwer (1965))

At the first Symposium on Energy Metabolism a small committee (K. L. Blaxter, Scotland; K. Nehring, Germany (DDR); W. Wöhlbier, Germany (DBR); E. Brouwer, Netherlands) was appointed to consider and to recommend constants and factors to be used in calculations of energy metabolism. After studying the problem a provisional set of constants and factors was sent for criticism to the participants of the symposium. No objections were raised. The committee therefore considers its main task to be finished and makes the following recommendations.

1   *Physical constants*
Atomic weights $C = 12.011$
                     $O = 15.999$
                     $H = 1.008$
                     $N = 14.007$
Molar volume O_2, N_2, CO_2, CH_4, $H_2 = 22.41$ l
Heat of combustion $CH_4 = 9.45$ kcal/l
                             $H_2 = 3.05$ kcal/l

From these figures the following values have been derived:

		Litre weight (g)	Mol. weight (g)	Carbon (g/l)	Combustion value (kcal/l)
O_2	31.998	1.428	—	—	—
CO_2	44.009	1.964	27.29	0.5360	—
CH_4	16.043	0.716	74.87	0.5360	9.45
N_2	28.014	1.250	—	—	—
H_2	2.016	0.090	—	—	3.05

2   *The amounts of oxygen consumed and of CO_2 and of heat
           produced when biological materials (protein, fat, carbohydrate)
           are oxidized*

It is clear that some of these values are not constants in the real
sense but averages. Agreed values are inserted in the following table:

Constants for protein, fat and carbohydrate when oxidized in the animal body

	Carbon (%)	Consumed on oxidation of 1 g		Set free on oxidation of 1 g			
		O_2 (g)	O_2 (l)	CO_2 (g)	CO_2 (l)	Heat (kcal)	Respiratory quotient
Protein†	52.00	1.366	0.957	1.520	0.774	4.40	0.809
Fats	76.70	2.875	2.013	2.810	1.431	9.50	0.711
Starch	44.45	1.184	0.829	1.629	0.829	4.20	1.00
Saccharose	42.11	1.122	0.786	1.543	0.786	3.96	1.00
Glucose	40.00	1.066	0.746	1.466	0.746	3.74	1.00

† The composition of protein is: N: 16%, C: 52%, kcal/g: 5.7.

3   *Formulae using the agreed values*

With the aid of the above constants the following formulae have
been computed.

(1) *Protein (P, g) from nitrogen (N, g)*:

$$P = 6.25 \times N$$

(2) *Carbon (C, g) in protein from nitrogen (N, g) in protein*:

$$C = 325 \times N$$

(3) *Fat (F, g) from carbon (C, g) in fat*:

$$F = 1.304 \times C$$

(4) *Energy balance (RE, kcal) from the carbon balance (C, g) and the nitrogen
balance (N, g)*:

$$RE = 12.388 \times C - 4.636 \times N$$

Alternatively the same can be calculated from the carbon balance (C,
g) and the protein balance (P, g):

$$RE = 12.388 \times C - 0.742 \times P$$

(5) *Heat production (M, kcal) from oxygen consumption (O_2, l), carbon
dioxide production (CO_2, l), methane production (CH_4, l) and urine N(N, g)*:

$$M = 3.866 \times O_2 + 1.200 \times CO_2 - 0.518 \times CH_4 - 1.431 \times N.$$

Alternatively the same can be calculated from oxygen consumption (O_2, l), carbonic acid production (CO_2, l), methane production (CH_4, l) and oxidized protein ($P = 6.25 \times$ urine-N, g):

$$M = 3.866 \times O_2 + 1.200 \times CO_2 - 0.518 \times CH_4 - 0.229 \times P.$$

It should be kept in mind that in using these formulae the CO_2 occurring in the urine as free CO_2 and as carbonate should be included.

4   *Special cases*

There are cases in which deviations from the recommended values are too great to permit use of these constants. This applies for instance to the calorific value of beet sugar which is definitely lower than that of starch and cellulose. Another example is the calorific value of fats containing a high proportion of short-chain fatty acids. In these and similar cases other constants can be used.

5   *Remarks*

The use of this set of constants and factors can never be compulsory. It is, however, desirable that all authors mention in their publications whether the foregoing constants and factors were used or whether one or more other constants were chosen.

Appendix IV

Analysis of heat flow through gradient layers

1   *The basic insulating layer*

The temperature distribution due to conduction of heat through a homogenous material is governed by the equation

$$\frac{\partial T}{\partial t} = \kappa \frac{\partial^2 T}{\partial x^2} \tag{A4.1}$$

where T is temperature, t is time, x is distance measured in the direction of heat flow (i.e. normal to an isothermal surface) and $\kappa = K/\rho c$ is the thermal diffusivity (thermal conductivity/density × specific heat). Carslaw & Jaeger (1947) give general solutions to this equation applicable to different physical problems. The solution for the temperature of a large slab of area A and thickness D, initially at zero temperature and starting at time $t = 0$ with a rate of heat flow Q into the surface $x = D$, whilst the surface $x = 0$ remains held at zero temperature is:

$$T = \frac{\dot{Q}x}{KA} - \frac{8\dot{Q}D}{KA\pi^2} \sum_{n=0}^{\infty} \frac{(-1)^n}{(2n+1)^2} \cdot e^{-\kappa(2n+1)^2\pi^2 t/4D^2} \cdot \sin\frac{(2n+1)\pi x}{2D} \tag{A4.2}$$

The temperature at the surface is obtained by substituting $x = D$:

$$T_F = \frac{\dot{Q}D}{KA}\left[1 - \frac{8}{\pi^2} \sum_{n=0}^{\infty} \frac{1}{(2n+1)^2} \cdot e^{-(2n+1)^2 t/\tau}\right] \tag{A4.3}$$

where

$$\tau = 4D^2/\kappa\pi^2 \tag{A4.4}$$

Since $\sum_{n=0}^{\infty} 1/(2n+1)^2$ is a series which rapidly converges to $\pi^2/8$ the square bracket rises assymptotically to unity.

The rate of rise (Fig. 5.12, Section 5.3.2c) is initially faster than a simple exponential, but for $t > \tau$ approximates very closely to that of the exponential $(1 - e^{-t/\tau})$.

2   *The metal foil covering the insulating layer*

The problem is the same as that for the insulating layer except that the boundary condition $T = 0$ at $x = 0$ is no longer applicable. If we assume that heat flow from this surface is governed by Newton's Law of Cooling, and we choose the heat flow coefficient so that the temperature finally attained at the inner surface of the foil is equal to the final surface temperature of the insulating layer, then using the subscript F to indicate properties of the foil, the boundary condition at $x = 0$ is

$$-K_F \frac{\partial T}{\partial x} = \frac{KT}{D} \tag{A4.5}$$

The solution for this boundary condition is also given by Carslaw & Jaeger (1947, p. 104). The surface temperature (T_F) of the foil is

$$T_F = \frac{\dot{Q} D_F}{K_F A} + \frac{\dot{Q} D}{KA} \left[1 - \sum_{n=1}^{\infty} \frac{2h(\alpha_n^2 + h^2) \cdot e^{-\kappa_F \alpha_n t}}{\alpha_n^2 \{ h + D_F(\alpha_n^2 + h^2) \}} \right] \tag{A4.6}$$

where $h = K/K_F D$, and a_n are the positive roots of the equation

$$a \tan(a D_F) = K/K_F D \tag{A4.7}$$

The roots of equation (A4.7) are tabulated by Carslaw & Jaeger (1947, p. 377). For an insulating layer covered by an aluminium foil, h is of the order of 1.5 m^{-1}, a_1 is 50–100 and very much less than a_2, a_3, etc., and the equation reduces to

$$T_F = \frac{\dot{Q}}{A} \left[-\frac{D_F}{K_F} + \frac{D}{K}(1 - e^{-a_1^2 \kappa_F t}) \right] \tag{A4.8}$$

The terms $\dot{Q} D_F / K_F A$ and $\dot{Q} D / KA$ represent the temperature gradients across the foil and insulating layer respectively. Equation (A4.8) is a simple exponential function with a time constant $\tau_F = 1/a_1^2 \kappa_F$. For aluminium $\kappa = 0.86 \times 10^{-4}$ and τ_F in most instances will be in the range 1–10 s.

3   *Reduced temperature gradient due to lateral heat flow in the thermocouple tape*

This problem was analysed by Bothe (1958). He considered the arrangement illustrated in Fig. 5.14 (Section 5.3.2e). The heavy line represents the metal thermocouple tape (thickness D', conductivity K') which is bent at right-angles to penetrate the insulating layer (thickness D, conductivity K). In order to estimate the temperature of the sensitive junction we assume that it is situated at the origin ($x = 0$) midway between two bridges (at $x = \pm L$). The following assumptions are made for simplicity:

(1) The lateral thermal resistance of the metal is negligible compared

with that of the insulating layer, so that the temperature gradient is zero at the bridges (at $x = \pm L$, $T = 0$).

(2) The physical properties of the metal are uniform (i.e. for the moment assume that it is all copper). Then the system is symmetrical and at $x = 0 \; dT/dx = 0$.

(3) $L \gg D$.

(4) Steady-state temperature conditions have been attained.

The lateral flow of heat per unit width of tape at any point along it is

$$-K'D'\frac{dT}{dx}.$$

We now consider a short element (dx) of the tape. The net lateral flow of heat out of the element along the tape in the x direction is

$$-K'D'\frac{d^2T}{dx^2}.dx$$

and the heat flow out of the element through the insulating layer is $KTdx/D$. If the heat flux per unit area impinging on the surface is $\dot{Q}/A$ then:

$$\frac{\dot{Q}}{A}.dx = \frac{KT}{D}.dx - K'D'\frac{d^2\,T}{dx^2}.dx$$

$$\therefore \frac{d^2T}{dx^2} - \frac{T}{l^2} + \frac{T_D}{l^2} = 0 \qquad (A4.8)$$

where $l = \sqrt{(DD'K'/K)}$ is a characteristic dimension (length) of the materials used in the construction of the gradient layer, and $T_D = \dot{Q}D/KA$ is the equilibrium temperature that is generated across the basic layer in the absence of any thermocouples (cf. equation (5.7)).

The solution of equation (A4.8) is

$$T = Ae^{x/l} + Be^{-x/l} + T_D \qquad (A4.9)$$

and inserting the boundary conditions assumed under (1) and (2) above gives

$$T = T_D\left[1 - \frac{e^{x/l} + e^{-x/l}}{e^{L/l} + e^{-L/l}}\right] \qquad (A4.10)$$

and at $x = 0$

$$\frac{T_{(x=0)}}{T_D} = \left[1 - \frac{2}{e^{L/l} + e^{-L/l}}\right] = 1 - \theta_1 \qquad (A4.11)$$

θ_1 is the fractional reduction in sensitivity due to lateral conduction of heat away from a measuring point situated at a distance L from a bridge.

4   *Direct heat flow through the metal bridges*

For estimation of direct heat flow through the metal bridges equation (A4.8) and its solution (A4.9) are still valid; but it is convenient to redefine the origin ($x = 0$) at the position of the bridge (with temperature gradient T_B) and to apply new boundary conditions:

The direct heat flow through a metal bridge of width D'' is

$$\dot{Q}_B = \frac{K'D'D''}{D} \cdot T_B \tag{A4.12}$$

but the heat flow into the bridge from the tape is $K'D'D''(\mathrm{d}T/\mathrm{d}x)_{x=0}$ and the heat received direct from the source is $\dot{Q}D'D''/A$

$$\therefore \frac{KD'D''T_B}{D} = KD'D'' \left(\frac{\mathrm{d}T}{\mathrm{d}x}\right)_{x=0} + \dot{Q}D'D''/A$$

$$\therefore \left(\frac{\mathrm{d}T}{\mathrm{d}x}\right)_{x=0} = \frac{T_B}{D} - \frac{\dot{Q}}{K'A} \tag{A4.13}$$

The other boundary condition is:

$$\text{as } x \to \infty, \ T \to T_D = \frac{\dot{Q}D}{K'A}$$

Inserting the boundary conditions into (A4.9) gives the general solution

$$T = \left(\frac{\dot{Q}}{K'A} - \frac{T_B}{D}\right) l e^{-x/l} + \frac{\dot{Q}D}{KA}$$

and at $x = 0$ this reduces to

$$T_B = \frac{\dot{Q}}{A}\left(\frac{l}{K'} + \frac{D}{K}\right) \Big/ \left(1 + \frac{l}{D}\right) \tag{A4.14}$$

Inserting this value into equation (A4.12) gives the heat flow through a metal bridge as

$$\dot{Q}_B = \frac{KD'D''}{DA}\left(\frac{l}{K'} + \frac{D}{K}\right) \Big/ \left(1 + \frac{l}{D}\right)$$
$$= D'' l \dot{Q}(l + D')/(l + D)A \tag{A4.15}$$

If there are n copper bridges per unit area the fraction (θ_2) of the total heat flux which passes through the copper bridges is

$$\theta_2 = \frac{\dot{Q}_B nA}{\dot{Q}} = D'' ln(l + D')/(l + D) \tag{A4.16}$$

5

Heat flow at the edges of platemeters

In platemeters the heat sensitive surface is covered at the edges and between air pathways by insulating material as illustrated in Fig. A4.1. At the edge of the pathway ($x = 0$) there is an abrupt transition between heat transfer and from the air and no heat transfer. The heat flow along the aluminium foil under the insulation is exactly analogous to the lateral heat flow along thermocouple tape and equation (A4.8) is again applicable except that for $x > 0$, $T_D = 0$. The boundary conditions in this instance are

as $x \to \infty$, $T \to 0$

as $x \to -\infty$, $T \to T_D$

The general solution may be expressed as

$$T = \frac{T_D}{2}\left[1 - \frac{\sqrt{x^2}}{x}\left(1 - e^{-\sqrt{x^2}/l}\right)\right] \tag{A4.17}$$

where $l = \sqrt{(DD'K'/K)}$ as before except that D' and K' now refer to the aluminium foil. Equation (A4.17) indicates that the temperature is $T_D/2$ at $x = 0$, and changes exponentially to zero under the insulation, and to T_D in the air pathway. For the region under the insulation ($x > 0$) the solution reduces to

$$T = \frac{T_D}{2}e^{-x/l} \tag{A4.18}$$

Consider unit length of air pathway of width w. The total heat flow is KwT_D/D. The heat flow through that part of the layer under the insulation is

$$\int_0^\infty (KT_D/2D)\,e^{-x/l} = KlT_D/2D$$

Fig. A4.1. A cross-section through the edge of a platemeter and diagram of the temperature gradient near the edge.

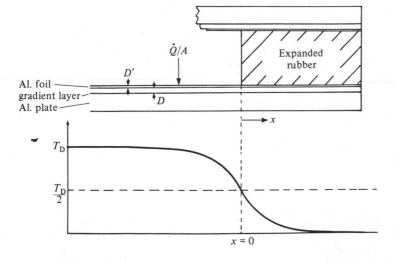

Therefore the proportion of the heat exchange that occurs under the insulation is $1/2w$.

Since l can be of the order of 1 cm and w 5 cm this can represent up to 10 % of the total.

Suppose that the layer extends for a distance L under the insulation. The proportion of the total heat escaping past the layer is then $1/2we^{-L/l}$.

Appendix V

Some physical properties of materials used in constructing calorimeters

	Density (ρ) kg m⁻³	Specific heat (c) J kg⁻¹ C⁻¹	Thermal conductivity (K) W m⁻¹ C⁻¹	Thermal diffusivity (κ) m² s⁻¹	Resistivity Ω m	Temperature coefficient of resistivity (a)
Aluminium	2.7×10^3	896	209	0.86×10^{-4}	2.83×10^{-8}	
Copper	8.89×10^3	385	385	1.12×10^{-4}	1.77×10^{-8}	0.0041
Constantan	8.88×10^3	410	22.6	0.62×10^{-4}	4.5×10^{-7}	0.006
Nickel		—	—	—	7.8×10^{-8}	
Tufnol	1.35×10^3	1460	0.25	1.27×10^{-7}	—	
Perspex Plexiglas	1.2×10^3	1460	0.25	1.43×10^{-7}	—	
Araldite (epoxy)	1.17×10^3	1460	0.25	1.46×10^{-7}	—	
Expanded polystyrene	1.5×10^1	1210	0.034	1.87×10^{-6}	—	
Water	1×10^3	4186	—	—	—	
Timber	5.3×10^2	1760	0.11	1.2×10^{-7}	—	

References

Abramson, E. (1943). 'Computation of results from experiments with indirect calorimetry'. *Acta Physiologica Scandinavica*, **6**, 1–19.

Adams, A. P., Vickers, M. D. A., Munroe, J. P. & Parker, C. W. (1967). 'Dry displacement gas meters'. *British Journal of Anaesthesia*, **39**, 174–83.

Alexander, G. (1961). 'Temperature regulation in the new-born lamb. II. A climatic respiration chamber for the study of thermoregulation, with an appendix on a correction for leaks and imperfect measurement of temperature and pressure in closed circuit respiration chambers'. *Australian Journal of Agricultural Research*, **12**, 1139–51.

Andrews, R. B. (1971). 'Net heart rate as a substitute for respiratory calorimetry'. *American Journal of Clinical Nutrition*, **24**, 1139–47.

Angely, L., Levart, E., Guiochon, G. & Peslerbe, G. (1969). 'General method to prepare standard samples for detector calibration in gas chromatographic analysis of gases'. *Analytical Chemistry*, **41**, 1446–9.

Armitage, G., Hervey, G. R. & Tobin, G. (1979). 'An automated long-term calorimeter for rats'. *Journal of Physiology* (*London*), **290**, 15–16P.

Armitage, G., Hervey, G. R., Rolls, B. J., Rowe, E. A. & Tobin, G. (1984). 'The effects of supplementation of the diet with highly palatable foods upon energy balance in the rat'. *Journal of Physiology* (*London*), **342**, 229–51.

Armsby, H. P. (1904). 'The respiration calorimeter at the Pennsylvania Experiment Station'. In *Annual Report of the Pennsylvania State College*, pp. 208–20.

Arnberg, B. T. (1962). 'Review of critical flowmeters for gas flow measurements'. *Journal of Basic Engineering*, **84**, 447–60.

d'Arsonval, M. A. (1886). 'Recherches de calorimetrie'. *Journal de l'Anatomie et de la Physiologie* (*Paris*), **22**, 113–61.

Arundel, P. A., Holloway, B. R. & Mellor, P. M. (1984). 'A low-cost modular oxygen-consumption device for small animals'. *Journal of Applied Physiology*, **57**, 1591–3.

Atwater, W. O. & Benedict, F. G. (1899). 'Experiments on the metabolism of energy and matter in the human body'. *United States Department of Agriculture, Office of Experimental Stations, Bulletin 69*.

Atwater, W. O. & Benedict, F. G. (1902). 'Experiments on the metabolism of energy and matter in the human body'. *United States Department of Agriculture, Office of Experimental Stations, Bulletin 109*.

312

Atwater, W. O. & Benedict, F. G. (1903). 'Experiments on the metabolism of energy and matter in the human body'. *United States Department of Agriculture, Office of Experimental Stations, Bulletin 136*.

Atwater, W. O. & Benedict, F. G. (1905). 'A respiration calorimeter with appliances for the direct determination of oxygen.' *Carnegie Institute of Washington, Publication No. 42*.

Atwater, W. O. & Rosa, E. B. (1899). 'Description of a new respiration calorimeter and experiments on the conservation of energy in the human body'. *United States Department of Agriculture, Office of Experimental Stations, Bulletin 63*.

Auchincloss, J. H., Gilbert, R. & Baule, G. H. (1970). 'Control of water vapour during rapid analysis of respiratory gases in expired air'. *Journal of Applied Physiology*, **28**, 245–7.

Aulic, L. H., Arnhold, H., Hauder, E. H. & Mason, A. D. (1983). 'A new open and closed circuit respiration chamber'. *Quarterly Journal of Experimental Physiology*, **68**, 351–7.

Bailey, C. B., Kitts, W. D. & Wood, A. J. (1957). 'A simple respirometer for small animals'. *Canadian Journal of Animal Science*, **37**, 68–72.

Baker, W. C. & Pouchot, J. F. (1983*a*). 'The measurement of gas flow. Part 1'. *Journal of the Air Pollution Control Association*, **33**, 66–72.

Baker, W. C. & Pouchot, J. F. (1983*b*). 'The measurement of gas flow. Part 2'. *Journal of the Air Pollution Control Association*, **33**, 156–62.

Balchum, O. J., Hartman, S. A., Slonim, N. B., Dressler, S. H. & Ravin, A. (1953). 'The permeability of the Douglas type bag to respiratory gases'. *Journal of Laboratory and Clinical Medicine*, **41**, 268–80.

Ballal, M. A. & Macdonald, I. A. (1982). 'An evaluation of the Oxylog as a portable device with which to measure oxygen consumption'. *Clinical Physics and Physiological Measurement*, **3**, 57–65.

Barnes, P. & Birdi, H. (1977). 'An assessment of the gapmeter lab kit'. *British Journal of Anaesthesia*, **49**, 277–9.

Barr, G. (1934). 'Two designs of flowmeter, and a method of calibration'. *Journal of Scientific Instruments*, **11**, 321–4.

Barratt, R. S. (1981). 'The preparation of standard gas mixtures. A review'. *The Analyst*, **106**, 817–49.

Barrett, J. F. & Robertson, J. D. (1937). 'A method for alcohol check experiments with the closed-circuit apparatus used in respiratory metabolism'. *Journal of Pathology and Bacteriology*, **45**, 555-60.

Bartels, H., Bucherl, E., Hertz, C. W., Rodewald, G. & Schwab, M. (1963). *Methods in Pulmonary Physiology*. London: Hafner.

Bate, G. C., D'Aoust, A. & Canvin, D. T. (1969). 'Calibration of infra-red CO_2 gas analysers'. *Plant Physiology*, **44**, 1122-6.

Beaver, W. L. (1973). 'Water vapor corrections in oxygen consumption calculations'. *Journal of Applied Physiology*, **35**, 928–31.

Bell, G. H., Davidson, J. N. & Scarborough, H. (1965). *Textbook of Physiology and Biochemistry, 6th Edition*. Edinburgh: Livingstone.

Benedict, F. G. (1909). 'An apparatus for studying the respiratory exchange'. *American Journal of Physiology*, **24**, 345–74.

Benedict, F. G. (1912). 'Ein Universalrespirationsapparat'. *Deutsches Archiv für Klinische Medizin*, **107**, 156–200.

Benedict, F. G. (1925). 'The control of gaseous metabolism apparatus'. *The Boston Medical and Surgical Journal*, **193**, 583–91.

Benedict, F. G. (1930). 'Helmet for use in clinical studies of gaseous metabolism'. *The New England Journal of Medicine*, **203**, 150–8.

Benedict, F. G. & Carpenter T. M. (1910). 'Respiration calorimeters for studying respiratory exchange and energy transformations of man'. *Carnegie Institution of Washington, Publication 123.*

Benedict, F. G. & Collins, W. E. (1920). 'A clinical apparatus for measuring basal metabolism'. *The Boston Medical and Surgical Journal*, **183**, 449–58.

Benedict, F. G. & Higgins, H. L. (1910). 'An adiabatic calorimeter for use with the calorimeter bomb'. *Journal of the American Chemical Society*, **32**, 461–7.

Benedict, F. G. & Lee, R. C. (1937). 'Lipogenesis in the animal body with special reference to the physiology of the goose'. *Carnegie Institution of Washington, Publication 489.*

Benedict, F. G. & Wardlaw, H. S. H. (1932). 'Some factors determining insensible perspiration in man'. *Archives of Internal Medicine*, **49**, 1019–31.

Beneken Kolmer, H. H. & Kreuzer, F. (1968). 'Continuous polarographic recording of oxygen pressure in respiratory air'. *Respiration Physiology*, **4**, 109–17.

Ben-Porat, M., Sideman, S. & Bursztein, S. (1983). 'Energy metabolism rate equation for fasting and postabsorptive subjects'. *American Journal of Physiology (Regulatory, Integrative, Comparative Physiology)*, **13**, R764–9.

Benzinger, T. H., Heubscher, R. G., Minard, D. & Kitzinger, C. (1958). 'Human calorimetry by means of the gradient principle'. *Journal of Applied Physiology (Supplement)*, **121**, S1–24.

Benzinger, T. H. & Kitzinger, C. (1949). 'Direct calorimetry by means of the gradient principle'. *Review of Scientific Instruments*, **20**, 849–60.

Bischoff, T. L. W. & Voit, C. (1860). *Die Gesetze der Ernährung des Fleischfressers.* Leipzig: C. F. Winter.

Blaxter, K. L. (1956a). 'Some observations on the digestibility of food by sheep and on related problems'. *British Journal of Nutrition*, **10**, 69–91.

Blaxter, K. L. (1956b). 'The nutritive value of feeds as sources of energy: a review'. *Journal of Dairy Science*, **39**, 1396–1424.

Blaxter, K. L. (1962). *The Energy Metabolism of Ruminants.* London: Hutchinson.

Blaxter, K. L. (1967). 'Techniques in energy metabolism studies and their limitations'. *Proceedings of the Nutrition Society*, **26**, 86–96.

Blaxter, K. L. (1978). 'Adair Crawford and calorimetry'. *Proceedings of the Nutrition Society*, **37**, 1–3.

Blaxter, K. L., Brockway, J. M. & Boyne, A. W. (1972). 'A new method for estimating the heat production of animals'. *Quarterly Journal of Experimental Physiology*, **57**, 60–72.

Blaxter, K. L., Graham, N. McC. & Rook, J. A. F. (1954). 'Apparatus for the determination of the energy of calves and of sheep'. *Journal of Agricultural Science (Cambridge)*, **45**, 10–18.

Blaxter, K. L. & Howells, A. (1951). 'The nutrition of the young Ayrshire calf. 2. A spirometer for the determination of the respiratory exchange of the calf'. *British Journal of Nutrition*, **5**, 25–9.

Blaxter, K. L. & Joyce, J. P. (1963). 'The accuracy and ease with which measurements of respiratory metabolism can be made with tracheostomized sheep'. *British Journal of Nutrition*, **17**, 523–37.

Blaxter, K. L. & Rook, J. A. F. (1953). 'The heat of combustion of the tissues of cattle in relation to their chemical composition'. *British Journal of Nutrition*, **7**, 83–91.

Bønsdorff Petersen, C. (1967). A description of an open circuit respiration apparatus for poultry'. In *Energy Metabolism of Farm Animals. Proceedings of the 4th Symposium, Warsaw, Poland. E.A.A.P. Publication No. 12*, eds. K. L. Blaxter, J. Kielanowski & G. Thorbek, pp. 437–45. Newcastle-on-Tyne: Oriel.

Booyens, J. & Hervey, G. R. (1960). 'The pulse rate as a means of measuring metabolic rate in man'. *Canadian Journal of Biochemistry & Physiology*, **38**, 1301–9.

Bothe, W. (1958). 'Some calculations concerning the gradient calorimeter'. *Journal of Applied Physiology*, **12**, S25–8.

Bradham, G. B. & Carter, H. T. (1972). 'An automatically compensating gradient-layer calorimeter for animal metabolic studies'. *Medical and Biological engineering*, **10**, 793–5.

Bradshaw, P. (1968). 'Thermal methods of flow measurements'. *Journal of Scientific Instruments (Journal of Physics E)*, **1**, 504–9.

Bradshaw, P. D. (1984). 'A/D and D/A conversion'. In *Designers Handbook of Integrated Circuits*, ed. A. B. Williams, pp. 8/1–48, New York: McGraw-Hill.

Braefield, A. E. & Llewellyn, M. J. (1982). *Animal Energetics*. Glasgow: Blackie.

Brain, T. J. S. & Scott, R. W. W. (1982). 'Survey of pipeline flowmeters'. *Journal of Physics. E: Scientific Instruments*, **15**, 967–80.

British Standards Institution (1979–83). *British standard methods for preparation of calibration gas mixtures, B.S. 4559*. London: B.S.I.

Brockway, J. M., McDonald, J. D. & Pullar, J. D. (1977). Calorimetric facilities at the Institute and the techniques employed to measure energy metabolism'. *Annual Report of the Rowett Research Institute*, **33**, 106–19.

Brody, S. (1945). *Bioenergetics and Growth*. New York: Reinhold (Reprinted in 1964, New York: Haffner).

Brouwer, E. (1958). 'On simple formulae for calculating the heat expenditure and the quantities of fat and carbohydrate metabolized in ruminants, from data on gaseous exchange and urine-N'. *Proceedings of the 1st Symposium on Energy Metabolism*. Rome: European Association for Animal Production, Publication No. 8.

Brouwer, E. (1965). 'Report of Sub-committee on Constants and Factors'. In *Energy Metabolism. Proceedings of the 3rd Symposium*, ed. K. L. Blaxter, London: Academic Press.

Brown, D., Cole, T. J., Dauncey, M. J., Marrs, R. W. & Murgatroyd, P. R. (1984) 'Analysis of gaseous exchange in open-circuit indirect calorimetry'. *Medical and Biological Engineering and Computing*, **22**, 333–8.

Burch, D. E., Gryvnak, D. A. & Williams, D. (1962). 'Total absorptance of carbon dioxide in the infrared'. *Applied Optics*, **1**, 759–65.

Burch, D. E., Singleton, E. B. & Williams, D. (1962). 'Absorption line broadening in the infrared'. *Applied Optics*, **1**, 359–63.

Burch, D. E. & Williams, D. (1962). 'Total absorptance of carbon monoxide and methane in the infrared'. *Applied Optics*, **1**, 587–94.

Bursztein, S., Saphar, R., Glaser, P., Taitelman, U., de Mytennaere, S. & Nedey, R. (1977). 'Determination of energy metabolism from respiratory functions alone'. *Journal of Applied Physiology*, **42**, 117–19.

Burton, A. C. (1935). 'Human calorimetry. II. The average temperature of the tissues of the body'. *Journal of Nutrition*, **9**, 261–79.

Byles, P. H. (1960). 'Observations on some continuously-acting spirometers'. *British Journal of Anaesthesia*, **32**, 470–5.

Cairnie, A. B. & Pullar, J. D. (1959). 'Temperature controller based on measurements of rate-of-change of temperature'. *Journal of Scientific Instruments*, **36**, 249–52.

Caldwell, F. T., Hammel, H. T. & Dolan, F. (1966). 'A calorimeter for simultaneous determination of heat production and heat loss in the rat'. *Journal of Applied Physiology*, **21**, 1665–71.

Cammell, S. B., Beever, D. E., Skelton, K. V. & Spooner, M. C. (1981). 'The construction of open-circuit calorimeters for measuring gaseous exchange and heat production in sheep and young cattle'. *Laboratory Practice*, **30**, 115–19.

Capstick, J. W. (1921). 'A calorimeter for use with large animals'. *Journal of Agricultural Science (Cambridge)*, **11**, 408–31.

Carlson, L. D., Honda, C. N., Saski, T. & Judy, W. V. (1964). 'A human calorimeter'. *Proceedings of the Society for Experimental Biology and Medicine*, **117**, 327–31.

Carpenter, T. M. (1923). 'An apparatus for the exact analysis of air in metabolism investigations with respiratory exchange chambers'. *Journal of Metabolic Research*, **4**, 1–25.

Carpenter, T. M. (1933*a*). 'The development of methods for determining basal metabolism of mankind'. *Ohio Journal of Science*, **33**, 297–313.

Carpenter, T. M. (1933*b*). 'Some considerations on precise analysis of air from respiration chambers'. *Journal of Biological Chemistry*, **101**, 595–601.

Carpenter, T. M. (1937). 'The constancy of the atmosphere with respect to carbon dioxide and oxygen content'. *Journal of the American Chemical Society*, **59**, 358–60.

Carpenter, T. M. & Fox, E. L. (1923). 'Alcohol check experiments with portable respiration apparatus'. *The Boston Medical and Surgical Journal*, **189**, 551–61.

Carpenter, T. M., Fox, E. L. & Sereque, A. F. (1929). 'The Carpenter form of the Haldane gas analysis apparatus. Changes made in the apparatus and details regarding its use'. *Journal of Biological Chemistry*, **83**, 211–30.

Carpenter, T. M., Lee, R. C. & Finnerty, A. E. (1930). 'Eine Apparat für die exakte und schnelle Analyse der Gase aus einer Respirationskammer'. *Sonderabdruck aus Wissenschaftliches. Archiv für Landwirtschaft, Abt B, Tierernahrung und Tierzucht*, **4**, 1–26.

Carter, K. B., Shaw, A., Richards, J. R., Boyd, I. A. & Harland, W. A. (1975). 'Calorimetry system for studying small animals following injury'. *Medical and Biological Engineering*, **13**, 551–7.

Cathcart, E. P. (1918). 'Method of estimating energy expenditure by indirect calorimetry'. *Journal of the Royal Army Medical Corps*, **31**, 339–52.

Cathcart, E. P. & Cuthbertson, D. P. (1931). 'The composition and distribution of the fatty substances of the human subject'. *Journal of Physiology (London)*, **72**, 349–60.

Chamberlain, E. A. C., Dixon, S. R. M. & Walters, P. L. (1954). 'A reciprocating pump for gas analysis apparatus'. *Journal of Scientific Instruments*, **31**, 66–7.

Charkey, L. W. & Thornton, P. A. (1959). 'A simple method for estimating oxygen consumption of small animals'. *Poultry Science*, **38**, 899–902.

Charlet-Lery, G. (1958). 'Méthode de mésure de courte duree des echanges gaseux'. In *Symposium on Energy Metabolism, Copenhagen, Denmark (EAAP Publication No. 8)*, eds. G. Thorbek & H. Aersøe, pp. 194–202. Rome: EAAP.

Charlet-Lery, G. & Laplace, J. P. (1976). 'Utilization of a tracheal cannula in the pig'. In *Energy Metabolism of Farm Animals. Proceedings of the Symposium on Energy Metabolism, Clermont-Ferrand, France (EAAP Publication No. 19)*, ed. M. Vermorel, pp. 339–42. Clermont-Ferrand: G. de Bussac.

Chase, J. D. (1964). 'Sintered plate flowmeter'. *Journal of Scientific Instruments*, **41**, 386–7.

Chasteau, V. A. L. & Cronje, J. S. (1974). 'An air-flow measuring system with high inherent accuracy for use on an open system respiration apparatus for small ruminants'. In *Energy Metabolism of Farm Animals. Proceedings of the 6th Symposium on Energy Metabolism, Hohenheim, B.R.D. (EAAP Publication No. 14)*, eds. K. H. Menke, H-J Lantzsch & J. R. Reichel, pp. 245–8. Rome: EAAP.

Christensen, B. G. (1946). A simple apparatus for simultaneous measurement of the standard metabolism and the respiratory quotient in small laboratory animals, supplemented with investigations on normal and hypophysectomized rats'. *Acta Physiologica Scandinavica*, 12, 11–26.

Christensen, C. C., Frey, H. M. M., Foenstelien, E., Aadland, E. & Refsum, H. E. (1983). 'A critical evaluation of energy expenditure estimates based on individual O_2 consumption/ heart rate curves and average daily heart rates'. *American Journal of Clinical Nutrition*, 37, 468–72.

Clarke, L. C. (1956). 'Monitor and control of blood and tissue oxygen tensions'. *Transactions of the American Society of Artificial Internal Organs*, 2, 41–9.

Clayton, D. L. (1970). 'A simple automated recording respirometer'. *Journal of Applied Physiology*, 28, 700–1.

Cline, W. H. & Watts, D. T. (1966). 'An automated basal metabolism apparatus for small animals'. *Proceedings of the Society for Experimental Biology and Medicine*, 122, 1264–6.

Collins, R. G., Musasche, V. W. & Howley, E. T. (1977). 'Preparation of matched reagents for use with the Scholander gas analyser'. *Journal of Applied Physiology*, 43, 164–6.

Collis, D. C. & Williams, M. J. (1959). 'Two-dimensional convection from heated wires at low Reynolds numbers'. *Journal of Fluid Mechanics*, 6, 357–84.

Colovos, N. F., Holter, J. B. & Hayes, H. H. (1968). 'Changes in the Carpenter modification of the Haldane gas analysis apparatus'. *Journal of Dairy Science*, 51, 1311–12.

Colvin, H. W., Wheat, J. D., Rhode, E. A. & Boda, J. M. (1957). 'A technique for measuring eructated gas in cattle'. *Journal of Dairy Science*, 40, 492–502.

Comte-Bellot, G. (1976). 'Hot-wire anemometry'. In *Annual Review of Fluid Mechanics*, Vol. 8, ed. M. Van Dyke, pp. 209–31. Palo Alto: Annual Reviews.

Consolazio, D. F. (1971). 'Energy expenditure studies in military populations using Kofranyi–Michaelis respirometers'. *American Journal of Clinical Nutrition*, 24, 1431–7.

Consolazio, C. F., Johnson, R. E. & Pecora, L. J. (1963). *Physiological Measurements of Metabolic Functions in Man*. New York: McGraw-Hill.

Cooper, E. A. & Pask, E. A. (1960). 'A bag for measuring respiratory volumes'. *The Lancet*, i, 369.

Corbett, J. L., Farrell, D. J., Leng, R. A., McClymont, G. L. & Young, B. A. (1971). 'Determination of energy expenditure of penned grazing sheep from estimates of carbon dioxide entry rate'. *British Journal of Nutrition*, 26, 277–91.

Corbett, J. L., Leng, R. A. & Young, B. A. (1967). 'Measurement of energy expenditure by grazing sheep and the amount of energy supplied by volatile fatty acids produced in the rumen'. In *Energy Metabolism of Farm Animals. Proceedings of the 4th Symposium on Energy Metabolism, Warsaw, Poland. (EAAP Publication No. 12)*, eds. K. L. Blaxter, J. Kielanowski & G. Thorbek, pp. 77–86. Newcastle-on-Tyne: Oriel Press.

Cormack, R. S. (1972). 'Eliminating two sources of error in the Lloyd–Haldane apparatus'. *Respiration Physiology*, 14, 382–90.

Coward, W. A. & Prentice, A. M. (1985). 'Isotope method for the measurement of carbon dioxide production in man'. *American Journal of Clinical Nutrition*, 41, 659–61.

Coward, W. A., Prentice, A. M., Murgatroyd, P. R., Davies, H. L., Cole, T. J., Sawyer, M., Goldberg, G. R., Haliday, D. & Macnamara, J. P. (1984). 'Measurement of CO_2 and water production rates in man using ^2H, ^{18}O-labelled H_2O; comparisons between calorimeter and isotope values'. In *Human Energy Metabolism: Physical Activity and Energy Expenditure Measurements in Epidemiological Research Based upon Direct and Indirect Calorimetry*, ed. van Es, pp. 126–8. Wageningen: Stichting Nederlands Institute voor de Voeding (Euro-Nut report; 5).

Coward, W. A., Roberts, S. B., Prentice, A. M. & Lucas, A. (1986). 'The $^2H_2$ $^{18}O_2$ method for energy expenditure measurements – clinical possibilities, necessary assumptions and limitations'. *Proceedings of the 7th Congress of the European Society for Parental and Enteral Nutrition, Clinical Nutrition & Medical Research.* Eds G. Dietze, A. Grumert, G. Kleinberger & G. Wolfram, pp. 169–77.

Cox, L. A., Almeida, A. P., Robinson, J. S. & Horsley, J. K. (1974). 'An electric respirometer'. *British Journal of Anaesthesia*, **46**, 302–10.

Crampton Smith, A. & Hahn, C. E. W. (1969). 'Electrodes for the measurement of oxygen and carbon dioxide tensions'. *British Journal of Anaesthesia*, **41**, 731–41.

Cresswell, E. (1961). Note on the effect of shearing, and of drenching with simulated rain, on the oxygen consumption and respiratory pattern of sheep'. *Empire Journal of Experimental Agriculture*, **29**, 162–4.

Cresswell, E. & Harris, L. E. (1961). 'An improved tracheal cannula for use in indirect calorimetry and other studies'. *The Veterinary Record*, **73**, 343–4.

Cruickshank, E. W. H. (1930). 'Automatic shaker for gas analysis'. *Journal of Physiology (London)*, **69**, 29–30P.

Cunningham, D. J. C. (1964). 'Claude Gordon Douglas'. *Biographical Memoirs of Fellows of the Royal Society*, **10**, 51–74.

Daines, M. E. (1969). 'The preparation of standard gas mixtures by a gravimetric technique'. *Chemistry and Industry*, **88** 1047–53.

Dale, H. E., Shanklin, M. D., Johnson, H. D. & Brown, W. H. (1967). 'Energy metabolism of the chimpanzee: a comparison of direct and indirect calorimetry'. *Journal of Applied Physiology*, **22**, 850–3.

Daniels, F. (1916). 'An adiabatic calorimeter'. *Journal of the American Chemical Society*, **38**, 1473–80.

Daniels, J. (1971). 'Portable respiratory gas collection equipment'. *Journal of Applied Physiology*, **31**, 164–7.

Dargol'tz, V. G. (1973). 'Analysis of the constants used in indirect calorimetry of birds'. *Soviet Journal of Ecology*, **4**, 328–33.

Dauncey, M. J., Murgatroyd, P. R. & Cole, T. J. (1978). 'A human calorimeter for direct and indirect measurements of 24-hour energy expenditure'. *British Journal of Nutrition*, **39**, 557–66.

Davies, N. J. H. & Denison, D. M. (1979). 'The measurement of metabolic gas exchange and minute volume by mass spectrometry alone'. *Respiration Physiology*, **36**, 261–7.

Davies, P. W. & Brink, F. (1942). 'Microelectrodes for measuring local oxygen tension in animal tissues'. *Review of Scientific Instruments*, **13**, 524–33.

Deighton, T. (1926). 'A new calorimeter for use with young farm animals'. *Journal of Agricultural Science (Cambridge)*, **16**, 376–82.

Deighton, T. (1939). 'A study of the metabolism of fowls. I. A calorimeter for the direct determination of the metabolism of fowls'. *Journal of Agricultural Science (Cambridge)*, **29**, 431–51.

Depocas, F. & Hart, J. S. (1957). 'Use of the Pauling oxygen analyser for measurement of oxygen consumption of animals in open-circuit systems and in a short-lag closed circuit apparatus'. *Journal of Applied Physiology*, **10**, 388–92.

Despretz, C. (1824). 'La cause de la chaleur animale'. *Journal de Physiologie Experimentale et de Pathologie*, **4**, 143–59.

Dewar, A. D. & Newton, W. H. (1948). 'The determination of total metabolism in the mouse'. *British Journal of Nutrition*, **2**, 123–41.

Diem, K. & Lentner, C. (1970). *Documenta Geigy, Scientific Tables*, 7th edn, Basle: Ciba-Geigy.

Douglas, C. G. (1911). 'A method for determining the total respiratory exchange in man'. *Journal of Physiology (London)*, **42**, xvii–xviii.

Douglas, C. G. (1956). 'Energy expenditure in man'. *Proceedings of the Nutrition Society*, **15**, 72–7.

Douglas, C. G. & Priestley, J. G. (1924). *Human Physiology. A Practical Course.* Oxford: Clarendon Press.

Dounis, E., Steventon, R. D. & Wilson, R. S. E. (1980). 'The use of a portable oxygen consumption meter (Oxylog) for assessing the efficiency of crutch walking'. *Journal of Medical Engineering and Technology*, **4**, 296–8.

Dowsett, E. E., Kekwick, A. & Pawan, G. L. S. (1963). 'An experimental approach to the mechanisms of weight loss. I.: A method of measuring metabolism in small animals'. *Metabolism*, **12**, 213–21.

Dressler, D. P., Mastio, G. J. & Allbritten, F. F. (1960). 'The clinical application of gas chromatography to the analysis of respiratory gases'. *Journal of Laboratory and Clinical Medicine*, **55**, 144–8.

Du Bois, E. (1927). *Basal Metabolism in Health and Disease.* 2nd edn. London: Bailliere, Tindall & Cox.

Duffield, F. A. (1926). 'A labour-saving device for use in gas analysis'. *Journal of Physiology (London)*, **61**, 29–30P.

Duggan, M. B., Tobin, G. & Milner, R. D. G. (1984). 'Calibration of a respiratory calorimeter for the study of children in Africa'. In *Human Energy Metabolism: Physical Activity and Energy Expenditure Measurements in Epidemiological Research Based upon Direct and Indirect Calorimetry*, ed. A. J. H. van Es, pp. 113–18. Wageningen: Stichting Nederlands Institute voor de Voeding (Euro-Nut report; 5).

Dulong, P. R. (1841). 'Memoire sur la chaleur animale'. *Annales de Chimie et de Physics*, 3e série, **1**, 440–55.

Durnin, J. V. G. A. & Edwards, R. G. (1955). 'Pulmonary ventilation as an index of energy expenditure'. *Quarterly Journal of Experimental Physiology*, **40**, 370–7.

Durnin, J. V. G. A. & Passmore, R. (1967). *Energy, Work and Leisure.* London: Heinemann.

Du Vigneaud, V. (1927). 'Some useful modifications of the Haldane gas-analysis apparatus'. *Journal of Laboratory and Clinical Medicine*, **13**, 175–80.

Dyer, C. A. (1947). 'A paramagnetic oxygen analyser'. *Review of Scientific Instruments*, **18**, 696–702.

EAAP (1958). *Symposium on Energy Metabolism, Publication No. 8*, eds., G. Thorbek & H. Aersøe. Rome: EAAP.

EAAP (1961–85). *Energy Metabolism of Farm Animals. Proceedings of Symposia.* Publication Nos. 10 (1961), 11 (1964), 12 (1967), 13 (1970), 14 (1973), 19 (1976), 26 (1979), 29 (1982), in Press (1985).

Eley, C., Goldsmith, R., Layman, D., Tan, G. L. E. & Walker, E. (1978). 'A respirometer for use in the field for the measurement of oxygen consumption. "The Miser", a miniature indicating and sampling electronic respirometer'. *Ergonomics*, **21**, 253–64.

Eley, C., Goldsmith, R., Layman, D. & Wright, B. M. (1976). 'A miniature indicating and sampling respirometer (Miser)'. *Journal of Physiology (London)*, **256**, 59–60P.

Ellliot, J. M. & Davison, W. (1975). 'Energy equivalents of oxygen consumption in animal energetics'. *Oecologia, (Berlin)*, **19**, 195–201.

Ellis, F. R. & Nunn, J. F. (1968). 'The measurement of gaseous oxygen tension utilizing paramagnetism: an evaluation of the "Sevomex" OA. 150 analyser'. *British Journal of Anaesthesia*, **40**, 569–78.

Engels, E. A. N., Inskip, M. W. & Corbett, J. L. (1976). 'Effect of change in respiratory quotient on the relationship between carbon dioxide entry rate in sheep and their energy metabolism'. In *Energy Metabolism of Farm Animals. Proceedings of the 7th Symposium on Energy Metabolism, Clermont-Ferrand, France (EAAP Publication No. 19)*, ed. M. Vermorel, pp. 339–42. Clermont-Ferrand: G de Bussac.

Enoch, H. Z. & Cohen, Y. (1981). 'Sensitivity of an infrared gas analyser used in the differential mode, to partial gas pressures of carbon dioxide and water vapour in the bulk air'. *Agricultural Meteorology*, **24**, 131–8.

van Es, A. J. H. (1967). 'Report to the sub-committee on constants and factors: constants and factors regarding metabolic water'. In *Energy Metabolism of Farm Animals. Proceedings of the 4th Symposium, Warsaw, Poland. E.A.A.P. Publication No. 12*, eds. K. L. Blaxter, J. Kielanowski & G. Thorbek, pp. 513–14. Newcastle-on-Tyne: Oriel.

Farrell, D. J. (1974). 'General principles and assumptions of calorimetry'. In *Energy Requirements of Poultry*, eds. T. R. Morris & B. M. Freeman, pp. 1–24. Edinburgh: British Poultry Science Ltd.

Farrell, D. J., Corbett, J. L. & Leng, R. A. (1970). 'Automatic sampling of blood and ruminal fluid of grazing sheep'. *Research in Veterinary Science*, **11**, 217–20.

Fatt, I. (1968). 'The oxygen electrode: some special applications'. *Annals of the New York Academy of Sciences*, **148**, 81–92.

Finucane, K. E., Egan, B. A. & Dawson, S. V. (1972). 'Linearity and frequency response of pneumotachographs'. *Journal of Applied Physiology*, **32**, 121–6.

Flatt, W. P., Waldo, D. R., Sykes, J. F. & Moore, L. A. (1958). 'A proposed method of indirect calorimetry for energy metabolism studies with large animals under field conditions'. In *Symposium on Energy Metabolism, Copenhagen, Denmark (EAAP Publication No. 8)*, eds. G. Thorbek & H. Aersøe, pp. 101–9. Rome: EAAP.

Fleisch, A. (1925). 'Der Pneumotachograph – ein Apparat zur Geschwindigkeitsregistrierung der Atemluft'. *Pflugers Archiv*, **209**, 713–22.

Fordham, M., Appenteng, K. & Goldsmith, R. (1978). 'The cost of work in medical nursing'. *Ergonomics*, **21**, 331–42.

Fowler, K. T. (1969). 'The respiratory mass spectrometer'. *Physics in Medicine and Biology*, **14**, 185–99.

Fowler, W. S., Blackburn, C. M. & Helmholtz, H. F. (1957). 'Determination of basal rate of oxygen consumption by open- and closed-circuit methods'. *Journal of Clinical Endocrinology*, **17**, 786–96.

Fox, H. C., Miller, D. S. & Payne, P. R. (1959). 'A ballistic bomb calorimeter for the rapid determination of some nutritive values of foodstuffs'. *Proceedings of the Nutrition Society*, **18**, ix–x.

Friesen, W. O. & McIlroy, M. B. (1970). 'Rapidly responding oxygen electrode for respiratory gas sampling'. *Journal of Applied Physiology*, **29**, 258–9.

Fry, D. L., Hyatt, R. E., McCall, C. B. & Mallos, A. (1957). 'Evaluation of three types of respiratory flowmeters'. *Journal of Applied Physiology*, **10**, 210–14.

Gagge, A. P. (1936). 'The linearity criterion as applied to partitional calorimetry'. *American Journal of Physiology*, **116**, 656–68.

Garby, L. (1985). 'Contribution to discussion'. In *Nutritional Adaptation in Man*, eds. Sir Kenneth Blaxter & J. C. Waterlow, p. 95. London: John Libby.

Garrow, J. S. & Hawes, S. F. (1972). 'The role of amino acid oxidation in causing "specific dynamic action" in man'. *British Journal of Nutrition*, **27**, 211–19.

Garrow, J. S., Murgatroyd, P., Toft, R. & Warwick, P. (1977). 'A direct calorimeter for clinical use'. *Journal of Physiology (London)*, **267**, 15–16P.

Garry, R. C., Passmore, R., Warnock, G. M. & Durnin, J. V. G. A. (1955). 'Studies on expenditure of energy and consumption of food by miners and clerks, Fife, Scotland, 1952'. *Medical Research Council, Special Report Series No. 289*. London: HMSO.

Gaudebout, C. & Blayo, M. C. (1975). 'Assessment of Scholander micromethod for gas concentrations versus weighing method'. *Journal of Applied Physiology*, **38**, 546–9.

Gentz, J., Persson, B., Hakkarainen, J. & Lindstedt, S. (1970). 'Apparatus measuring oxygen consumption and expired CO_2 and $^{14}CO_2$ in small subjects'. *Journal of Applied Physiology*, **29**, 731–9.

Geppert, J. & Zuntz, N. (1888). 'Ueber die Regulation der Athmung.' *Archiv für die Gesamte Physiologie des Menschen und der Tiere*, **42**, 188–245.

Gillbe, C. E., Heneghan, C. P. H. & Branthwaite, M. A. (1981). 'Respiratory mass spectrometry during general anaesthesia'. *British Journal of Anaesthesia*, **53**, 103–9.

Godal, A., Belenky, D. A., Standaert, T. A., Woodrum, D. E., Grimsrud, L. & Hodson, W. A. (1976). 'Application of the hot-wire anemometer to respiratory measurements in small animals'. *Journal of Applied Physiology*, **40**, 275–7.

Golden, F. StC., Hampton, I. F. G. & Smith, D. (1979). 'Cold tolerance in long-distance swimmers'. *Journal of Physiology (London)*, **290**, 48–9P.

Golden, F. StC., Hampton, I. F. G. & Smith, D. (1980). 'Lean long distance swimmers'. *Journal of the Royal Naval Medical Service*, **66**, 26–30.

Goldsmith, R., O'Brien, C., Tan, G. L. E., Smith, W. S. & Dixon, M. (1978). 'The cost of work on a vehicle assembly line'. *Ergonomics*, **21**, 315–23.

Goldsmith, R., Tan, G. L. E., Walker, E. & Wright, B. M. (1976). 'Validation of a miniature indicating and sampling respirometer (Miser)'. *Journal of Physiology (London)*, **256**, 102–3P.

Gothard, J. W. W., Busst, C. M., Branthwaite, M. A., Davies, N. J. H. & Denison, D. M. (1980). 'Applications of respiratory mass spectrometry to intensive care'. *Anaesthesia*, **35**, 890–5.

Grad, B. (1952). 'A simple method for the measurement of the oxygen consumption and heart rate of rats'. *Endocrinology*, **50**, 94–9.

Grafe, E. (1909). 'Ein kopfrespirationsapparat'. *Deutsches Archiv für Klinische Medizin*, **95**, 529–42.

Grootenhuis, P. (1949). 'The flow of gases through porous metal compacts'. *Engineering*, **167**, 291–2.

Gray, R. & McCracken, K. J. (1976). 'A system of centralised gas analysis and data capture to service several open and closed circuit respiration chambers'. In *Energy Metabolism of Farm Animals. Proceedings of the 7th Symposium, Vichy, France. E.A.A.P. Publication No. 19*, ed. M. Vermorel, pp. 335–8. Rome: EAAP.

Grenvik, A., Hedstrand, U. & Sjogren, H. (1966). Problems in pneumotachography'. *Acta Anaesthesiologica Scandinavica*, **10**, 147–55.

Haldane, J. S. (1892). 'A new form of apparatus for measuring the respiratory exchange of animals'. *Journal of Physiology (London)*, **13**, 419–30.

Haldane, J. S. (1898). 'Some improved methods of gas analysis'. *Journal of Physiology (London)*, **22**, 465–80.

Haldane, J. S. (1912). *Methods of Air Analysis*. London: Griffin (2nd edn 1918; 3rd edn 1920; 4th edn. 1935).

Haldi, J., Wynn, W. & Breding, H. (1961). 'Apparatus for measuring oxygen consumption of small animals'. *Journal of Applied Physiology*, **16**, 923–5.

Hallback, I., Karlsson, E. & Ekblom, B. (1978). 'Comparison between mass spectrometry and Haldane technique in analysing O_2 and CO_2 concentrations in air gas mixtures'. *Scandinavian Journal of Clinical and Laboratory Investigation*, **38**, 285–8.

Hamilton, L. H. & Kory, R. C. (1960). 'Application of gas chromatography to respiratory gas analysis'. *Journal of Applied Physiology*, **15**, 829–37.

Hammell, H. T. & Hardy, J. D. (1963). 'A gradient type of calorimeter for measurement of thermoregulatory processes in the dog'. In *Temperature, its Measurement and Control in Science and Industry*, *Vol.* III, ed.-in-chief C. M. Herzfield, Part 3, ed. J. D. Hardy, pp. 31–42. New York: Reinhold.

Hampton, I. F. G., Stainthorpe, S. & Tobin, G. (1982). 'Determination of oxygen consumption in man over extended periods of time'. *Journal of Physiology (London)*, **330**, 8–9P.

Hardy, J. D. (1949). 'Heat transfer'. In *Physiology of Heat Regulation and the Science of Clothing*, ed. L. H. Newburgh, pp. 78–108. Philadelphia: Saunders.

Hardy, J. D. & Du Bois, E. F. (1938a). 'The technic of measuring radiation and convection'. *Journal of Nutrition*, **15**, 461–75.

Hardy, J. D. & Du Bois, E. F. (1938b). 'Basal metabolism, radiation, convection and vaporization at temperatures of 22 to 35 °C'. *Journal of Nutrition*, **15**, 477–97.

Hardy, J. D., Milhorat, A. T. & Du Bois, E. F. (1941). 'Basal metabolism and heat loss of young women at temperatures from 22 to 25 °C'. *Journal of Nutrition*, **21**, 383–404.

Harrison, M. H., Brown, G. A. & Belyavin, A. J. (1982). 'The "Oxylog": an evaluation'. *Ergonomics*, **25**, 809–820.

Havstad, K. M. & Malechek, J. C. (1982). 'Energy expenditure by heifers grazing crested wheatgrass of diminishing availability'. *Journal of Range Management*, **35**, 447–50.

Hawkins, A. E. (1958). 'A self adjusting unit for the control of an animal calorimeter'. *Journal of Scientific Instruments*, **35**, 440–2.

Heitmann, H., Buckles, R. G. & Laver, M. B. (1967). 'Blood PO_2 measurements: performance of microelectrodes'. *Respiration Physiology*, **3**, 380–95.

Henderson, Y. (1918). 'Applications of gas analysis. IV The Haldane gas analyser'. *Journal of Biochemistry*, **33**, 31–8.

Heneghan, C. P. H., Gillbe, C. E. & Branthwaite, M. A. (1981). 'Measurement of metabolic gas exchange during anaesthesia. A method using mass spectrometry'. *British Journal of Anaesthesia*, **53**, 73–6.

Hervey, G. R. (1973). 'Extended Fortran for the "8" family'. *Decus Europe: Proceedings of the 9th Seminar*. Digital Europe.

Hervey, G. R. & Tobin, G. (1982). 'The Signal Gas Blender as a means of calibrating an indirect calorimeter'. *Journal of Physiology (London)*, **330**, 6–7P.

Heusner, A. A., Roberts, J. C. & Smith, R. E. M. (1971). 'Circadian patterns of oxygen consumption in *Peromyscus*'. *Journal of Applied Physiology*, **30**, 50–5.

Hill, A. V. & Hill, A. M. (1914). 'A self-recording calorimeter for large animals'. *Journal of Physiology (London)*, **48**, xiii–xiv.

Hill, D. W. (1959) 'Rapid measurement of respiratory pressures and volumes'. *British Journal of Anaesthesia*, **31**, 352–8.

Hill, D. W. & Powell, T. (1967). Cross-sensitivity effects in non-dispersive infrared gas analysers using condenser microphone detectors'. *Journal of Scientific Instruments*, **44**, 189–94.

Hill, R. W. (1972). 'Determination of oxygen consumption by use of the paramagnetic oxygen analyser'. *Journal of Applied Physiology*, **33**, 261–3.

Holman, S. W. (1895). Calorimetry: 'Methods of cooling correction'. *Proceedings of the American Academy of Arts and Science*, **31**, 245–54.

Holtkamp, D. E., Ochs, S., Pfeiffer, C. C. & Heming, A. E. (1955). 'Determination of the oxygen consumption of groups of rats'. *Endocrinology*, **56**, 93–104.

Howland, R. J. & Newman, K. (1985). 'A high precision automatic closed-circuit respirometer for small animals'. *Journal of Applied Physiology*, **58**, 1031–3.

Huebscher, R. G., Schutrum, L. F. & Parmelee, G. V. (1952). 'A low-inertia low-resistance heat flow meter'. *Transactions of the American Society of Agricultural Engineers*, **58**, 275–86.

Humphrey, S. J. E. & Wolff, H. S. (1977). 'The Oxylog'. *Journal of Physiology (London)*, **267**, 12P.

Hunt, L. M. (1969). 'An analytic formula to instantaneously determine total metabolic rate for the human system'. *Journal of Applied Physiology*, **27**, 731–3.

Hyde, C. G. & Mills, F. E. (1932). *Gas Calorimetry*. London: Ernest Benn.

Jacobsen, S., Johansen, O. & Garby, L. (1982). '5.8 m^3 human heat-sink calorimeter with online data acquisition, processing and control'. *Medical and Biological Engineering and Computing*, **20**, 29–36.

Janac, J., Catsky, J. & Jarvis, P. G. (1971). 'Infra-red gas analysers and other physical analysers'. In *Plant Photosynthetic Production: Manual of Methods*, ed. Z. Sestak, J. Catsky & P. G. Jarvis, pp. 111–97. The Hague: Dr W. Junk N.V. Publ.

Jequier, E. & Schutz, Y. (1983). 'Long term measurements of energy expenditure in humans using a respiration chamber'. *American Journal of Clinical Nutrition*, **38**, 989–98.

Johnson, R. E., Robbins, F., Schilke, R., Mole, P., Harris, J. & Wakat, D. (1967). 'A versatile system for measuring oxygen consumption in man'. *Journal of Applied Physiology*, **22**, 377–9.

Jones, H. B. (1950). 'Respiratory system: nitrogen elimination'. In *Medical Physics*, vol. 2, ed. O. Glasser, pp. 855–71. Chicago: The Year Book Publishers, Inc.

Jones, H. M. (1929). 'An accuracy check for metabolism apparatus'. *Journal of Laboratory and Clinical Medicine*, **14**, 750–3.

Kappagoda, C. T., Stoker, J. B. & Linden, R. J. (1974). 'A method for the continuous measurement of oxygen consumption'. *Journal of Applied Physiology*, **37**, 604–7.

Kelly, C. F., Bond, T. E. & Heitman, H. (1963). 'Direct "air" calorimetry for livestock'. *Transactions of the American Society of Agricultural Engineers*, **6**, 126–8.

Kelly, N. C. (1977). *The Intake and Utilization by Sheep of the Energy and Nitrogenous Constituents of Diets Based on Grass Silage*. PhD Thesis. University of Glasgow, pp. 82–95.

Kendall, P. P., McLean, J. A. & Smith, J. F. (1971). 'An integrating gas sampling system for open-circuit calorimetry'. *Journal of Physiology (London)*, **219**, 4–5P.

Kerslake, D. McK. (1972). *The Stress of Hot Environments*. Cambridge: Cambridge University Press.

Kilday, M. V. (1950). 'A quantitative study of the carbon monoxide formed during the absorption of oxygen by alkaline pyrogallol'. *Journal of the Research of the National Bureau of Standards*, **45**, 43–59.

King, L. V. (1914). 'On the convection of heat from small cylinders in a stream of fluid'. *Philosophical Transactions of the Royal Society, A.*, **214**, 373–432.

Kinney, J. M., Morgan, A. P., Domingues, F. J. & Gildner, K. J. (1964). 'A method for continuous measurement of gas exchange and expired radioactivity in acutely ill patients'. *Metabolism*, **13**, 205–11.

Klauer, F., Turowski, E. & Wolff, V. (1941). 'Determination of oxygen in gas mixtures by physical methods'. *Zeitschrift für Angewandte Chemie*, **54**, 494–6.

Kleiber, M. (1935). 'The California apparatus for respiration trials with large animals'. *Hilgardia*, **9**, 1–70.

Kleiber, M. (1961). *The Fire of Life*. New York: John Wiley.

Klein, P. D., James, W. P. T., Wong, W. W., Irving, C. S., Murgatroyd, P. R., Cabrera, M., Dallosso, H. M., Klein, E. R. & Nichols, B. L. (1984). 'Calorimetric validation of the doubly-labelled water method for the determination of energy expenditure in man'. *Human Nutrition: Clinical Nutrition*, **38C**, 95–106.

Kofranyi, E. & Michaelis, H. F. (1940). 'Ein tragbarer Apparat zur bestimmung des Gasstoffwechsels'. *Arbeitsphysiologie*, **11**, 148–50.

Koller, D. & Samish, Y. B. (1964). 'A null-point compensating system for simultaneous and continuous measurement of net photosynthesis and transpiration by controlled gas-stream analysis'. *Botanical Gazette*, **125**, 81–8.

Kreuzer, F., Rogeness, G. A. & Bornstein, P. (1960). 'Continuous recording of respiratory air oxygen tension'. *Journal of Applied Physiology*, **15**, 1157–8.

Krogh, A. (1913). 'A bicycle ergometer and respiration apparatus for the experimental study of muscular work'. *Skandinavisches Archiv für Physiologie*, **30**, 375–94.

Krogh, A. (1916). *The Respiratory Exchange of Animals and Man*. London: Longmans Green & Co.

Krogh, A. (1920). 'The calibration, accuracy and use of gas meters'. *Biochemical Journal*, **14**, 282–9.

Krogh, A. & Rasmussen, O. (1922). 'Ueber bestimmung des Energieumsatzes bei Patienten'. *Weiner Klinische Wochenschrift*, **35**, 803–6.

Laulanié, F. (1894). 'De la marche des alterations de l'air dans l'asphyxie en vase clos'. *Archives de Physiologie Normale et Pathologique*, **26**, 845–59.

Lawton, R. W., Prouty, L. R. & Hardy, J. D. (1954). 'A calorimeter for rapid determination of heat loss and heat production in laboratory animals'. *Review of Scientific Instruments*, **25**, 370–7.

Lefèvre, J. (1911). *Chaleur Animal et Bioenergetique*. Paris: Masson et Cie.

Legg, B. J. & Parkinson, K. J. (1968). 'Calibration of infra-red gas analysers for use with carbon dioxide'. *Journal of Scientific Instruments (Journal of Physics E)*, **1**, 1003–6.

Lehmann, C., Muller, F., Munk, I., Senator, H. & Zuntz, N. (1893). 'Untersuchungen an zwei hungernden Menschen'. *Virchow's Archiv für Pathologische Anatomie und Physiologie*, **131**, Supplement p. 51.

Lehrer, E. & Ebbinghaus, E. (1950). 'Ein Apparat zur Sauerstoffmessung in Gasgemischen auf magnetischer Grundlage'. *Zeitschrift für Angewandte Physik*, **2**, 20–4.

Levy, A. (1964). 'The accuracy of the bubble meter method for gas flow measurements'. *Journal of Scientific Instruments*, **41**, 449–53.

Levy, L. M., Bernstein, L. M. & Devor, D. (1958). 'Scholander apparatus for accurate estimation of carbon dioxide in small samples of expired air'. *Journal of Applied Physiology*, **13**, 309–12.

Lewis, R. C., Iliff, A. & Duval, A. M. (1943). 'The comparative accuracy of the closed circuit bedside method and the open circuit chamber procedure for the determination of basal metabolism'. *Journal of Laboratory and Clinical Medicine*, **28**, 1238–45.

Lewis, R. C., Kinsman, G. M. & Iliff, A. (1937). 'The basal metabolism of normal boys and girls from two to twelve years old, inclusive'. *American Journal of Diseases of Children*, **53**, 348–428.

Lifson, N., Gordon, G. B. & McClintock, R. (1955). 'Measurement of total carbon dioxide production by means of $D_2{}^{18}O$'. *Journal of Applied Physiology*, **7**, 704–10.

Lifson, N., Gordon, G. B., Visscher, M. B. & Nier, A. O. C. (1949). 'The fate of utilized molecular oxygen and the source of the oxygen of respiratory carbon dioxide studied with the aid of heavy oxygen'. *Journal of Biological Chemistry*, **180**, 803–11.

Lilienthal, J. L., Zierler, K. L. & Folk, B. P. (1949). 'A simple volumeter for measuring the oxygen consumption of small animals'. *Bulletin of the John Hopkins Hospital*, **84**, 238–44.

Lilly, J. C. (1950). 'Respiratory system: methods; gas analysis'. In *Medical Physics*, vol. II, ed. O. Glasser, pp. 845–55. Chicago: Year Book Publishers.

Lister, G., Hoffman, J. I. E. & Rudolph, A. M. (1974). 'Oxygen uptake in infants and children: a simple method for measurement'. *Paediatrics*, **53**, 656–62.

Livingstone, S. D. (1968). 'Calculation of mean body temperature'. *Canadian Journal of Physiology and Pharmacology*, **46**, 15–7.

Lloyd, B. B. (1958). 'A development of Haldane's gas-analysis apparatus'. *Journal of Physiology (London)*, **143**, 5–6P.

Long, C. L., Carlo, M. A., Schaffel, N., Schiller, W. S., Blakemore, W. S., Spencer, J. L. & Broell, J. R. (1979). 'A continuous analyser for monitoring respiratory gases and expired radioactivity in clinical studies'. *Metabolism*, **28**, 320–32.

Lukin, L. (1963). 'Instability of the "steady state" during exercise'. *Internationale Zeitschrift für Angewandte Physiologie Einschliesslich Arbeitsphysiologie*, **20**, 45–9.

Lund, D. D. & Gisolfi, C. V. (1974). 'Estimation of mean skin temperature during exercise'. *Journal of Applied Physiology*, **36**, 625–8.

Lundy, H., McLeod, M. G. & Jewitt, T. R. (1977). 'An automated multicalorimeter system: experiments on laying hens'. *British Poultry Science*, **19**, 173–86.

Lunge, G. & Ambler, H. R. (1934). *Technical Gas Analysis*. London: Gurney and Jackson.

Lunn, J. N., Molyneux, L. & Pask, E. A. (1965). 'A device for the measurement of ventilation in young children under general anaesthesia'. *Anaesthesia*, **20**, 135–44.

Lusk, G. (1928). *The Elements of the Science of Nutrition*. 4th edn. Reprinted 1976, New York: Johnson Reprint Corp.

MacLagan, N. F. & Sheahan, M. M. (1950). 'The measurement of oxygen consumption in small animals by a closed circuit method'. *Journal of Endocrinology*, **6**, 456–62.

MacLaury, D. W. & Johnson, T. H. (1968). 'Method for measuring oxygen consumption'. *Poultry Science*, **47**, 1377–9.

McLean, J. A. (1971). 'A gradient layer calorimeter for large animals'. *Journal of the Institution of Heating and Ventilating Engineers*, **39**, 1–8.

McLean, J. A. (1972). 'On the calculation of heat production from open-circuit calorimetric measurements'. *British Journal of Nutrition*, **27**, 597–600.

McLean, J. A. (1974). 'Improved air-sampling system for open-circuit calorimetry'. *Laboratory Practice*, **23**, 14–15.

McLean, J. A. (1987). Analysis of gaseous exchange in open-circuit calorimetry. *Medical and Biological Engineering and Computing*. **25**, 239–40.

McLean, J. A. (1986). 'The significance of carbon dioxide and methane measurements in the estimation of heat production'. *British Journal of Nutrition*. **55**, 631–3.

McLean, J. A. & Davidson, J. (1978). 'An analogue output device for recording flow rate'. *Laboratory Practice*, **27**, 289–90.

McLean, J. A., Downie, A. J., Jones, C. D. R., Stombaugh, D. P. & Glasbey, C. A. (1983a). 'Thermal adjustments of steers (*bos taurus*) to abrupt changes in environmental temperature'. *Journal of Agricultural Science (Cambridge)*, **100**, 305–14.

McLean, J. A., Downie, A. J., Watts, P. R. & Glasbey, C. A. (1982). 'Heat balance of steers (*Bos taurus*) in steady-temperature environments'. *Journal of Applied Physiology*, **52**, 324–32.

McLean, J. A., Stombaugh, D. P., Downie, A. J. & Glasbey, C. A. (1983*b*). 'Body heat storage in steers (*Bos taurus*) in fluctuating thermal environments'. *Journal of Agricultural Science (Cambridge)*, **100**, 315–22.

McLean, J. A. & Watts, P. R. (1976). 'Analytical refinements in animal calorimetry'. *Journal of Applied Physiology*, **40**, 827–31.

McLean, J. A. & Young, B. A. (1985). 'A new calibration method for oxygen consumption'. In *Human Energy Metabolism: Physical Activity and Energy Expenditure Measurements in Epidemiological Research based upon Direct and Indirect Calorimetry*. ed. A. J. H. van Es, pp. 41–3. Wageningen. Stichting Nederlands Instituut voor de Voeding. (Euro-Nut. report No. 5).

MacLeod, M. G., Lundy, H. & Jewitt, T. R. (1985). 'Heat production by the mature male turkey (*Meleagris gallopavo*): preliminary measurements in an automated, indirect, open-circuit multi-calorimeter system'. *British Poultry Science*, **26**, 325–33.

Machta, L. & Hughes, E. (1970). 'Atmospheric oxygen in 1967 to 1970'. *Science*, **168**, 1582–4.

Magnus-Levy, A. (1907). In *Metabolism and Practical Medicine, Vol. 1. The Physiology of Metabolism*, ed. C. Von Noorden, English translation edn I. Walker Hall. pp. 185–282. London: Heinemann.

Malhotra, M. S., Sen Gupta, J. & Rai, R. M. (1963). 'Pulse count as a measure of energy expenditure'. *Journal of Applied Physiology*, **18**, 994–6.

Margaria, R. (1933). 'A gas analysis apparatus'. *Journal of Scientific Instruments*, **10**, 242–6.

Marston, H. R. (1948). 'Energy transactions in the sheep'. *Australian Journal of Scientific Research, Series B (Biological Sciences)*, **1**, 93–129.

Maxfield, M. E. & Smith, P. E. (1967). 'Abbreviated methods for measurement of oxygen consumption in work physiology'. *Human Factors*, **9**, 587–94.

Mendelsohn, E. (1964). *Heat and Life*. Cambridge, Mass: Harvard University Press.

Merrill, A. L. & Watt, B. K. (1955). 'Energy value of foods'. *Agricultural Handbook No. 74*, Washington: United States Department of Agriculture.

Miller, B. G., Kirkwood, J. K., Howard, K., Tuddenham, A. & Webster, A. J. F. (1981). 'A self-compensating, closed-circuit respiration calorimeter for small mammals and birds'. *Laboratory Animals*, **51**, 313–17.

Miller, D. S. & Payne, P. R. (1959). 'A ballistic bomb calorimeter'. *British Journal of Nutrition*, **13**, 501–8.

Miller, J. E., Carroll, A. J. & Emerson, D. E. (1965). 'Preparation of primary standard gas mixtures for analytical instruments'. *Report of Investigations 6674*, Washington: United States Department of the Interior, Bureau of Mines.

Misson, B. H. (1974). 'An open-circuit respirometer for metabolic studies on the domestic fowl: establishment of standard operating conditions'. *British Poultry Science*, **15**, 287–97.

Mitchell, H. H. & Hamilton, T. S. (1932). 'The effect of the amount of feed consumed by cattle on the utilisation of its energy content'. *Journal of Agricultural Research*, **45**, 163–91.

Mitchell, D. & Wyndham, C. H. (1969). 'Comparison of weighting formulas for calculating mean skin temperature'. *Journal of Applied Physiology*, **26**, 616–22.

Mitchell, D., Wyndham, C. H., Atkins, A. R., Vermeulan, A. J., Hofmeyer, H. S., Strydom, N. B. & Hodgson, T. (1968). 'Direct measurement of the thermal responses of nude resting men in dry environments'. *Pflugers Archiv*, **303**, 324–43.

Møllgard, H. & Anderson, A. C. (1917). 'Respiration apparatus: its significance and use in experiments with dairy cattle'. *Beretning fra Forsøgslaboratoriet Kobenhaven*, **94**, 1–180.

Monteith, J. L. (1954). 'Error and accuracy in thermocouple psychrometry'. *Proceedings of the Physical Society*, **67**, 217–26.

Monteith, J. L. (1972). 'Latent heat of evaporization in thermal physiology. *Nature New Biology*, **236**, 96.

Montoye, H. J., van Huss, W. D., Reineke, E. P. & Cockrell, J. (1958). 'An investigation of the Müller-Franz calorimeter'. *Arbeitsphysiologie*, **17**, 28–33.

Moors, P. J. (1977). 'A closed-circuit respirometer for measuring the average daily metabolic rate of small mammals'. *Oikos*, **28**, 304–8.

Morey, N. B. (1936). 'An analysis and comparison of different methods of calculating the energy value of diets'. *Nutrition Abstracts and Reviews*, **6**, 1–12.

Morrison, P. R. (1947). 'An automatic apparatus for the determination of oxygen consumption'. *Journal of Biological Chemistry*, **169**, 667–79.

Morrison, P. R. (1951). 'An automatic manometric respirometer'. *Review of Scientific Instruments*, **22**, 264–7.

Moses, L. E. (1947). 'Determination of oxygen consumption in the albino rat'. *Proceedings of the Society for Experimental Biology and Medicine*, **64**, 54–7.

Mount, L. E. (1968). *Climatic Physiology of the Pig*. London: Arnold.

Mount, L. E., Holmes, C. W. & Start, I. B. (1967). 'A direct calorimeter for the continuous recording of heat loss from groups of growing pigs over long periods'. *Journal of Agricultural Science (Cambridge)*, **68**, 47–55.

Mountcastle, V. B. (1974). *Medical Physiology*, 13th edn. St Louis: C. V. Mosby.

Müller, E. A. & Franz, H. (1952). 'Energieverbrauchsmessungen bei beruflicher Arbeit mit einer verbesserten Respirations-Gasuhr'. *Arbeitsphysiologie*, **14**, 499–504.

Murgatroyd, P. R. (1984). 'A 30 m³ direct and indirect calorimeter'. In *Human Energy Metabolism: Physical Activity and Energy Expenditure Measurements in Epidemiological Research based upon Direct and Indirect Calorimetry*, ed. A. J. H. van Es, pp. 46–8. Wageningen: Stichting Nederlands Instituut voor de Voeding. (Euro-Nut report No. 5).

Murlin, J. R. & Burton, A. C. (1935). 'Human calorimetry. I. Description of a respiration calorimeter'. *Journal of Nutrition*, **9**, 233–59.

Nelson, G. O. (1971). *Controlled Test Atmospheres. Principles and Techniques*. Ann Arbor: Ann Arbor Science.

Nielsen, B. (1966). 'Regulation of body temperature and heat dissipation at different levels of energy – and heat production in man'. *Acta Physiologica Scandinavica*, **68**, 215–27.

Nijkamp, H. J. (1969). 'Determination of the urinary energy – and carbon output in balance trials'. *Zeitschrift für Tierphysiologie, Tierernahrung und Futtermittelkunde*, **25**, 1–9.

Nijkamp, H. J. (1971). 'Some remarks and recommendations concerning bomb calorimetry'. *Zeitschrift für Tierphysiologie, Tierernahrung und Futtermittelkunde*, **27**, 115–21.

Noyons, A. K. M. (1927). *The Differential Calorimeter and the Determination of Human Basal Metabolism*. Louvain: Reûe Fonteyn.

Noyons, A. K. M. (1937). 'Méthode d'enregistrement continu de la teneure en CO_2 et en O_2 des gaz respiratoires au moyen du diaferomètre thermique, servant à l'étude du métabolisme des tissus des animaux et de l'homme'. *Annales de Physiologie et de Physiochimie Biologique*, **13**, 909–35.

Nunn, J. F. (1977). *Applied Respiratory Physiology*, 2nd edn, p. 241. London: Butterworths.

Nunn, J. F., Bergman, N. A., Coleman, A. J. & Casselle, D. C. (1964). 'Evaluation of the Servomex paramagnetic analyser'. *British Journal of Anaesthesia*, **36**, 666–73.

Nunn, J. F. & Ezi-Ashi, T. I. (1962). 'The accuracy of the respirometer and ventigrator'. *British Journal of Anaesthesia*, **34**, 422–32.

Orsini, D. & Passmore, R. (1951). 'The energy expended carrying loads up and down stairs: experiments using the Kofranyi-Michaelis calorimeter'. *Journal of Physiology (London)*, **115**, 95–100.

Otis, A. B. (1964). 'Quantitative relationships in steady state gas exchange'. In *Handbook of Physiology. Vol. I. Section 3. Respiration*, eds. W. O. Fenn & H. Rahn, pp. 681–98. Washington: The American Physiological Society.

Owens, F. N., Cooper, D. P., Goodrich, R. D. & Meiske, J. C. (1969). Bomb calorimetry of high-moisture materials'. *Journal of Dairy Science*, **52**, 1273–77.

Ower, E. & Pankhurst, R. C. (1977). *The Measurement of Air Flow*. Oxford: Pergamon Press.

Palmes, E. D. & Park, C. R. (1947). 'Thermal regulation during early acclimatization to work in a hot dry environment'. *United States Army Medical Department Field Research Laboratory, Report 2–17–1*, Fort Knox, Kentucky.

Parkinson, K. J. & Legg, B. J. (1971). 'A new method for calibrating infrared gas analysers'. *Journal of Physics E: Scientific Instruments*, **4**, 598–600.

Parkinson, K. J. & Legg, B. J. (1978). 'Calibration of infra-red analysers for carbon dioxide'. *Photosynthetica*, **12**, 65–7.

Passmore, R. (1967). 'Energy balances in man'. *Proceedings of the Nutrition Society*, **26**, 97–101.

Passmore, R. & Draper, M. H. (1965). 'Energy Metabolism'. In *Newer Methods of Nutritional Biochemistry*, Vol. 2, ed. A. A. Albanese, pp. 41–83. New York: Academic Press.

Patterson, R. (1979). 'The design and application of a mass spectrometer-based metabolic function analyser'. In *The Medical and Biological Application of Mass Spectrometry*, eds. J. P. Payne, J. A. Bushman & D. W. Hill, pp. 119–140. London: Academic Press.

Paul, A. A. & Southgate, D. A. T. (1978). *The Composition of Foods*. 4th edn. London: H.M.S.O.

Pauling, L., Wood, R. E. & Sturdivant, J. H. (1946). 'An instrument for determining the partial pressure of oxygen in a gas'. *Journal of the American Chemical Society*, **63**, 795–8.

Payne, J. P., Bushman, J. A. & Hill, D. W. (1979). *The Medical and Biological Application of Mass Spectrometry*. London: Academic Press.

Perkins, J. F. (1954). 'Plastic Douglas bags'. *Journal of Applied Physiology*, **6**, 445–7.

Perry, R. H., Green, D. W. & Maloney, J. O. (1984). *Perry's Chemical Engineers' Handbook*. Revised 5th edn, pp. 5–20. New York: McGraw-Hill.

Peters, J. P. & van Slyke, D. D. (1932). *Quantitative Clinical Chemistry*, Vol. II Methods. Baltimore: Williams & Wilkins.

Peters, J. P. & van Slyke, D. D. (1932). *Quantitative Clinical Chemistry*, Vol. 1, 2nd edn. London: Balliere, Tindall & Cox.

Pettenkofer, M. (1862). 'Ueber die Respiration'. *Annalen der Chemie und Pharmacie*, **123**, Suppl. 2, 1–52.

Pettenkofer, M. & Voit, C. (1866). Untersuchungen über den Stoffverbrauch des normalen Menchen'. *Zeitschrift für Biologie*, **2**, 478–573.

Pickwell, G. V. (1968). 'An automatically compensating heat-exchange calorimeter for metabolic studies'. *Medical and Biological Engineering*, **6**, 425–31.

Pols, W. M. (1962). 'Pneumotachography'. *Acta Anaesthesiologica Scandinavica*, Supplement IX.

Poole, G. W. & Maskell, R. C. (1975). 'Validation of continuous determination of respired gases during steady-state exercise'. *Journal of Applied Physiology*, **38**, 736–8.

Pritchard, R., Guy, J. J. & Connor, N. E. (1977). *Industrial Gas Utilization. Engineering Principles and Practice*, pp. 363–94. Epping (Essex): Bowker.

Prouty, L. R., Barret, M. J. & Hardy, J. D. (1949). 'A simple calorimeter for the continuous determination of heat loss and heat production in animals'. *Review of Scientific Instruments*, **20**, 357–62.

Pullar, J. D. (1969). 'Methods of calorimetry. A. Direct'. In *Nutrition of Animals of Agricultural Importance, Part 1*, ed. D. P. Cuthbertson, Vol. 17, pp. 471–90 of *International Encyclopedia of Food and Nutrition*. Oxford: Pergamon Press.

Quattrone, P. D. (1965). 'Improved gradient-layer animal calorimeter'. *Review of Scientific Instruments*, **36**, 832–8.

Ramanathan, N. L. (1964). 'A new weighting system for mean surface temperature of the human body'. *Journal of Applied Physiology*, **19**, 531–3.

Raymond, W. F., Canaway, R. J. & Harris, C. E. (1957). 'An automated adiabatic bomb calorimeter'. *Journal of Scientific Instruments*, **34**, 501–3.

Regnault, V. & Reiset, J. (1849). 'Recherches chimiques sur la respiration des animaux des diverses classes'. *Annales de Chimie et de Physique. Series III*, **26**, 299–519.

Richards, T. W. (1909). 'Recent investigations in thermochemistry'. *Journal of the American Chemical Society*, **31**, 1275–83.

Richards, T. W., Henderson, L. J. & Forbes, G. S. (1905). 'Elimination of thermometric lag and accidental loss of heat in calorimetry'. *Proceedings of the American Academy of Arts and Science*, **41**, 3–19.

Riche, J. A. & Soderstrom, G. F. (1915). 'The respiration calorimeter of the Russell Sage Institute of Pathology in Bellevue Hospital'. *Archives of Internal Medicine*, **15**, 805–28.

Riendeau, R. P. & Consolazio, C. F. (1959). 'A new method of calibrating the Müller-Franz respiration gas meter'. *Journal of Applied Physiology*, **14**, 154–6.

Ritzman, E. G. & Benedict, F. G. (1929). 'Simplified techniques and apparatus for measuring energy requirements of cattle'. *New Hampshire Agricultural Experiment Station Bulletin 240*.

Roth, P. (1922). 'Modifications of apparatus and improved technic adaptable to the Benedict type of respiration apparatus. III. Graphic method for the estimation of the metabolic rate'. *The Boston Medical and Surgical Journal*, **186**, 491–8.

Rubner, M. (1894). 'Die Quelle der Thierschen warme'. *Zeitschrift für Biologie*, **30**, 73–142.

Rubner, M. (1902). *The Laws of Energy Consumption In Nutrition*. English Translation, 1982, New York: Academic Press.

Samish, Y. B. (1978). 'Measurement and control of CO_2 concentration in air is influenced by the desiccant'. *Photosynthetica*, **12**, 73–5.

Scheid, P., Slama, H. & Piiper, J. (1971). 'Electronic compensation of the effects of water vapor in respiratory mass spectrometry'. *Journal of Applied Physiology*, **30**, 258–60.

Schoeller, D. A. (1983). 'Energy expenditure from doubly labeled water: some fundamental considerations in humans'. *American Journal of Clinical Nutrition*, **38**, 999–1005.

Schoeller, D. A. & van Santen, E. (1982). 'Measurement of energy expenditure in humans by doubly labeled water method'. *Journal of Applied Physiology*, **53**, 955–9.

Schoeller, D. A. & Webb, P. (1984). 'Five-day comparison of the doubly labeled water method with respiratory gas exchange'. *American Journal of Clinical Nutrition*, **40**, 153–8.

Schoffelen, P. F. M., Saris, W. H. M., Westerterp, K. R. & ten Hoor, F. (1984). 'Evaluation of an automated indirect calorimeter for measurement of energy balance in man'. In *Human Energy Metabolism: Physical Activity and Energy Expenditure Measurements in Epidemiological Research based upon Direct and Indirect Calorimetry*, ed. A. J. H. van Es, pp. 51–4. Wageningen: Stichting Nederlands Instituut voor de Voeding (Euro-Nut report No. 5).

Scholander, P. F. (1947). 'Analyser for accurate estimation of respiratory gases in one-half cubic centimetre samples'. *Journal of Biological Chemistry*, **167**, 235–50.

Schulze, K., Kairam, R., Stefanski, M., Sciacca, R. & James, L. S. (1981). 'Continuous measurement of minute ventilation and gaseous metabolism of newborn infants'. *Journal of Applied Physiology*, **50**, 1098–1103.

Senftleben, H. (1930). 'Magnetische Beeinflussung des Warmeleitvermogens paramagnetischer Gase'. *Physikalische Zeitschrift*, **31**, 961–3.

Severinghaus, J. W. (1960). 'Respiratory system: methods; gas analysis'. In *Medical Physics*, vol. III, ed. O. Glasser, pp. 550–9. Chicago: Year Book Publishers.

Severinghaus, J. W. (1963). 'High-temperature operation of oxygen electrode giving fast response for respiratory gas sampling'. *Clinical Chemistry*, **9**, 727–33.

Severinghaus, J. W. (1968). 'Measurements of blood gases: PO_2 and PCO_2'. *Annals of the New York Academy of Sciences*, **148**, 115–32.

Scharpey-Schafer, E. (1932). 'A metabolism apparatus for small animals'. *Quarterly Journal of Experimental Physiology*, **22**, 95–100.

Shephard, R. J. (1955). 'A critical examination of the Douglas bag technique'. *Journal of Physiology (London)*, **127**, 515–24.

Sherebrin, M. H. & Burton, A. C. (1964). 'Method of using a compensated gradient to improve time response of an animal calorimeter'. *Review of Scientific Instruments*, **35**, 1668–73.

Sibbald, I. R. (1963). 'A comparison of two bomb calorimeters'. *Poultry Science*, **42**, 780–1.

Simmons, L. F. G. (1949). 'A shielded hot-wire anemometer for low speeds'. *Journal of Scientific Instruments*, **26**, 407–11.

Sinclair, T. R., Buzzard, G. H. & Knoerr, K. R. (1976). 'A pressure method for frequent differential calibration of CO_2 infrared analyser'. *Photosynthetica*, **10**, 188–92.

Smith, E. (1955). 'A new apparatus for long term measurements of oxygen consumption in small mammals'. *Proceedings of the Society for Experimental Biology and Medicine*, **89**, 499–501.

Smith, W. D. A. (1963). 'Uptake of nitrous oxide during the induction of Anaesthesia. Part I: Apparatus and methods'. *British Journal of Anaesthesia*, **35**, 224–36.

Smothers, J. L. (1966). 'A volumetric respirometer for small animals'. *Journal of Applied Physiology*, **21**, 1117–18.

Snellen, J. W. (1966). 'Mean body temperature and the control of thermal sweating'. *Acta Physiologica et Pharmacologica Neerlandica*, **14**, 99–174.

Snellen, J. W. & Baskind, P. (1969). 'The air-conditioned system, and operation and performance of the human calorimeter at the Human Sciences Laboratory'. *Chamber of Mines of South Africa, Research Report No. 24/69.*

Snellen, J. W., Chang, K. S. & Smith, W. (1983). 'Technical description and performance characteristics of a human whole-body calorimeter'. *Medical and Biological Engineering and Computing*, **21**, 9–20.

Snellen, J. W., Mitchell, D. & Wyndham, C. H. (1970). 'Heat of evaporation of sweat'. *Journal of Applied Physiology*, **29**, 40–4.

Sodal, I. E., Bowman, R. R. & Filley, G. F. (1968). 'A fast-response oxygen analyser with high accuracy for respiratory gas measurement'. *Journal of Applied Physiology*, **25**, 181–3.

Sonden, K. & Tigerstedt, R. (1895). 'Untersuchungen uber de Respiration und dem Gesammstoff weschel des Menschen'. *Archiv für Physiologie*, **6**, 1–224.

Sparkes, D. W. (1968). 'A standard choked nozzle for absolute calibration of air flowmeters'. *The Aeronautical Journal of the Royal Aeronautical Society*, **72**, 335–8.

Spinnler, G., Jequier, E., Favre, R., Dolivo, M. & Vannotti, A. (1973). 'Human calorimeter with a new type of gradient layer'. *Journal of Applied Physiology*, **35**, 158–65.

Staub, N. C. (1961). 'A simple small oxygen electrode'. *Journal of Applied Physiology*, **16**, 192–4.

Stock, M. J. (1975). 'An automatic, closed-circuit oxygen consumption apparatus for small animals'. *Journal of Applied Physiology*, **39**, 849–50.

Stokes, M. R. (1977). *The relationship between the pattern of fermentation in the rumen and energy metabolism in sheep*. PhD Thesis, University of Glasgow, pp. 91–101.

Stratford, B. S. (1964). 'The calculation of the discharge coefficient of profiled choked nozzles and the optimum profile for absolute air flow measurement'. *Journal of the Royal Aeronautical Society*, **68**, 237–45.

Strite, G. H. & Yacowitz, H. (1956). 'A simplified method for estimating the rate of oxygen consumption of young chicks'. *Poultry Science*, **35**, 142–4.

Strømme, S. B. & Hammel, H. T. (1968). 'A simple method for calibration of gas meters'. *Journal of Applied Physiology*, **25**, 184–5.

Swift, R. W. (1933). 'New details of technic in air analysis'. *Journal of Laboratory and Clinical Medicine*, **18**, 731–8.

Tamas, I. A., Urquhart, N. S. & Ozbun, J. L. (1971). 'Calibration of infrared analysers for CO_2: mathematical model and experimental results'. *Applied Spectroscopy*, **25**, 191–5.

Thomas, C. C. (1911). 'The measurement of gases'. *Journal of the Franklin Institute*, **172**, 411–60.

Thomson, E. F. (1979). 'Description of simplified open-circuit respiration equipment for cattle'. *Laboratory Practice*, **28**, 1315–17.

Thorbek, G. & Neergaard, L. (1970). 'Construction and function of an open circuit respiration plant for cattle'. In *Energy Metabolism of Farm Animals. Proceedings of the 5th Symposium on Energy Metabolism, Zurich, Switzerland* (EAAP Publication No. 13) eds. A. Schurch & C. Wenk, pp. 243–6. Zurich: Juris Verlag.

Tiemann, J. (1969). 'Ein Respirationskalorimeter zur bestimmung des Transformationsgrades ver Verbrennung im menschlichen Organismus'. *Elektromedizin*, **14**, 135–45.

Tissot, J. (1904). 'Nouvelle méthode de mesure et d'inscription du débit et des mouvements respiratoires de l'homme et des animaux'. *Journal de Physiologie et de Pathologie Générale*, **6**, 688–700.

Tobin, G. & Hervey, G. R. (1984). 'Accuracy in indirect calorimetry'. In *Human Energy Metabolism: Physical Activity and Energy Expenditure Measurements in Epidemiological Research based upon Direct and Indirect Calorimetry*, ed. A. J. H. van Es, pp. 32–6. Wageningen: Stichting Nederlands Instituut voor de Voeding (Euro-Nut report No. 5).

Torda, T. A. & Grant, G. C. (1972). 'Test of a fuel cell oxygen analyser'. *British Journal of Anaesthesia*, **44**, 1108–12.

Trusell, F. & Diehl, H. (1963). 'Efficiency of chemical desiccants'. *Analytical Chemistry*, **35**, 674–7.

Tschegg, E., Sigmund, A., Veitl, V., Schmid, P. & Irsigler, K. (1979). 'An isothermic, gradient-free, whole-body calorimeter for long-term investigations of energy balance in man'. *Metabolism*, **28**, 764–70.

Turner, H. G. & Thornton, R. F. (1966). 'A respiration chamber for cattle'. *Proceedings of the Australian Society for Animal Production*, **6**, 413–19.

Turney, S. Z. & Blumenfeld, W. (1973). 'Heated Fleisch pneumotachometer: a calibration procedure'. *Journal of Applied Physiology*, **34**, 117–21.

Verdin, A. (1973). *Gas Analysis Instrumentation*. London: Macmillan.

Vermorel, M., Bouvier, J-C., Bonnet, Y. & Fauconneau, G. (1973). 'Construction et fonfonctionment de 2 chambres respiratoires du type "circuit ouvert" pour jeunes bovins'. *Annales de Biologie Animale Biochimie Biophysique*, **13**, 659–81.

Verstegen, M. W. A., van der Hel, W., Koppe, R. & van Es, A. J. H. (1971). 'An indirect calorimeter for the measurement of the heat production of large groups of animals kept together'. *Mededelinge Landbouwhogeschool Wageningen*, **71**, 1–13.

Visser, J. & Hodgson, T. (1960). 'The design of a human calorimeter for the determination of body heat storage'. *South African Mechanical Engineer*, **9**, 243–60.

Wainman, F. W. & Blaxter, K. L. (1958a). 'Closed-circuit respiration apparatus for the cow and steer'. In *Symposium on Energy Metabolism* (EAAP Publication No. 8) eds. G. Thorbek & H. Aersøe, pp. 80–4. Rome: EAAP.

Wainman, F. W. & Blaxter, K. L. (1958b). 'Developments in the design of closed-circuit respiration apparatus for animals weighing up to 70 kg'. In *Symposium on Energy Metabolism*, (EAAP Publication No. 8) eds. G. Thorbek & H. Aersøe, pp. 85–95. Rome: EAAP.

Wainman, F. W. & Paterson, D. (1963). 'A note on the collection of urine from male cattle and sheep'. *Journal of Agricultural Science (Cambridge)*, **61**, 253–4.

Waring, J. J. & Brown, W. O. (1965). 'A respiration chamber for the study of energy utilisation for maintenance and production in the laying hen'. *Journal of Agricultural Science Cambridge*, **65**, 139–46.

Watts, D. T. & Gourley, D. R. H. (1953). 'A simple apparatus for determining basal metabolism of small animals in student laboratory'. *Proceedings of the Society for Experimental Biology and Medicine*, **84**, 585–6.

Watts, P. R. (1976). 'A system for automatic collection and removal of urine from male farm animals'. *British Journal of Nutrition*, **36**, 295–7.

Weast, R. C. ed. (1979). *Handbook of Chemistry and Physics*. Boca Raton, Fa.: C.R.C. Press.

Webb, P. (1985). *Human Calorimeters*. New York: Praeger.

Webb, P. & Annis, J. F. (1968). 'Cooling required to suppress sweating during work'. *Journal of Applied Physiology*, **25**, 489–93.

Webb, P., Annis, J. F. & Troutman, S. J. (1972). 'Human calorimetry with a water cooled garment'. *Journal of Applied Physiology*, **32**, 412–18.

Webb, P. & Troutman, S. J. (1970). 'An instrument for continuous measurement of oxygen consumption'. *Journal of Applied Physiology*, **28**, 867–71.

Webster, A. J. F. (1967). 'Continuous measurement of heart rate as an indicator of the energy expenditure of sheep'. *British Journal of Nutrition*, **21**, 769–85.

Webster, W. M. & Cresswell, E. (1957). 'A new technique in indirect calorimetry'. *The Veterinary Record*, **69**, 526–7.

Weil, J. V., Sodal, I. E. & Speck, R. D. (1967). 'A modified fuel cell for the analysis of oxygen concentration of gases'. *Journal of Applied Physiology*, **23**, 419–22.

Weir, J. B. de V. (1949). 'New methods for calculating metabolic rate with special reference to protein metabolism'. *Journal of Physiology (London)*, **109**, 1–9.

Weissman, C., Askanazi, J., Milic-Emili, J. & Kinney, J. M. (1984). 'Effect of respiratory apparatus on respiration'. *Journal of Applied Physiology*, **57**, 475–80.

Weissman, C., Damask, M. C., Askanazi, J., Rosenbaum, S. H. & Kinney, J. M. (1985). 'Evaluation of a non-invasive method for the measurement of metabolic rate in humans'. *Clinical Science*, **69**, 135–41.

Wenger, C. B. (1972). 'Heat of evaporation of sweat: thermodynamic considerations'. *Journal of Applied Physiology*, **32**, 456–9.

Wennberg, L. A. (1974). 'A new formula for estimating oxygen consumption in man and animals'. *European Journal of Applied Physiology*, **34**, 65–8.

Whitelaw, F. G. (1974). 'Measurement of energy expenditure in the grazing ruminant'. *Proceedings of the Nutrition Society*, **33**, 163–72.

Whitelaw, F. G., Brockway, J. M. & Reid, R. S. (1972). 'Measurement of carbon dioxide production in sheep by isotope dilution'. *Quarterly Journal of Experimental Physiology*, **57**, 37–55.

Widdowson, E. M. (1955). 'Assessment of the energy value of human foods'. *Proceedings of the Nutrition Society*, **14**, 143–54.

Willard, H. N. & Wolf, G. A. (1951). 'A source of error in the determination of basal metabolic rates by closed-circuit technic'. *Annals of Internal Medicine*, **34**, 148–62.

Williams, A. B. (1984). *Designers Handbook of Integrated Circuits*. New York: McGraw-Hill.

Wilmore, J. H., Davis, J. A. & Norton, A. C. (1976). 'An automated system for assessing metabolic and respiratory function during exercise'. *Journal of Applied Physiology*, **40**, 619–24.

Winslow, C-E. A., Herrington, L. P. & Gagge, A. P. (1936a). 'A new method of partitional calorimetry'. *American Journal of Physiology*, **116**, 641–55.

Winslow, C-E. A., Herrington, L. P. & Gagge, A. P. (1936b). 'The determination of radiation and convection exchanges by partitional calorimetry'. *American Journal of Physiology*, **116**, 669–84.

Wolff, H. S. (1958b). 'The accuracy of the integrating motor pneumotachograph (Imp)'. measurement of energy expenditure by indirect calorimetry'. *Quarterly Journal of Experimental Physiology*, **43**, 270–83.

Wolff, H. S. (1958b). 'The accuracy of the integtrating motor pneumotachograph (Imp)'. *Journal of Physiology (London)*, **141**, 36–7P.

Wright, B. M. (1955). 'A respiratory anaemometer'. *Journal of Physiology (London)*, **127**, 25P.

Yamamoto, S. & Ogura, Y. (1985). 'Variations in heart rate and relationship between heart rate and heat production of breeding Japanese Black Cattle'. *Japanese Journal of Livestock Management*, **20**, 109–18.

Yoshida, I., Shimada, Y. & Tanaka, K. (1979). 'Evaluation of a hot-wire respiratory flowmeter for clinical applicability'. *Journal of Applied Physiology*, **47**, 1131–5.

Young, A. C. & Carlson, L. D. (1952). 'Thermal gradient calorimeter'. *Arctic Aeromedical Laboratory, Ladd Air Force Base, Alaska, Report, Project No. 22–1301–0002*.

Young, A. C., Carlson, L. D. & Burns, H. L. (1955). 'Regional heat loss by temperature gradient calorimetry'. *Arctic Aeromedical Laboratory, Ladd Air Force Base, Alaska, Report No. 2, Project No. 8–7951*.

Young, B. A. (1970). 'Application of the carbon dioxide entry rate technique to measurement of the energy expenditure by grazing cattle'. In *Energy Metabolism of Farm Animals. Proceeding of the 5th Symposium on Energy Metabolism, Zurich, Switzerland* (EAAP Publication No. 13) eds. A. Schurch & C. Wenk, pp. 237–411. Zurich: Juris Verlag.

Young, B. A. & Corbett, J. L. (1972). 'Maintenance energy requirement of grazing sheep in relation to herbage availability'. *Australian Journal of Agricultural Research*, **23**, 57–76.

Young, B. A., Fenton, T. W. & McLean, J. A. (1984). 'Calibration methods in respiratory calorimetry'. *Journal of Applied Physiology*, **56**, 1120–5.

Young, B. A., Kerrigan, B. & Christopherson, R. J. (1975). 'A versatile respiratory pattern analyser for studies of energy metabolism of livestock'. *Canadian Journal of Animal Science*, **55**, 17–22.

Young, B. A., Leng, R. A., Whitelaw, R. G., McClymont, G. L. & Corbett, J. L. (1967). 'Estimation of energy expenditure from measurements of carbon dioxide entry rate'. In *Energy Metabolism of Farm Animals. Proceedings of the 4th Symposium on Energy Metabolism, Warsaw, Poland*, eds. K. L. Blaxter, J. Kielanowski & G. Thorbek, pp. 435–6. Newcastle-upon-Tyne: Oriel Press.

Young, B. A. & Webster, M. E. D. (1963). 'A technique for the estimation of energy expenditure in sheep'. *Australian Journal of Agricultural Research*, **14**, 867–73.

Zuntz, N. (1905). 'Ein nach dem Prizip von Regnault und Reiset gebaulter Respirationsapparat'. *Archiv für Anatomie und Physiologie*, Supplement 431–5.

Zuntz, N., Loewy, A., Muller, F. & Caspari, W. (1906). *Hohenklima und Bergwanderungen*. Berlin: Deutsches Verlagshaus Bong & Co.

Index